DESIGN FOR NORTHERN CLIMATES

1988 INTERNATIONAL GRAND PRIZE WINNER
$10,000.00
IN THE BUILDING DESIGN, TECHNOLOGY AND
URBAN INFRASTRUCTURE COMPETITION

SPONSORED BY
THE INTERNATIONAL WINTER CITIES CORPORATION
THE AMERICAN INSTITUTE OF ARCHITECTS
THE ROYAL ARCHITECTURAL INSTITUTE OF CANADA
THE ALBERTA ASSOCIATION OF ARCHITECTS
THE CITIES of SAPPORO, JAPAN and TROMSO, NORWAY

DESIGN FOR NORTHERN CLIMATES

■

COLD-CLIMATE PLANNING AND ENVIRONMENTAL DESIGN

■

VLADIMIR MATUS

VNR VAN NOSTRAND REINHOLD COMPANY
New York

Library of Congress Catalog Card Number 87-29572
ISBN 0-442-26463-1

Printed in the United States of America

Designed by Ann Gold

Van Nostrand Reinhold Company Inc.
115 Fifth Avenue
New York, New York 10003

Van Nostrand Reinhold Company Limited
Molly Millars Lane
Wokingham, Berkshire RG11 2PY, England

Van Nostrand Reinhold
480 La Trobe Street
Melbourne, Victoria 3000, Australia

Macmillan of Canada
Division of Canada Publishing Corporation
164 Commander Boulevard
Agincourt, Ontario M1S 3C7, Canada

16 15 14 13 12 11 10 9 8 7 6 5 4 3 2 1

Library of Congress Cataloging-in-Publication Data

Matus, Vladimir.
Design for northern climates.

Bibliography: p.
Includes index.
1. Architecture—Cold regions. 2. Architecture and climate—Cold regions. I. Title.
NA2542.C75M38 1988 720'.911'3 87-29572
ISBN 0-442-24643-1

To the City of Hamilton

CONTENTS

FOREWORD

Climate has little effect on the architecture of most North American cities. Although the architecture of Santa Fe suggests hot, arid lands, and a few small towns in Quebec and the Maritime Provinces have architecture that mitigates the effects of cold weather, the urban face of most communities from Dallas to Edmonton is identical. The designs of Kansas City and Winnipeg, for instance, do not indicate their very different winters and summers. And individuals living in Atlanta and Minneapolis/St. Paul rely on air conditioners or furnaces to cope with extreme temperatures.

Within the last decade, however, the concept of a livable winter city has begun to penetrate Canada, the United States, northern Japan, and North and Central Europe. The subject has been raised at conferences, in books, and by associations founded to explore climate-sensitive environments.

Vladimir Matus, in *Design for Northern Climates,* offers design solutions for architects and city planners working in Denver, Des Moines, Detroit, as far north as Regina, Calgary, and beyond. This text makes the concept of a livable winter city credible.

Matus's design solutions are based largely on the sun as the greatest natural source of heat. He describes how entire cities as well as individual homes can make the most efficient use of the sun's energy. He explores the effects of the sun on individuals' psyches—how solar and human rhythms interact. Our ancient and cultural prejudices are examined to explain our attachment to southern values, which often stands in the way of a scientific approach to design solutions in the north.

Design for Northern Climates offers methods of attaining a complete ecological description of a given area. Matus's innovative solutions for describing land forms and his descriptions of how land form affects the plan, shape, and location of buildings suitable for cold climate are invaluable tools for every urban developer and builder.

Solar Envelopes, Skydomes, Solar Pyramids, and Solar Templates are a few of the practical innovations discussed in the text. Ordinary concerns, such as the purpose and function of windows, are also addressed. A case study of a low-density residential neighborhood puts Matus's theories into practice.

Matus includes guidelines for creating habitable shelters, neighborhoods, and high-density cities. We discover here how large buildings can be designed to use and misuse solar and other heat, the effect tall buildings have on wind, light, and the general comfort of pedestrians.

Matus offers a wealth of information in the form of charts, diagrams, and checklists. A complete bibliography on winter living completes this comprehensive guide to design for northern climates.

William C. Rogers
Consultant, Minnesota International Center

PREFACE

Design for Northern Climates is the product of over two decades of research and study. The fundamental concepts of this publication are based on three major areas of research:

1. the physical and mental dimensions of individuals
2. living and working conditions of harsh environments, especially in the Arctic
3. the architectural and urban-planning response to the energy crisis of the mid-1970s

In the late fifties, worldwide research was conducted into the stress that results from an excess or lack of environmental stimuli. I was directly involved in designing experimental facilities for social and sensory deprivation studies. These experiments confirmed that a certain type, range, and intensity of environmental stimuli are vital to human physical and mental health. In addition, it was found that these stimuli can be manipulated to affect human perception of space and time.

For example, one set of experiments involved distortion of time perception. Subjects were led to believe that the skylight in a small windowless room was real. However, what appeared to be a skylight was in fact a light fixture—a complex system of fluorescent and incandescent lights and color filters. By manipulating the intensity and color of light, it was possible to simulate natural light as it changes through the day. These changes affected the subjects' perception of time, and their subjective assessment of time could be altered by deliberate changes in the external environment.

Ten years later, in the late sixties, while working on a variety of research-planning projects sponsored by the Department of Indian Affairs and Northern Development, I studied living and working conditions in the Arctic. I was thus able to comprehend more fully the magnitude of environmental stress that can result when a certain combination of environmental stimuli are absent from or present in one's surroundings.

This experience led to the realization that humans in harsh environments must be accommodated in environments that simulate as closely as possible the original rich, stimulating conditions of the human race's hypothetical, ancestral home, East Africa. These ideal conditions will be referred to as our "primordial climate" or "thermal paradise" in the ensuing chapters.

The second realization was that all human-made ecological systems, such as houses and communities, located in extremely cold regions should be as autonomous as possible. Their capacity to utilize ambient energies should be maximized in order to minimize dependence on distant energy sources. The third and rather disheartening realization was that a direct, rational response to the harsh climate would most likely fail. Interviews with arctic dwellers continually revealed that cultural prejudices, a universal attachment to southern values, and a southern spatial and architectural vocabulary may be the most difficult obstacles to overcome before introducing and implementing genuinely northern solutions.

Another decade passed and the energy crisis hit the Western world. The all-encompassing journalistic phrase *energy crisis* served to cloud the true nature of the problem. In reality it was a supply crisis involving a single resource—a problem repeated and overcome throughout human history. In fact, any living organism that is completely dependent on only one type of energy source has always been vulnerable. Taking into account well-documented macroenvironmental dynamics, any highly specialized organism is in danger of extinction if its survival is contingent upon the availability of a single energy source.

The relatively short energy crisis in the seventies had a significant impact on the evolution of new concepts concerning human settlements, and it helped expand our understanding of the mechanics of the natural environment. Like everyone else at the time, architects, followed by planners, searched to make houses and communities more energy efficient and thus less dependent on distant and often unreliable sources. The original challenge at its simplest was to maintain the existing quality of life while lowering costs. It soon became apparent, however, that not only was this an attainable goal, but that, at a much lower cost, the quality of life could even be significantly improved.

The initial response to the sudden scarcity of resources was an attempt to increase efficiency. All human-made systems were scrutinized in an effort to eliminate waste. It became evident, however, that within the chain of interdependent natural and human-made systems, the total elimination of by-products and wastes is neither possible nor necessarily desirable. The concept of throughput, which broadened the narrow input-output focus and incorporated environmental issues into the energy-efficiency agenda, was therefore introduced. More efficient links between interdependent systems were now sought; the waste of one system would fuel another. Energy efficiency came to be directly linked with overall economic performance; cost calculations, previously consisting of construction, maintenance, and upgrading, now became "life-cycle costing," consisting of inception, development, a productive period, obsolescence, and the cost of burial, or even better, the cost of recycling.

In the planning and designing of human settlements, costs were also measured in relation to gains or losses in the quality of life. This inevitably led to investigations concerning the nature of well-being. Since a lack of well-being is more obvious and manifests itself in a variety of health problems and social pathologies, one way to assess a built environment is by the absence or incidence of such problems.

The other possibility is to compare the ideal or primordial environment, suited to human physical and mental dimensions, with the built environment. This presupposes knowledge of what constitutes ideal conditions and the capability of expressing it in quantitative terms. Using comfort parameters as a yardstick for the evaluation of built environments became a new and exciting possibility. For example, the human comfort zone, usually described as a particular combination of air temperature, air movement, and humidity, was compared with the air temperature, air movement, and humidity inside and outside of buildings in a given environment. Having specific goals for maintaining human comfort makes planning for livability during all seasons possible. In northern latitudes, where the harshness of the climate requires that special attention be paid to the cold season, the "livable winter city" concept was born and became increasingly popular among planning and design professionals.

The City of Hamilton Planning Department was one of the first in Canada to address energy and planning issues directly; when the Planning Department received seed money from the Ontario Ministry of Housing, serious research began. This grant initiated a period characterized by the systematic development of new planning methods that, under the directorship of Victor Abraham, lasted over a decade. At the very beginning, it appeared that for the purpose of creating practical planning tools for a municipality, it would suffice simply to adopt and adapt the already well-known theories, principles, and techniques developed by Victor Olgyay and Ralph Knowles. These pioneering ideas were finally amended, modified, and expanded to such an extent that a new plannning and design system resulted.

The occasional papers and reports published by the City of Hamilton Planning Department attracted domestic and later international academic and professional interest. An international network, which included approximately three hundred participants, was established, leading to an intensive exchange of information. The original purpose of the planning and design method was to assist local municipal planners. International interest came as a surprise. The interest from countries with hot climates and in the southern hemisphere was quite unexpected. It became obvious that, regardless of the context within which they were described, these methods and techniques were universal in their application.

After an initial printing of only 200 to 400 copies of each report, the municipality needed outside funds to publish additional copies. Although the end of the energy crisis terminated much of the cash flow from external sources, interest in and demand for the reports continued, along with the suggestion that the work deserved wider circulation. Thus, the idea was born of producing a book to summarize and organize the experimental planning techniques developed under the auspices of the Hamilton Planning Department and the sponsorship of the Ontario Ministry of Housing, and to include some of my research work at the Czechoslovak Academy of Sciences, the University of Toronto, and the results of all the largely unpublished research work that has been in progress since the mid-seventies.

Although the ultimate responsibility for the content and format of the book lies with me, I cannot emphasize enough that it is the product of a collective effort by those who were directly or indirectly responsible for the actual research. I have received help and encouragement from so many people, that it is impossible for me to express my gratitude to each one individually. They include federal, provincial, and municipal officials; members of university faculties in Canada, the United States, and other countries; practising professionals, researchers, developers, my colleagues and friends in Hamilton, the researchers and staff of the Royal Botanical Gardens in Hamilton, and the researchers and staff of the Canadian Climate Centre, Environment Canada, in Downsview. Thus, I owe my sincere thanks to all the people who, during the time of my research and investigations, as well as during the preparation of the manuscript, offered their encouragement, advice, and technical assistance.

Special gratitude is due to my mentor, Victor Abraham, Director of Planning, who as an architect and planner with international experience, kept me on the right track with his knowledge and managerial skills. Because the Planning Department is the smallest one per capita in Canada, it required extraordinary administrative skills to sustain my research for such an extended period of time—particularly in the absence of external financial support during the recent recession.

I also wish to express my appreciation to Kewal Bhargava, who established the graphic style and prepared all of the art for this book. I would like to extend my sincere gratitude to Mac Carson, who, when he was Commissioner of Social Services, first gave me the idea of writing a book, offered advice, and was on hand in moments of disorientation and panic.

I am indebted to many members and friends of the Livable Winter City Association: Jack Royle, Toronto, the founder of the Canadian Livable Winter City movement; Dr. William C. Rogers, University of Minnesota; and Arni Fullerton, architect and planner, Edmonton, Alberta, who provided both inspiration and direct encouragement for which I am most grateful.

I would like to thank George Banz, architect and planner, Toronto, for his enthusiastic interest in my work and for his lively and inspired conversations. I also owe thanks to Simon Miles, President of the Ontario Conservation Council, not only for his continuing interest, but also for his critical review of my concepts. I am indebted to Dr. Cezar Heine, Inuvik, Northwest Territories, and Dr. Robin Eastwood, Clarke Institute of Psychiatry, Toronto, for providing me with the most recent research materials related to environmental health issues.

Again, I must apologize for my inability to provide detailed credits to all those with whom I was in communication. I would like to express my sincere thanks to all of my reviewers for their encouragement and appreciation of my work. Sincere gratitude is extended to Professor F. Curtis, University of Regina; Professor M. Hough, University of Toronto; S. Mendiratta, Mohawk College; J. Hix, Architect, Toronto; Professor M. J. Macalik, University of Nova Scotia; Professor D. Schramm, University of Wisconsin; R. Robbie, Architect, Toronto; Professor A. Waterhouse, University of Toronto; Dr. W.R.D. Sewell, University of Victoria; Professor A.F. Burghardt, McMaster University; Professor W. Smith, Nova Scotia College of Art and Design; Professor J.M. Wolfe, McGill University; Dr. P. Broberg, Landskronagruppen, Sweden; K. Greenberg, Director of Urban Design, Toronto; W. MacKay, Director, Ministry of Municipal Affairs; T. Boutilier, Planner, Kitchener; C. Williams, RWDI, Guelph. I am also indebted to Dr. Audrey Douglas, Toronto, for her critical review and for her constructive suggestions regarding the original manuscript.

I would also like to extend my sincere thanks to Donna Rossler and her associates at Van Nostrand Reinhold who devoted much energy and thought to my work and whose valuable suggestions are greatly appreciated.

Finally and most important, I would like to express sincere thanks to my long-suffering wife, Alina, for endless hours at the word processor, and for her patience and moral support. She contributed extensively to the preparation and editing of the original manuscript. She tamed the stubborn computer and put up with many obstacles and difficulties surrounding the preparation of the manuscript. Without her dedication and help, this book would not have been possible.

INTRODUCTION

The intent of *Design for Northern Climates* is to help design professionals maximize indoor and outdoor thermal comfort through physical design, particularly in regions with extended winters.

A long-recognized role of the architect is to create a comfortable indoor climate. Urban planners, however, have only recently begun to consider the impact of environmental stress—of which weather is the major component—on urban planning and design. Thus, it is of paramount importance to understand and anticipate the impact of the external climate, and of the sun in particular, on building construction, and conversely, the impact of building construction on the internal and external climate. This is particularly crucial in northern latitudes where environmental stress plays such an important role in human physical and psychological well-being.

Historically, two major factors account for the apparent lack of consistency in the attention paid to local environmental characteristics in the design process: economic as well as cultural and societal factors.

Because of the low cost and easy availability of energy, western industrialized society was able to shield itself from the whims of the weather by simply adjusting the thermostat. The shape, form, and orientation of buildings became irrelevant, and the process of selecting building materials and construction methods was reduced to simple load-bearing considerations. Shape and orientation ceased to be important to human comfort within a building. In addition, the low cost of energy led to an unprecedented development in long-distance transportation networks, which enabled the west to depend on resources that were not necessarily available in the immediate vicinity. Wires, pipes, cargo boats, cargo trains, and transport trucks temporarily eliminated the necessity of searching for local alternatives.

The second circumstance is related to societal and cultural factors. Historically, protective shelters were the product of local climate, materials available locally, the level of technical skills and craftsmanship, and lifestyle or survival techniques. Once a successful regional solution was developed, it was used to ensure the comfort and welfare of subsequent generations; the process of satisfying basic human needs became a need in itself. It is therefore easy to understand why a Turk is generally most happy in a Turkish-style house, and the English insist on English-style houses.

When a culturally programmed group resettles in an area whose environment is significantly different from the original environment, the satisfaction of secondary, or cultural, needs can work effectively against the satisfaction of primary human needs. One is unlikely to find a Turkish-style house in the Northwest Territories or in Alaska, but it is equally difficult to find many houses or communities there that are designed and constructed in direct response to local environmental circumstances. Many areas in northern latitudes were settled by immigrants from the south, who had definite requirements for the home and community. These fossilized southern houses, streets, and public spaces keep appearing despite the high cost of energy, municipal services, and the cost (more difficult to measure) of human psychological and physical discomfort. Culture is a double-edged sword: it reinforces survival in a predictable environment, but it can impede progress in times of change. Thus, one of the most difficult tasks for any designer is to provide an optimal environment for human activities within the constraints of the predominant cultural norms.

Because of the close relationship between environment and human physical and psychological comfort, any successful design must consider the external climate (the sun in particular) and how it affects the performance of the building envelope and the internal and external environment. Although the external climate is normally the starting point for determining the design, envelope, general layout, and construction of a building, the actual construction instantly generates changes in the external climate. This interaction makes the designer responsible, to some extent, for the internal and external climate.

Human-centered, climate-sensitive interior design, building design, urban design, and urban planning are absolutely dependent on a synergy of professional disciplines. Both the architect and urban planner must cooperate to assess how proposed buildings will affect the surrounding climate and how this climate will affect the building design.

Blanket regulations or solutions are ineffective, since no two buildings, building sites, streets, city corners, city blocks, or neighborhoods are the same. Most communities have a wide range of structures that are unique in detail and that coexist with the topography, climate, streets, and spaces to form highly complex relationships. Just as the physical form of built-up areas is complex and unique, so is the urban microclimate, a composite by-product of the intricate geometry of horizontal and vertical surfaces, the shape and orientation of buildings, street alignment, the pattern of urban open spaces, and the heat and pollution generated by human activities. Each wall, roof, building, pavement, street, courtyard, or park generates its own conditions, which are a product of the interaction between the given urban entity, its physical surroundings, and the broader regional weather pattern. In order to preserve and enhance indoor and outdoor comfort, climate-sensitive design must consider these interactions, particularly in northern latitudes.

The control of climatic elements through a variety of design interventions must always be measured against a number of social, engineering, and economic constraints. Only a unified approach to the resolution of conflict yields solutions that will ultimately be in the best interest of the community as a whole. As a result, it becomes critical to the design process that a practical methodology and tools that provide effective guidance are made available. The methodology described in the ensuing chapters will provide the designer with a way to prepare a cost-efficient, climate-sensitive layout for an individual shelter, subdivision, neighborhood, village, or city. The climate-sensitive planning and design approach described in this book is not intended to replace conventional practices but rather to add to them and to focus the designer's attention on areas that have traditionally been neglected.

An environmentally sensitive design process requires a wide range of meteorological information, and these different sets of data require different methods of interpretation. Thus, a format for tabulating selected weather components on a regional scale—a synopsis of mesoclimatic elements—was developed to help the designer understand the dynamic weather system governing a given area. The broader meteorological properties of a region must be superimposed over the local topography, since even relatively slight changes in altitude, slope orientation, and slope gradient can generate significant differences in radiation pattern, temperature, airflow, and the general microclimate. In high latitudes and/or high altitudes especially, major differences in climate (including ground temperature, moisture distribution, and wind patterns) can result from minor topographical variations and must be carefully considered in the process of deciding land-use allocation, densities, transportation systems, road patterns, the cost of municipal services, and building design. On a smaller scale, general weather system modifications originating from individual buildings and building complexes must be considered.

Design for Northern Climates is divided into four parts.

The first part comprises chapters 1 and 2, which place human dwellings within the context of a dynamic environment. The complexity of the terrestrial environment is explored as are the challenges of surviving potentially deadly surroundings.

The second part, chapters 3, 4, and 5, introduce a planning and design methodology for low-density residential neighborhoods. Guidelines are provided for designing according to sun movement, considering solar access as well as shadow impact. A neighborhood case study and an analysis of its results conclude this part.

The third part, chapters 6 and 7, is devoted to high-density urban environments. This part examines the genesis of a unique weather system, called *an urban heat island,* and the design challenges of densely developed urban areas with high-rise buildings in regions dominated by a low sun and long shadows.

The fourth part includes the final chapter, which is devoted to *Solar Templates,* a simple and indispensible tool that is described within the context of a northern environment (the information can be applied, however, to any latitude where the sun shines, be it northern or southern).

Information presented herein has an empirical bias; geometry is approached in a deliberately concrete way. Moreover, in an effort to make this book a practical reference manual for the design professional, all mathematical and scientific formulas are reduced to a bare minimum. Readers without a mathematical background should be able to follow the development of the subject, although a basic knowledge of math and geometry will be helpful. The aim is not to determine exact performance via engineering-style calculations, but to maximize local opportunities with intelligent and sensitive design.

PART 1

THE SHELTER

CHAPTER 1

ENERGY AND NATURAL SYSTEMS

Biology leaves us with the impression that when the level of environmental stress becomes intolerable, certain living organisms elicit one of three major responses. The ultimate response is extinction. Others continue to live, but undergo accelerated mutation as a means of adapting to change. Still others respond with attempts to replicate their original environment, essentially denying the existence of changed conditions. The latter is the most demanding alternative and is the response of humans—their passport to the universe.

With this formula for survival, humans are able to inhabit practically any environment on the surface of the earth. It is more difficult, however, to duplicate primordial conditions in harsh environments, which requires a great deal of ingenuity, energy, and resources. A long process involving experimentation is necessary for developing a successful strategy to inhabit hostile zones.

LIVING SYSTEMS AND TERRESTRIAL ENVIRONMENTS

The search for new and better means of accommodating humans in harsh climates must begin with an improved understanding of the terrestrial environment. The observable environment is a product of the interaction of gravity, topography, water, and the gaseous atmosphere. The climatic system's propulsive mechanism is the sun in conjunction with the spinning movement of the orbiting earth. The cyclic exposure to the sun's radiation makes the diurnal/nocturnal and the seasonal or annual cycles the most visible phenomena on earth. Polar regions are subjected primarily to seasonal cycles, whereas the region between the tropic of Cancer and the tropic of Capricorn displays minimal or no seasonal variation (fig. 1-1).

As a result of these two cycles, the elements of the terrestrial environment are continuously seeking equilibrium, and this equilibrium is continously being disturbed. This is manifested in natural phenomena, such as air, that is heated by the sun and rises; the resultant vacuum is replaced by another mass of air rushing in; on the way, this moving air causes waves, rearranges sand dunes, and erodes the shore. A graphic representation of this dynamic reality is next to impossible (fig. 1-2).

LIVING SYSTEMS AND ENVIRONMENTAL STRESS

All living organisms are programmed to survive. Their ability to monitor the environment continuously is a key element of their success. A simplified picture of an individual living system would show the presence of an

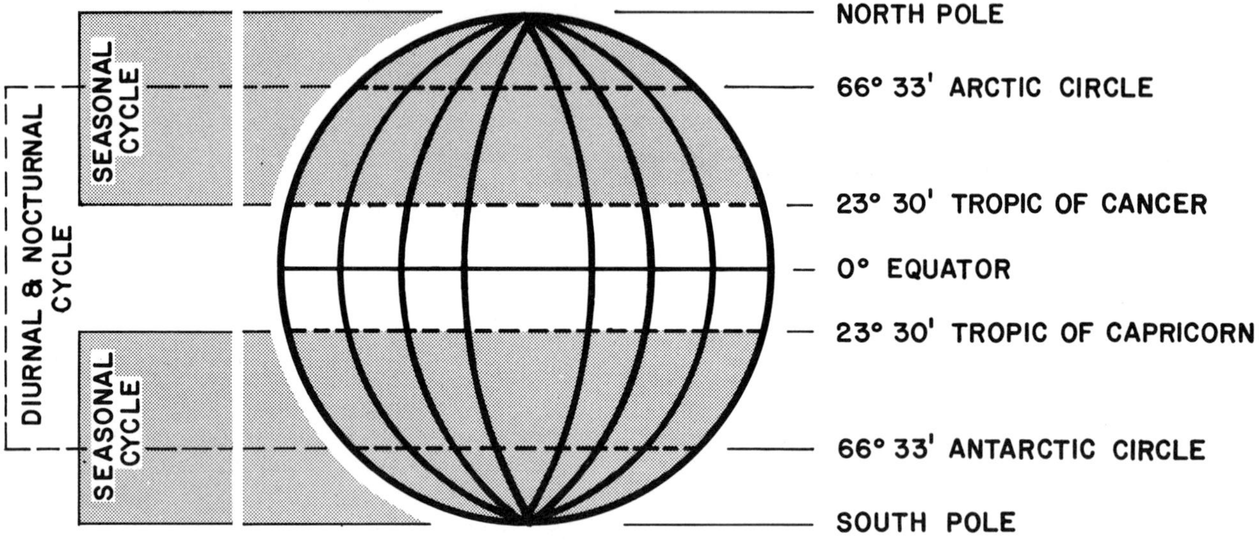

1-1. Seasonal and diurnal/nocturnal variation.

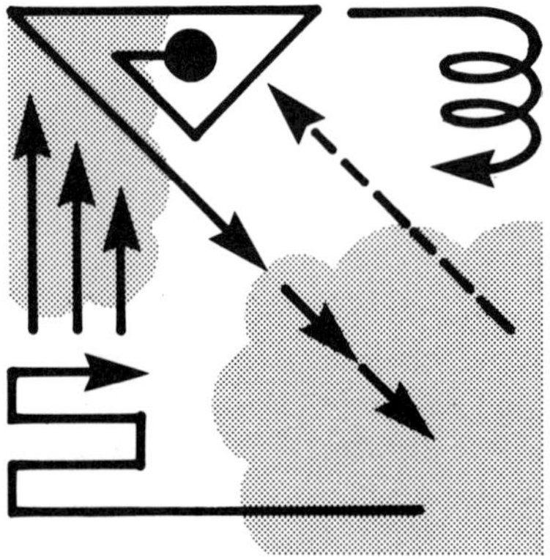

1-2. Elements of the terrestrial environment are continuously engaged in an array of conflicts.

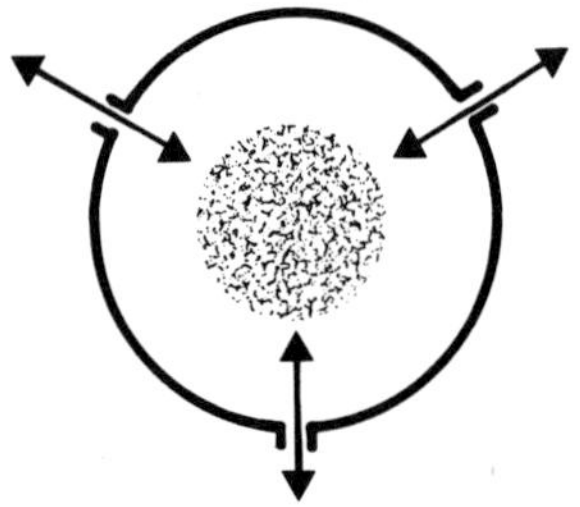

1-3. A living system has an envelope equipped with sensors to monitor the outside environment (information input) and gates for trading energy and matter. Metabolism takes place within a protected environment.

outer, usually porous, skin that is equipped with sensors to monitor the outside environment (information input); it would possess openings (gates) for trading energy and matter. Metabolism, based on a continuous flow of energy and matter, takes place within the inner, protected environment (fig. 1-3). The susceptibility of this individual system to environmental variation stressors is determined by

1. the surface to volume ratio (depending on the form, the same surface area can host different volumes)
2. the physical properties of the skin:
   - its filtering, reflecting, and absorbing capacity
   - its general strength
   - its elasticity (limits, tolerances)
   - its fatigue, life expectancy
   - its ability to regenerate and heal
3. the level of metabolism (the rate at which it devours energy)

In essence, the art of survival lies in maintaining an internal equilibrium in an unstable world. Internal equilibrium implies stability and stability implies predictability, which implies survival of the system over time. The logic circuit is thus completed and the system's survival programming is realized.

The external environment is an inexhaustible source of life-supporting and life-destroying forces. The possible impact of environmental stress on an individual living system can be assessed in one of two ways.

1. Where deformation is confined to its elasticity limits, a system is able to restore its original state—no permanent harm is done (fig. 1-4).
2. Where deformation is not confined to its elasticity limits, environmental forces extend beyond the extreme limits of the system's elasticity and cause a permanent deflection (scars); the system is unable to resist or cope with the stress and restore its lost

balance; environmental forces become destructive (fig. 1-5).

In general, a number of defensive and remedial options are available to an individual living system both prior to and after an environmental attack. The following techniques can be used *prior* to an environmental attack:

1. *anticipation*—a continuous monitoring of the outside environment (the gathering and processing of information and data)
2. *prevention*—adjusting and modifying the form, reinforcing the envelope (fortification), storing energy, boosting metabolism and inner resistance
3. *avoidance*—changing location (if the system is mobile)

After an environmental attack takes place and causes a permanent deformation, a limited number of healing, repairing, corrective, restoring, and upgrading actions are available. Measures that are undertaken before an attack are less costly and less complex than those that must be applied afterward.

Survival Strategies

Living systems in pursuit of survival must favor and concentrate on options related to the preattack situation. Thus, in order to secure a continuity in the exchange of material, energy, and information necessary for survival, certain mobile living systems perpetually seek more advantageous areas. In other words, *escape* from hostile or destructive environmental forces and anticipated catastrophes (a migratory response to environmental stress) has been adopted as a life-style (fig. 1-6).

Where a system has either limited or no mobility, however, the stability of its internal environment can be secured by adopting a different strategy of prevention—shape adaptation and adjustment. A variety of living-system shapes reflect the particular conditions of a given environmental area or region. In addition, celestial mechanics make these environmental conditions changeable within a given area. Certain species therefore change shape according to specific local conditions and seasonal variations—a form-shaping response to environmental stress. But solutions are locale specific; blueprints have a limited use in other domains, and lateral movement is either risky or impossible (fig. 1-7).

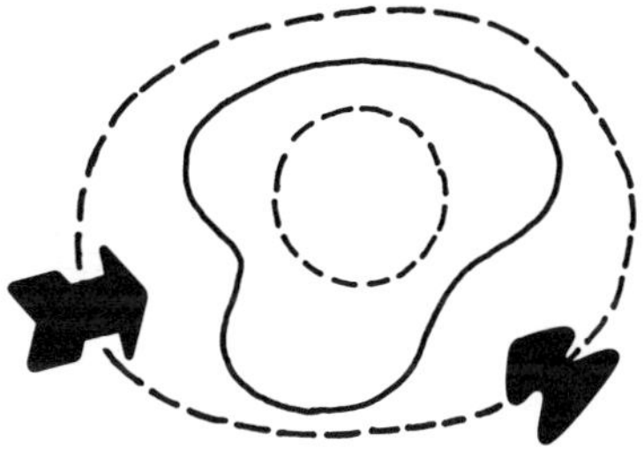

1-4. If deformation of a living system is within elasticity limits, no permanent harm is done.

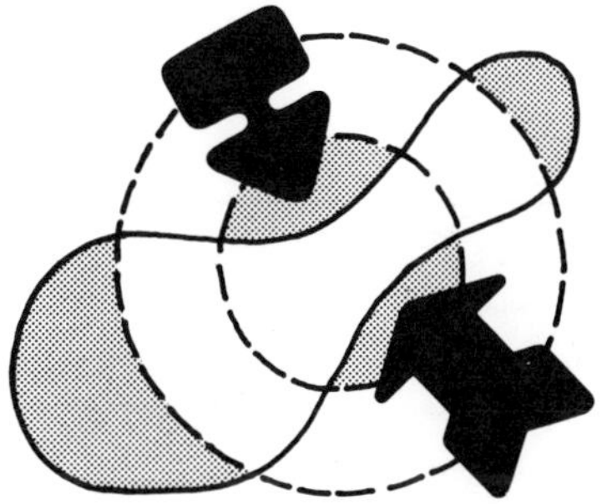

1-5. If deformation of a living system exceeds elasticity limits, a permanent dent results.

1-6. A migratory response to hostile environmental forces.

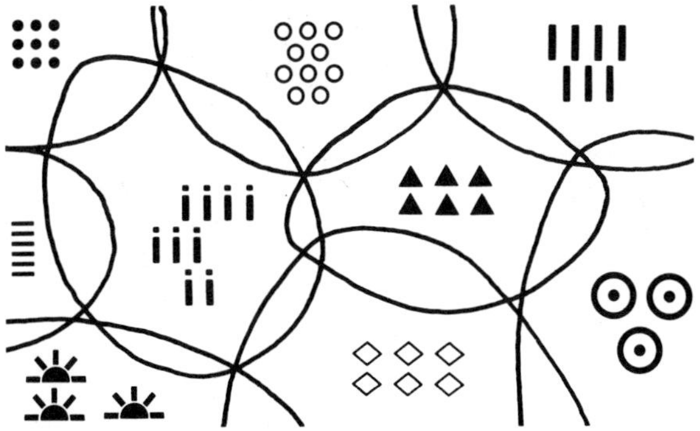

1-7. The shape adaptation of a living system with limited mobility.

HUMAN RESPONSE TO ENVIRONMENTAL STRESS

Anthropologists argue convincingly that humans evolved in a warm climate, and their departure from this thermal paradise introduced many challenges to their survival. Humans (unlike plants and animals who *adapt* to change) developed enlarged brains. They were then able to create a more secure internal environment. Their first response to seasonal change, a response already practiced by some animals, involved the avoidance of unpleasant times and places. Humans moved in response to adverse weather, seeking more hospitable areas (fig. 1-8).

Humans also had to adjust to the diurnal/nocturnal cycle. Thermal comfort at night was secured by erecting simple, portable structures, or by taking advantage of natural shelters, such as caves and large trees.

This nomadic movement or seasonal migration had a definite pattern, one that followed the path of edible animals and that was repeated every year. By following the path of "good times and good places," humans used the land as a solar collector. The art of survival lay in the successful harnessing and conversion of energy from living, solar batteries—plants and animals.

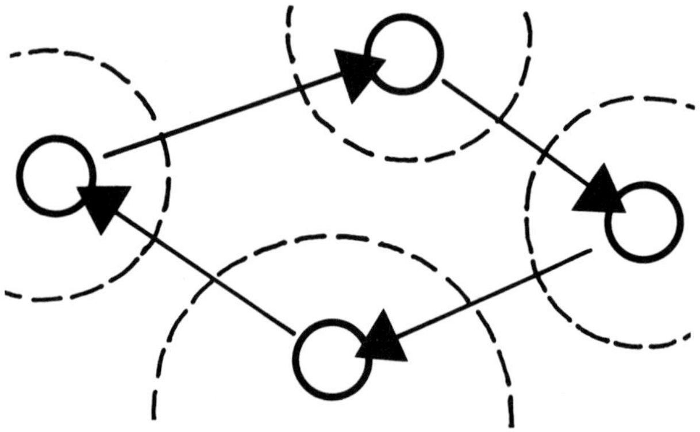

1-8. A nomadic response by humans to hostile environmental forces—avoidance of unpleasant times and places.

Although this mode of existence might appear sufficiently effective and even seem to introduce variety to an otherwise routine existence, statistics indicate that one square mile can support only one nomad in mild climates and not even one in northern latitudes. In the High Arctic, for example, 50,000 square miles were needed to support one Inuit family. Despite an exhausting daily struggle, a nomadic existence could be threatened by weather, and survival remained a gift.

Nomadic shelters, designed and constructed to cope with diurnal variations or with short periods of adverse weather, were relatively simple machines. They were put together from a variety of elements that were found on the spot and were sometimes combined with standard parts, such as furs, ropes, and poles, that were carried along. These early shelters performed remarkably well despite their simplicity.

Step by step and over a long period of time, new survival methods and strategies evolved. Smaller and smaller areas of land were used more intensely. As agriculture developed, this more limited territory was modified to produce a human-made, more efficient solar collector. One square mile of land was now capable of supporting up to one hundred people. A sequence of human activities that directly affected natural processes replaced escape as the primary response to environmental stress (fig. 1-9).

With the introduction of agriculture, however, a new set of problems arose. As humans were anchored to one spot, it became necessary to construct an efficient machine for mitigating environmental variations. Such a machine should be capable of maintaining the primordial climate regardless of outside variations. Thus, the skills necessary to create and maintain an indoor temperature that remained within the narrow range of the human comfort zone were developed hand in hand with the skills associated with supplying food. With the ability to construct a permanent "all seasons" shelter which

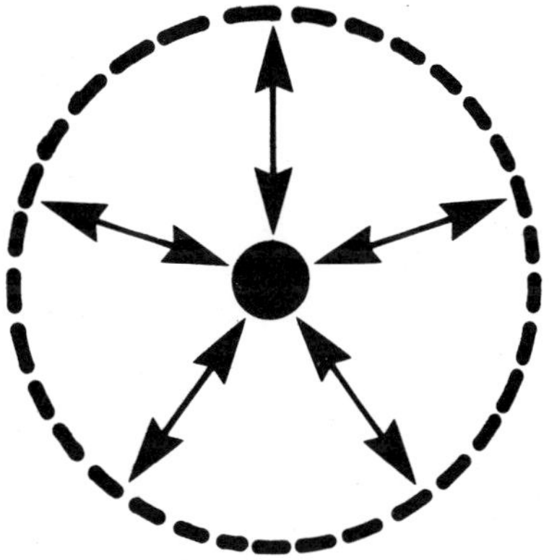

1-9. An agricultural response to the environment consisted of a sequence of human activities that altered natural processes.

imitated the environmental conditions that best suited individuals' biological make-up, came the ability to shake the chains of environmental dependence.

The establishment of a permanent home is necessary for the colonization and exploitation of less hospitable or hostile environments. Thus, humanity has a vested interest in the construction of shelters; the art of construction is refined by generations, and its recipes are embedded in the collective memory.

Shelter as an Autonomous System

Shelter can be viewed as an intermediary between occupants and the natural environment. It can also be seen as a human-made system that surrounds its creator and so might be described as an extension of a human being. Its similarities to an individual living system are remarkable. Its protective envelope or skin, comprises a roof and walls and is equipped with sensors —doors, windows—to facilitate the exchange of information, energy, and matter between the inside of the dwelling and its surroundings.

The internal equilibrium, a replica of humanity's legendary thermal paradise, is vitally important not only for the human occupant but also for the survival of the shelter itself. The internal equilibrium of a stationary system is maintained by

1. introducing energy through a powerful energy source
2. adjusting its shape and construction to local environmental conditions

A shelter is also similar to a living system in that it is susceptible to seasonal and diurnal variations. The study and evaluation of this susceptibility can be conveniently broken down into

1. the surface to volume ratio (the same surface area can host different volumes)
2. the general form/shape
3. the type of material and construction
4. the physical properties of the shell (the exterior and all exposed surfaces)
   - general strength
   - reflecting, absorbing, and filtering capacity
   - elasticity (limits, tolerances and so on)
   - fatigue, decay, life expectancy
   - the cost of upgrading and maintenance
5. the level of internal activity
6. the amount of energy pumped in

Maintaining thermal equilibrium is of paramount importance to the shelter, its occupants, and the community as a whole. The method chosen to secure that bubble of proverbial thermal paradise therefore has profound cultural side-effects.

Shelter as an Energy-devouring System

In rare times when a powerful and easily accessible high-density energy source is available, an alternative to the form-shaping response emerges. Both local environmental peculiarities and the rhythm of diurnal and annual variations can be offset by increased energy and matter (fig. 1-10). The capacity to accelerate the metabolistic processes of devouring and converting energy and matter results in a *Star Wars*-like combat—a head-on collision between energies—but it also allows such systems to function anywhere regardless of local stress characteristics; seasonal variations can be ignored. In addition, the shape of the individual system does not have to be a result of local environmental stress (fig. 1-11).

The survival strategy that depends on securing internal equilibrium by the input of energy is simple and comprehensive. While, on the one hand, this mode of existence can be associated with significant control over the environment and ultimate freedom, the OPEC oil crisis in the mid-seventies, on the other hand, reminds us that any variation or interruption in the flow of energy can leave the system open to attack and can affect world economies. This path is also conducive to conquest, domination, and environmental rape.

The advent of the Industrial Revolution and its power-generating systems gradually enabled us to move toward this sort of "environmental independence." Recently, though, this move has been effected with unparalleled speed. Bulldozers could be used to make all sites equivalent, and oversized heating and cooling systems could transform every house into an individual cell of thermal paradise.

As a consequence of this newly acquired freedom, humanity enjoyed unprecedented delights: it no longer had to derive the shape and construction of its houses from the peculiarities of the local environment; human shelter could now be, and often was, inspired by a picture-book, romanticized past.

This trend promptly became part of the cultural mainstream. With the help of current accounting and tax systems, a human dwelling became a standardized, tradable piece of merchandise. The subsequent buying and selling of houses became a convenient and relatively quick method of creating personal assets and building up one's social status. In the world of frenzied house trading, the resale value is all-important.

As individual dwellings plug into a powerful energy-feeding system, they are able to respond to environmental variations at the touch of a button. The relationship between a house and its natural surroundings ceases to be a problem.

The architect and planner dissolved their marriage and pursued their independent courses. The art of

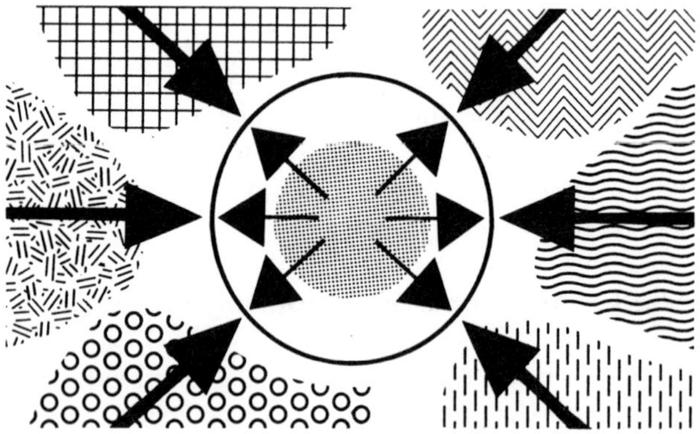

1-10. An energy-devouring response, or the mitigation of environmental stress with an increased amount of energy.

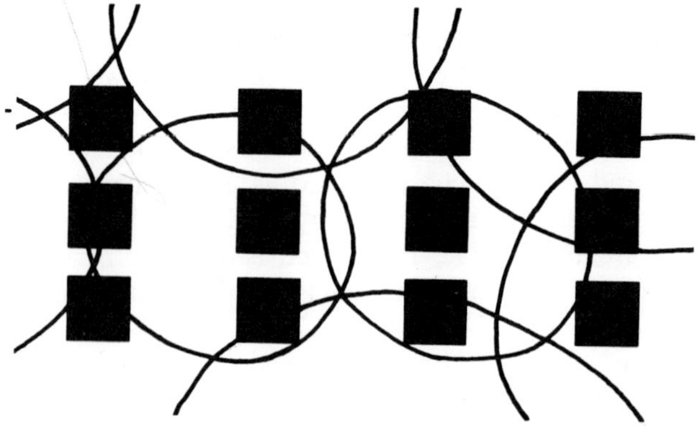

1-11. Energy-devouring systems can be located with no concern for local environmental forces.

building human settlements swiftly degenerated into a set of rules that anyone could follow with ease. Ultimately, there was no real need for a professional—the architect became primarily a draftsman, and the planner a bureaucrat. Teams of bureaucrats, laymen, and paraprofessionals rapidly cloned suburbs throughout the Western Hemisphere.

TOWARD AN ENVIRONMENTAL FIT

In the 1970s, it became apparent that the pursuit of environmental independence—the devouring of energy—was not necessarily the best course to follow. Since the high-energy input mode of response to seasonal and diurnal variation is based on seemingly abundant, accessible, and inexpensive high-density energy sources, the human shelter is extremely vulnerable to changes in its supply system, such as the availability and cost of oil or a power failure.

In addition, a response to cyclic environmental forces that is based on the performance of complex mechanical systems has its limits, such as peak loading, maintenance, susceptibility to human error, and acts of sabotage.

These conditions may make the costs of the energy necessary to maintain thermal equilibrium prohibitively high. This realization led to a reevaluation of the other possible strategy: shape adaptation and adjustment. Within the framework of this strategy, the architect's task was to develop a building form that could adapt to natural forces solely by virtue of its construction and shape. The urban planner's task was to develop a spatial framework within which buildings could maintain their inner equilibrium at the lowest possible cost. Such a spatial framework would automatically provide a predictable relationship between buildings and their natural or human-made surroundings.

It has to be stressed that the relationship between the natural environment, shelter, and human beings is in a state of constant flux and continuous readjustment. If habitation is taken to be the interaction of these three basic components, then a shelter becomes an intricate system for the satisfaction of physical and psychological needs.

By understanding human needs, it becomes possible to change or to modify them and to design houses and communities that serve these needs in harsh climates.

CHAPTER 2

THE IMPACT OF CLIMATE ON INDIVIDUALS

HUMAN HEALTH AND THE ENVIRONMENT

Planners and designers should be especially interested in the impact environmental forces have on individuals' physical and mental health, their personalities and lifestyles. Conventional climate-sensitive design is based on five environmental components: hours of sunshine, air temperature, air movement, humidity, and precipitation. Research in other scientific fields, namely in medicine and psychology, is producing evidence that other environmental factors contribute to the well-being of individuals; these include cosmic and terrestrial radiation, the earth's magnetism, the rate of change of barometric pressure, and the earth's movement in relation to the sun. Under certain circumstances, these components become stressors that act on the somatic and psychic level. The following is a brief review of the relationship between environmental light, rhythms of the sun, human health, and behavior.

Circadian Rhythm

Organs and systems in the human body exhibit a variety of rhythms, the duration of which vary from one cycle per second (the heartbeat) to one cycle per year (the secretion of serotonin). That a human being is a circadian organism is key to understanding the interactions between solar and human rhythms. The circadian rhythm displays twenty-four hour cycles in which the day phase exceeds the night phase by approximately 25 percent.

These phases are regulated by two major oscillators. The day activity and body temperature cycles are controlled by strong oscillators; the weak oscillators regulate the resting and sleeping cycles and are influenced by the strong oscillators. For example, under normal circumstances, the body temperature determines the timing and duration of sleep. Both strong and weak rhythms are self-pacing, but they do submit to external control. In the absence of such a control, they proceed autonomously and their timing and phasing become out-of-synch.

The most powerful and effective synchronizer is the twenty-four-hour light/dark cycle. In regions where the sun's rhythm is grossly distorted or nonexistent, the human organism suffers from a disturbance in its circadian rhythm. The uncoupling of weak and strong components produces the physiological setting for a whole list of pathological phenomena, including irritability, disorientation or confusion, various aches and pains, time and distance disorientation, digestive malfunctioning, a disruption in mental efficiency, an increase in mental disorders, disturbed sleep patterns, a weakened resistance to infection, an increased susceptibility to drugs (for example, an alcohol hangover), a disturbance of the mechanism that allows one to adapt easily to change, and an inability to interact socially.

The human organism always turns toward the predominant available synchronizer. Humans as social beings will therefore follow the social clock in the absence of sun. Another potential synchronizer is artificial light. However, artificial light and social activity share the same limitation as timetables: they can be manipulated by human will. Their capacity to entrain the human circadian rhythm is consequently questioned.

Seasonal Rhythm

Winter dominates seasonal variations as we travel from the tropic circles to the poles. The light-dark ratio is distorted to the point where a new phenomenon emerges—the unique polar day and polar night.

It is essential to keep in mind that constant light suppresses the weak component of circadian rhythm, the one associated with the rest and sleep phase. Similarly, the continuous absence of light counteracts the activity phase; the stress of polar day (a season characterized by continuous light) is not compensation for the stress of polar night (a season characterized by the continued absence of light). As a result, the nothern dweller is under year-round stress, which demands different methods of adjustment.

In regions characterized by a balanced four-season cycle, pronounced environmental stress is exhibited during the fall and winter. Statistical data confirms that a large number of physical and mental disorders have roots in environmental stressors that dominate the last two phases of the annual rhythm. The actual manifestation and virulence of any environmental stressor is, of course, dependent on local conditions, such as an accelerated rate of change in temperature or barometric pressure.

Since humans are continuously migrating and filling less and less hospitable niches on the earth, the type of environmental stress to which different groups are subjected will differ from region to region. Whereas plants and animals cope with new circumstances primarily through the mechanics of evolution, humans continuously recreate the thermal conditions of humanity's lengendary womb.

One can formulate a scale of environmental conditions ranging from utmost hospitality (where there is little or no need for protection) to utmost hostility (where total shielding is necessary). Between these two extremes lies a whole spectrum of specific situations that require special attention and unique solutions.

Planners and builders of settlements located in extremely hot or cold climates must develop a respect for the distinctive nature of a given locality; they must be aware of the available range of solutions and understand their consequences. An understanding of local environment must begin with a description of the three global coordinates: latitude, longitude, and altitude.

Latitude

Regions located between the tropic of Cancer and the tropic of Capricorn display relatively mild seasonal variations. Beyond the boundaries of these two tropics' bands, seasons are gradually dominated by low temperatures. As one moves toward the poles, the cold season becomes more pronounced, and roughly from the 60° parallel onward, the winter season should generally be considered the dominant season, interrupted only by a short, albeit occasionally pleasant, summer (see fig. 1-1).

The daily light-dark ratio becomes increasingly altered, ultimately resulting in polar day and in polar night.

Longitude

Although regions located at the same latitude share the same patterns of sunlight, the regions may be affected by different environmental stressors. For example, in Spain, at 43°N latitude, one would enjoy the sunny beaches of San Sebastian; in France, one could escape the harshness of winter in Nice and Marseilles; in Italy one would visit Rome, Livorno, and Prato; proceeding eastward, one would encounter sunny tourist destinations like Split in Yugoslavia and Varna in Bulgaria; and in the U.S.S.R., one could visit Sochi and Suchumi, long-established resorts that still reflect the opulence of Imperial Russia.

Although located at 43°N latitude, its longitude removes Hamilton, Ontario, Canada from the list of southern resorts. Barring the profound differences in their sunpath geometries, the closest European counterpart to Hamilton would be Oslo at 60°N latitude, a cold northern environment with northern lights, frost, and snow. Indeed, Hamilton at the best of times experiences 150 days between the last spring frost and the first fall frost, the most reliable time being mid-July when the weather is comparable with vacation resorts in Europe of the same latitude. The weather during the remainder of the year is usually less than comfortable, particularly for those susceptible to pneumonia, arthritis, or rheumatism.

In essence, environmental conditions determined by latitude are modified by ocean currents, the proximity of large bodies of water, the movement of air masses, and continental topography.

Altitude

The third and last global coordinate indicates altitude. Zero is the elevation at sea level; moving upward, the temperature drops at a Standard Lapse Rate of 2°C per 300 meters (3.5°F per 1,000 ft.). The following example illustrates the extent of the impact of altitudinal variation on environmental change: in Ecuador the city of Manta, at S 00–59–00 latitude, W 80–42–00 longitude, and 10–0m (32.8 ft.) altitude, is located on the equator at sea level. The yearly average temperature is 25.5°C (77.9°F); but nearby, the top of Mariscal Sucre Mountain (S 00–08 and W 78–30–00), with its peak at 2,866 meters (1,777 miles), is glaciated throughout the year.

This brief description of the relationship between global coordinates and environmental variations suggests the need to identify their interaction with the human organism.

CULTURE AND THE LOCAL ENVIRONMENT

Environmental variations influence many aspects of a particular culture. On a basic level, it is environmental stress that determines how much time we spend indoors, the kinds of activities we perform, and even the way we raise our children. For example, encouraging children to spend time outdoors establishes an independence from adult supervision. Free movement in open spaces permits expressive and volatile behavior. On the other hand, if environmental conditions keep people confined to their homes, familial and social relations become all-important. In the more restrictive indoor environment, children are usually discouraged from displaying too much emotion and encouraged to accept uncontrollable events calmly.

Early restrictions, or the lack of restrictions, have a profound impact on shaping the adult personality, behavioral patterns, as well as social conduct and interaction. The link between general environmental conditions and culture is most clearly manifested in the indigenous design of buildings and settlements and in the form and use of urban spaces.

A change in the climate, in the availability of natural resources, in the ecosystem, or in geographic location upsets the equilibrium between the natural environment and a given culture. In the past, changes usually took place over longer periods of time. This allowed time to adjust to new situations, to alter life-style, and to develop locale-specific techniques for survival (especially in the construction of buildings and communities).

It was the accelerated rate of change that resulted from advanced technology introduced during the Industrial Revolution that severed the long-established

bond between a given environment and a particular way of life. Consequently, the application of inappropriate building designs and construction techniques became widespread because of an abundance of inexpensive energy sources.

Despite expectations to the contrary, however, access to inexpensive energy sources and the luxuries this afforded, such as enclosed transportation and easy access to more hospitable regions, could not fully mitigate the impact of the physical environment on human physical and mental health.

THE DYNAMICS OF HUMAN NEEDS

The energy and skills necessary to cope with a sudden change in the direction and magnitude of environmental stress have profound side effects. The construction of new fortifications affects humans not so much in the physical sense as in the psychological and social senses. The best means of understanding the dynamics of these socio-psychological factors is to study them from the lowest level of evolution.

A prerequisite for survival in cold and harsh environments is the development of design and construction skills that can be used to mitigate environmental stress and duplicate a stable, primordial climate. Once they succeed, these skills become an indispensable part of a survival strategy for all members of the group. The group must also ensure that the skills and knowledge are passed on to the next generation.

Throughout history, humans have developed a very complicated way of adapting to the environment. Shelters were originally built for survival—to satisfy primary needs. The satisfaction of these biological needs was eventually realized, not by confronting the physical environment directly, but as a matter of habit, convention, rules, laws, or even fashion. From the earliest stage of human evolution, the most basic needs were satisfied in a way typical of a particular group. The challenge was not to build just any kind of protection against the elements, but to construct this protection using traditional patterns accepted by the community or tribe. Consequently, methods of meeting human habitation needs were based on the consensus of a group or cultural area and standardized in the early stages of development, even when the environment demanded a change in methods or materials of construction. A break from tradition was seen as a threat to survival.

The Iroquois, for example, lived in a long house—a single dwelling that housed family groups—built from materials (such as branches) that were readily available. The prairie Indians had few trees and a more nomadic life-style than the Iroquois. They constructed a portable teepee from locally available skins. The Inuit developed a hemispherical snow house for protection against wind and cold.

When a group exhausted particular building materials, the shape of the structure and the method of construction often copied the original as closely as possible; the shelter now had to satisfy not only a primary need, but also a secondary need. For example, ancient Greek construction was originally done in wood. When the Greeks began to build monumental structures from stone, they continued to use principles of post and beam construction, including fastening devices (nails and pegs) carved into the stone.

As a result, this petrified process of satisfying human needs became a need in itself; thus, historically, a member of the Iroquois band needed a house of the Iroquois type, and an Inuit needed a house of the Inuit type. Beneath this secondary need, the primary or basic need (protection against the rain, snow, wind, and sun) was concealed. All offspring were systematically and consciously programmed to accept this value system.

In a complex social system, basic biological imperatives become overshadowed by cultural needs. These emotionally determined imperatives, which evolved from the biological requirements of an individual or family, became the source of municipal by-laws, building codes, regulations governing the training and certification of craftspeople, and an array of new requirements that shape human behavior and social order. Secondary needs may even work *against* primary needs. For example, fashion promotes shoes that cause blisters, clothes that provide inadequate protection against the cold, furniture that can injure the spine, and houses that do not maintain homoiothermic conditions. Satisfying secondary needs makes it difficult to introduce architectural designs and concepts of urban planning that respond directly to the harsh northern climates.

EVOLUTION OF THE HUMAN SHELTER

Human shelter was developed primarily to mitigate periodic and incidental climatic stress, to reproduce the thermal conditions of the primordial climate, and to protect against a variety of dangers and threats from the natural surroundings, such as predators and enemies. The final shape, size, construction, and internal layout of a human shelter is determined by the regional climate, learned and inherited forms, experiences, and a variety of social factors. In a broader sense, the shelter itself is a product of the human capacity to regulate its survival strategy through given and continuously changing circumstances. Its evolution is driven by the principle of seeking the best possible satisfaction of needs.

The simplest forms of shelter include fences (either upright or sloped), walls without a roof, and roofs without walls. A completely enclosed shelter is, of course, the most efficient system for creating a desirable climate in colder regions. These simple shelters were originally constructed only as shields against extreme environmental stress, they were not necessarily designed to satisfy other human needs.

The Arandas, for example, use their hutches exclusively as protection against the wind and rain; their fire is located outdoors. However, in other locations, prolonged periods of bad weather can force individuals to cook, eat, bathe, socialize, manufacture, repair, and maintain tools and clothing, as well as perform administrative and management functions within their shelters. Indoor cooking, in turn, creates new needs. Whereas the Arandas, who cook outdoors, have simple kitchen utensils and stones for grinding, the early Aztecs, Iroquois, and Incas had not only pots and pans but fireplaces, chairs, and tables, so they could eat comfortably. In addition, their eating space was separate from living areas, and they had storerooms for food. With an escalation in environmental stress comes the evolution of a simple shelter into an intricate system of habitation.

From a global perspective, it is clear that the northern form of habitation was determined by the need for protection against cold, wind, rain, and snow. In the south, a form of protection against other kinds of extremes such as excessive heat and sunshine, was developed. The classic Roman or Arabic house, with its internal court full of shadows and a fountain, is an excellent example. Similarly, ancient Egyptian houses were exposed to the north, the direction of the prevailing cold winds. Their flat roofs, sheltered by a variety of wooden structures, were utilized for social and recreational activities during the day as well as after sunset. The humming fountains of the south have their counterpart in the flickering flames of northern fireplaces.

From time to time a change in temperature, food scarcity, social or economic upheaval could seriously disturb an established equilibrium. A period of tension usually follows and is accompanied by the mobilization of available mental and physical forces to restore balance. One example of accelerated evolution due to a shift in environmental stress is the metamorphosis of the medieval English manor as effectively described by James Burke in the television series, *Connections*.

A Medieval English Manor

An environmental stimulus may be helpful, if not necessary, for the advancement of survival techniques. Consequently, resettling or migrating into harsh weather regions or being exposed to adverse climates could be viewed as advantageous. The benefits of a rather severe environmental assault, one close to the limits of tolerance, can be illustrated by describing the consequences of a climatic minicatastrophe that descended on Europe in the midthirteenth century. After centuries of relative warmth, a cold wave that lasted almost two hundred years profoundly altered living and working patterns in Europe. These two centuries were characterized by short summers and long, severely cold winters.

Innovation was necessary to survive this new type of winter. The well-documented metamorphosis of the medieval English manor is an interesting example of the innovation that resulted from environmental stress. The early feudal system was a self-sufficient organization of pooled resources. In the center of the manor house was a "great hall," a large room with an open fire and a hole in the roof, that was used for sleeping, food preparation, and eating and allowed for no privacy.

The first reaction to the advancing cold was to introduce a chimney, which kept a fire going, used heat more efficiently, and released smoke and sparks into the environment at a safe distance. The design and construction of chimneys were continually refined, since a well-functioning chimney requires a good working knowledge of draft principles. The knowledge gained through these refinements had dramatic impact on the arts of glass and iron making, fundamental skills in the Industrial Revolution of the eighteenth century.

Chimney construction was a challenge to craftspersons, who needed to design structures that could withstand thermal shocks, were airtight, and could prevent fire from reaching the wood constructions. The chimney, and later a group of chimneys, became the structural and functional spine of the house.

Environmental stress had an impact on house construction as well. Open patios disappeared and buildings were closed off and built stronger to withstand harsher storms, rain, wind, and snow.

The newly acquired thermal comfort gradually became an improvement over the internal environment of the warmer period. A cluster of chimneys allowed the house to be divided into a number of separate, heated rooms. Ceilings were lowered to decrease the internal volume of air to be heated. More sophisticated food preparation took place in a separate kitchen. Heated bedrooms offered privacy. Romance and love poetry began to flourish in the cold darkness of northern Europe. New construction techniques made indoor bathing possible, which led to an increased appreciation of a clean human body.

Reading and paperwork were popular indoor activities during the long winter and had a positive impact on general management and economic development. The invention of the button and knitting is also attributed to this two-century cold spell, and the attention paid to clothing resulted in garments that retained heat better. The newly acquired thermal comfort meant smaller rooms and more intense contact between individuals. This, along with improved hygiene led to better manners and a more refined life-style in warm and commodious houses.

The paintings of that period convey a sense of interior comfort. Plaster, first used only as a sealant around chimneys, was gradually applied to all walls, which were painted in rich, warm colors. Later, wood paneling was introduced to improve further the thermal comfort of occupants. Blankets, originally hung to block drafts, became rich wall tapestries, improving aesthetics and the indoor climate. An improvement in the design and craftsmanship of furniture and other everyday objects is also evident in paintings of the period. Footstools were used to keep feet off the floor, warm water was stored in jars for washing hands and feet, towels hung on racks. Windows were backed by internal wooden shutters, and both were divided into individually openable segments, so that ventilation, illumination, and

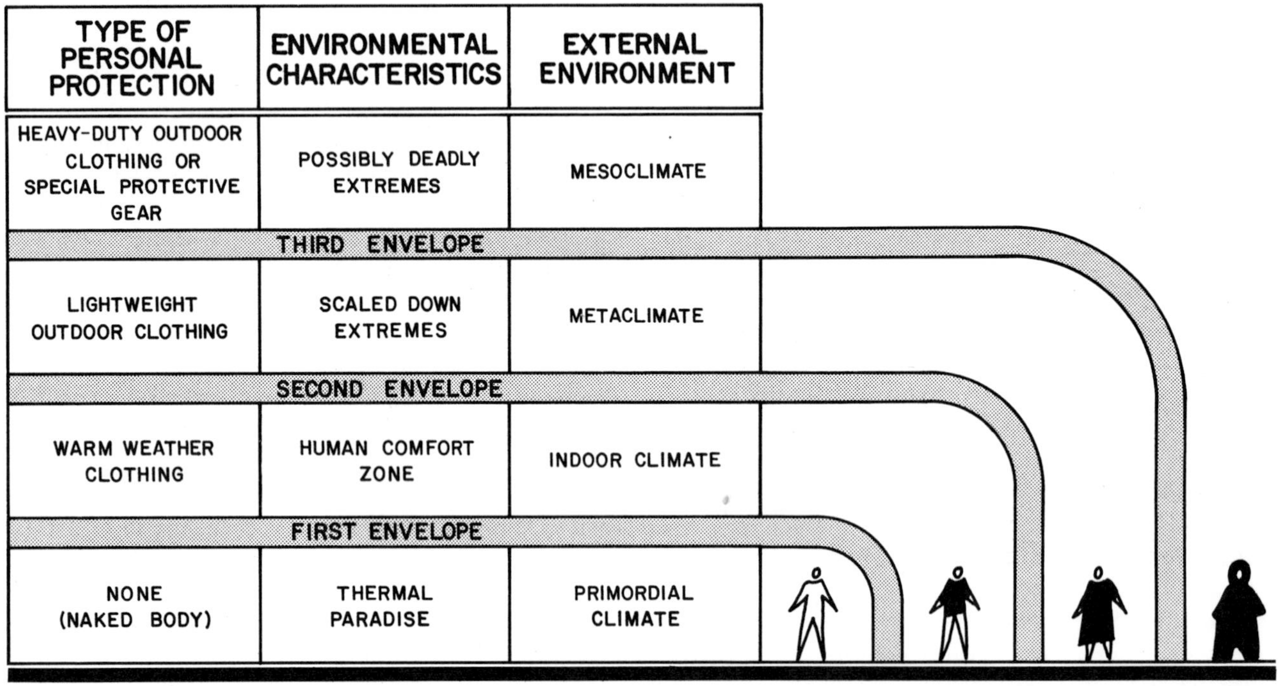

2-1. The human defense system against harsh climate.

thermal insulation could be adjusted to suit the particular needs and desires of the occupants.

The story of environmental stress and the metamorphosis of the ancient manor dwelling is one example of how intensive environmental stress can trigger a set of responses that accelerate both social and technological evolution and ultimately improve human welfare.

The Individual House

There is no single blueprint for an acceptable northern house. Any attempt to develop typology along latitude lines is doomed to fail. Only houses built in areas displaying similar environmental variations would share similar design principles: the same variations, not the same latitude, determine the design solution. The design response to harsh climates is locale-specific, although methodologies and approaches are transferable.

The design of the northern house must serve the same basic needs as its southern counterpart. In addition, the northern house must counteract the effects of a harsh external environment. In principle, the external environment must be transformed, within the house, to become the primordial human environment.

In fact, the primordial thermal environment is one that allows individuals to exist comfortably without clothes (fig. 2-1). Except for the bathroom, which is warmer, we keep our houses and workplaces slightly cooler than humanity's legendary paradise, since cultural conditioning precludes the pursuit of a reasonable assessment of the cost of manufacturing, distributing, and maintaining indoor clothing as compared with the cost of raising indoor temperatures a few degrees and thus eliminating the need for any clothing indoors. However, for the purposes of this analysis, it is assumed that the desirable indoor environment is equivalent to the primordial thermal environment, although it is actually equivalent to the primordial thermal environment with an allowance made for clothes worn indoors.

For most of the world, the roof and walls of the conventional house is a satisfactory solution. In moderately cold climates, a well-insulated envelope and carefully designed fenestration is satisfactory. In extremely hostile environments, however, other solutions are emerging. In one, the primordial climate is separated from the outside environment not only by a wall but by a generous space with its own unique climate. For example, if a glass box were attached to a house, and particularly if it were located on the sunny side, and the adjacent wall and floor had heat-absorbing and storing capacity, then the box would become a sun space with its own climate. The internal climate of the sun space would serve as a buffer betwen the indoor and outdoor climates.

Such a concept might evolve even further: a glass dome could be constructed to cover the entire house (fig. 2-2). Depending on the severity of the mesoclimate (the climate encompassing a limited geographical area), the variations in the metaclimate (the climate within a town covered by an envelope) might not fall below freezing. For example, grass might remain green throughout the winter. Pursuing this line of investigation further, we discover that according to the laws of physics, the larger the volume enclosed, the greater the buffering capacity: the metaclimate has a tendency to resemble outside conditions less and less. A sufficiently large, enclosed volume of space develops its own weather pattern, including air movement, condensation, and even cloud formation. The metaclimatic variations and weather pattern are propelled by two independent impetuses: the metabolistic heat as well as other by-products of human activity and variations in the mesoclimate.

Large volumes, such as would be required to enclose whole cities (fig. 2-3), also have an enormous load-

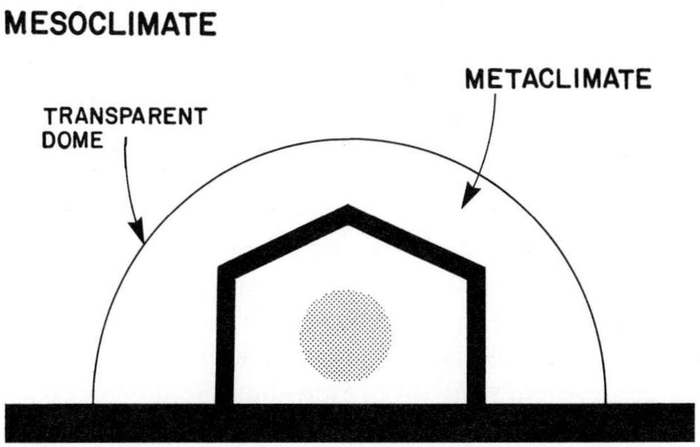

2-2. A membrane structure can be extended to cover an entire house.

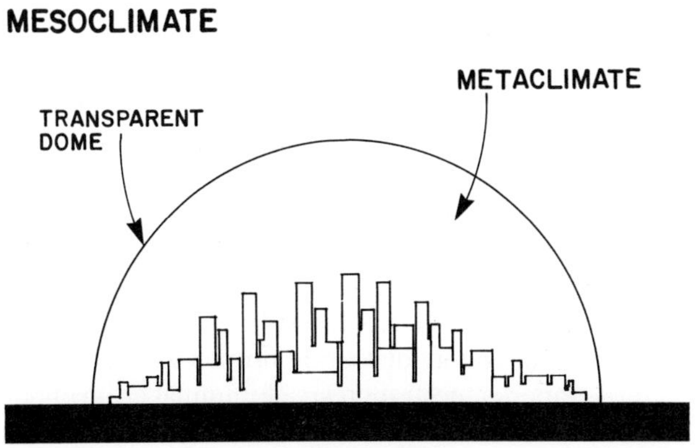

2-3. A membrane structure can be extended to cover an entire city.

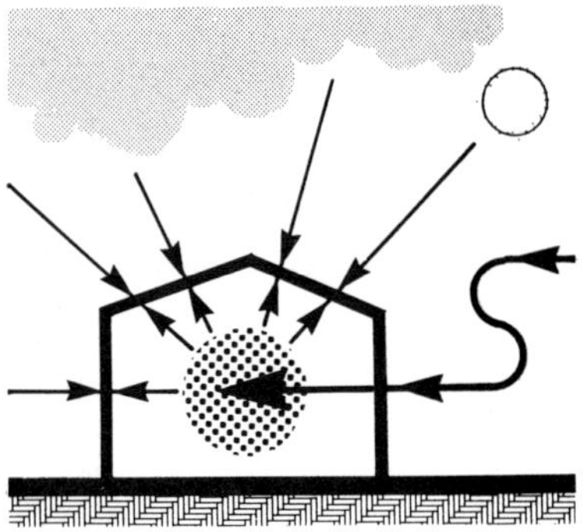

2-4. A conventional house.

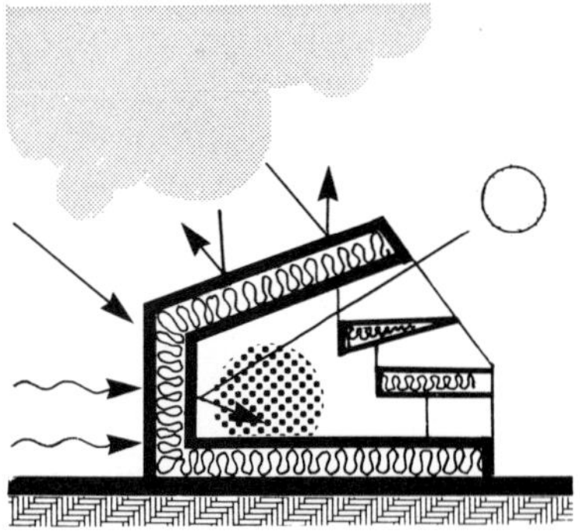

2-5. A passive solar house.

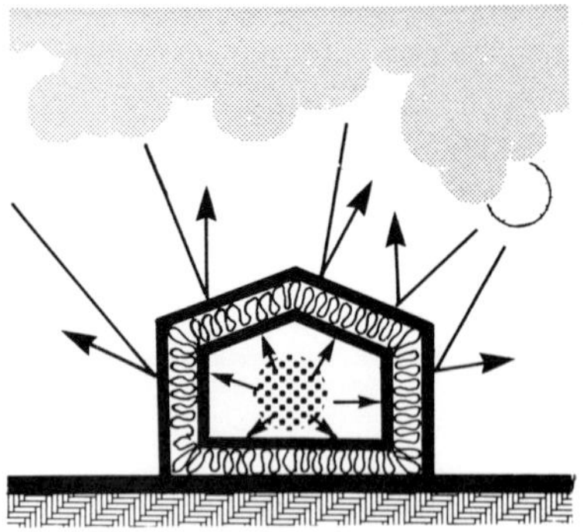

2-6. A superinsulated house.

carrying capacity. A temperature differential of a few degrees can assist in holding, or even hold, a lightweight enclosure in place. Any subsequent increase in volume would make the dome airborne. In such an extreme case, the engineering task would be to design anchors rather than weight-bearing foundations.

Examples of generic forms of metaclimatic buffers are an unheated sun space beside a southern wall, a double-shell house, or a roofed street. There are no major engineering or economic problems in implementing metaclimatic concepts on a large scale, but resentment of these unprecedented methods of human accommodation may be expected, since individuals are frequently unwilling to break from tradition.

Cultural conditioning is one of the most difficult obstacles to cost-efficient, comfortable housing designs in northern regions. As previously stated, the prime function of a house or building is to maintain at minimum cost a thermal milieu within the limits of the human comfort zone. Larger buildings with a lower surface to volume ratio offer thermal comfort at less expense. Large buildings, such as apartment buildings and row houses, or at least duplexes, are most cost effective. Experience shows, however, that most cultures place greater value on the single-family house.

Current housing production can be divided roughly into three groups.

1. CONVENTIONAL HOUSES are heated by energy imported from distant locations (fig. 2-4). They are very popular but very expensive to maintain and so are most appropriate in areas with inexpensive and accessible energy. An efficient delivery system or central distribution grid is a prerequisite. These houses are southern transplants held together by oversized, energy-devouring furnaces; site location, orientation, fenestration are of minimal or no importance. Thus, the design and construction of conventional houses are not much affected by the environment.
2. SOLAR HOUSES, heated by the sun, function only in particular environmental niches (fig. 2-5). Designs are locale specific and are usually not transferable. The design of solar houses is still in the experimental stage; the unpredictable nature of its prime energy source makes the performance of the solar house equally unpredictable. There is an urgent need to incorporate new technologies and inventions in the areas of thermal storage and solar harnessing systems. Consequently, solar houses are popular primarily among inventors and innovators. A thorough investigation of the mesoclimatic profile, topography, site location orientation, fenestration, and internal layout are prerequisites for successful designs. Solar heating systems work best in conjunction with superinsulated houses.
3. SUPERINSULATED HOUSES (gaining in popularity) are warmed by internal (metabolistic) heat and have either no or a low dependency on energy delivery systems or on an energy distribution grid (fig. 2-6). A superinsulated house does not require a thorough investigation of the mesoclimatic profile, topography site planning, fenestration, orientation, or internal layout. A superbly executed prototype is almost environmentally independent and thus easily transferred to other locations. A prerequisite for an effective design is the availability of high-tech materials, excellent insulation, advanced construction techniques, and well-trained construction crews and building trades. It should be noted, however, that because the structure must be airtight, internal air pollution and an associated danger of radiation can result. In addition, the limited or virtually nonexistent need for direct sunlight may lead to a reduction in window size, which would be a regressive step in light of the newly acquired knowledge about the importance of sunlight to individuals' well-being.

A virtually ignored problem results from our failure to recognize that human thermal comfort (the proper air temperature, air movement, and humidity) is only a component of the primordial climate. Another component is the diurnal/nocturnal rhythm.

Conventional technology permits the elimination or scaling down of environmental extremes to fit the human comfort zone. During months of uninterrupted daylight, building envelopes may be programmed to block light for twelve hours and thus respond to the human circadian rhythm. During prolonged periods without sunlight, however, "day" is created by capturing and storing energy during the period of light, by harnessing ambient energy during the dark period, or by using energy brought from a remote location. At present, this is extremely costly.

The need to approximate the cyclic diurnal/nocturnal rhythm of the primordial environment is an engineering challenge that has been addressed sporadically in medical journals. A solution may ultimately come from an unexpected direction, such as from wonder drugs, or mirrors launched in geosynchronous orbit. The most common response is a reliance on the "social clock," the rhythm of activities that includes work, leisure, entertainment, eating, and sleeping. But this is a poor substitute for the powerful and original solar synchronizer.

In 1968 I completed a series of design studies at the University of Toronto related to living and working in the High Arctic. A sketch of one of the experimental structures is shown in figure 2-7. Its spherical shape was chosen because

1. it has a favorable surface/volume ratio
2. of the aerodynamic properties of a sphere
3. a shell can be constructed from identical, prefabricated elements, since wall and roof elements are the same
4. there is a minimum foundation area / minimum permafrost disturbance

The upper hemisphere above the "roof garden" was divided into two parts. One was fixed; it functioned as a reflector of the low arctic sun and delivered light into the living areas by way of an internal court. The other part was retractable, a heavily insulated blanket that functioned as an "eyelid"; it opened and closed automatically to simulate the diurnal/nocturnal rhythm. During the arctic night, the eyelid would be closed permanently and a surrogate sun suspended above the courtyard would become the time giver. The intensity and color of this internal "sun" would be programmed to change throughout the day as does the natural sun in moderate climates (fig. 2-8).

In order to have humans respond to a surrogate synchronizer, light must be of a very high intensity, an extremely expensive proposition. A holistic accounting system, however, considers potential savings from improved health and satisfaction. Without considering individuals' well-being, orderly migration to harsh climatic zones is virtually impossible. The absence of pleasant, habitable homes and communities would limit the population in these extreme regions to tran-

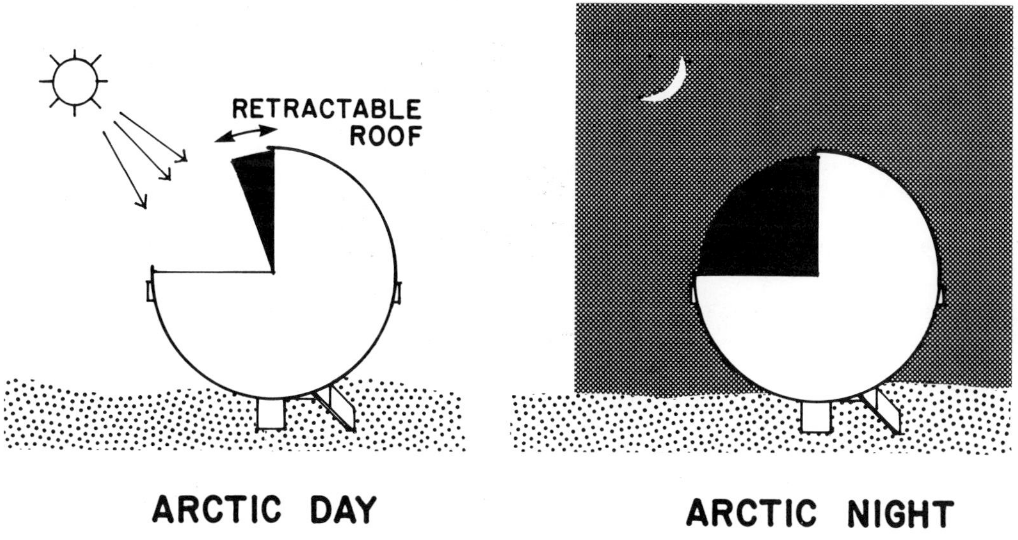

2-7. Spherical house design for the High Arctic.

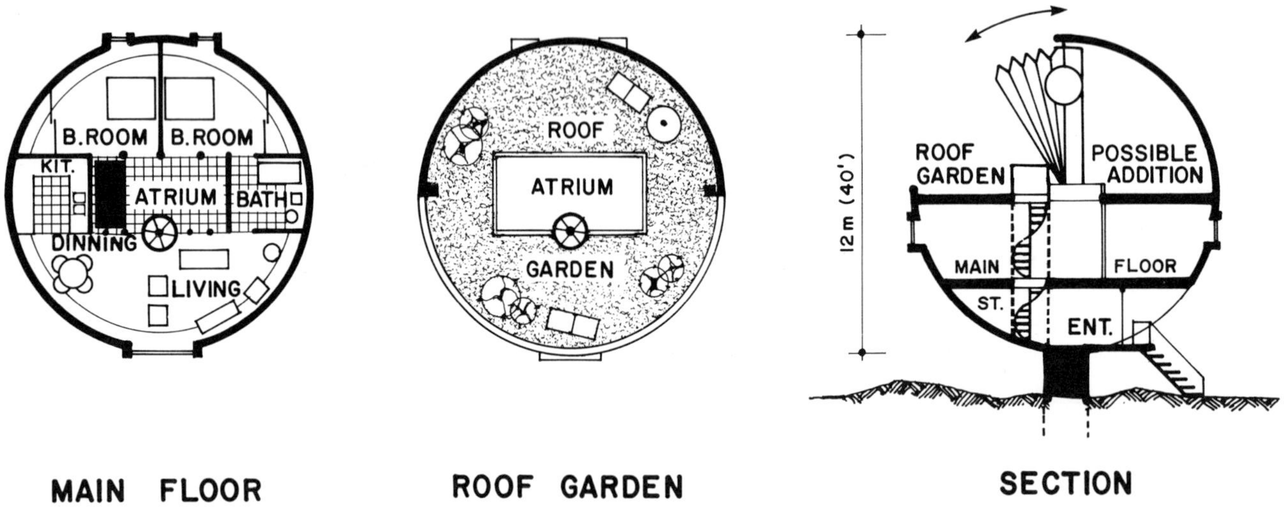

2-8. Plans and section of the experimental arctic house.

sients and a handful who possess a unique capacity to adapt.

Notwithstanding the unsolved engineering challenges, there are a number of new life-support engineering systems, materials, building components, and design concepts on the drafting boards, some of which are already entering the market. Three areas of innovation in particular may significantly improve comfort levels and revolutionize the design and cost efficiency of buildings meant for cold climates: intelligent building components, superwindows, and integrated mechanical systems.

Intelligent Building Components

In principle and by definition, "intelligent" systems have the capacity to sense a change in the environment, consult a data base, choose and execute an action. For example, an intelligent skylight would respond to conditions in the internal operating environment and operate according to programmed instructions. It would also monitor the external environment and close at any sign of precipitation. The amount and quality of daylight would be automatically controlled to suit the occupants' needs. In addition, the skylight could be equipped with a sun-tracking mirror, and the harnessed light may be conditioned by concentration, diffusion, radiant heat filtration, and so on.

Superwindows

Prototypes of windows have been developed recently that provide three times the insulation of conventional double glazing. Windows with ten times the insulation of double-glazed windows should be available in the near future. This level of performance can be achieved by improving existing technology with the use of infrared—a reflective, low-emissive coating that directs heat back to the interior—and the evacuation or filling of spaces between multilayered, laser-sealed glass panels with a high molecular weight gas such as argon.

A window with an R value higher than thirty would have a profound, or even revolutionary, impact on building design. The perimeter of a conventional building could cease to be a zone that responds to external variations (fig. 2-9). Internal heating or cooling sources would not have to be located under windows or on perimeter walls, and thus, significant savings could be realized from shorter distribution lines.

A superinsulated, intelligent window could allow heat to move in either direction according to the season or the time of day; it could change its insulation and light infiltration properties as needed. Because such a window would practically eliminate heat loss (it always gains more than it loses), it would allow the designer to take advantage of eastern, western, and even northern sunlight. A new set of rules for fenestration may therefore be developed that alter conventional layouts and design criteria for buildings in cold climates.

Integrated Mechanical Systems

Technological trends are toward the progressive integration of all mechanical services and life-support devices into a single system. It is anticipated that one piece of equipment with a central control module will be developed and will consist of space heating, an air to air exchanger, gray water heat recovery, and food refrigeration. The central control module would balance solar and internal metabolistic heat input against heat requirements and losses. All excess heat would be stored and utilized when needed. Since every building is a system that must breathe (take in and emit air), a metaclimatic buffer around the building may significantly increase the efficiency of life-support systems.

Future generations of integrated systems may also incorporate waste recycling and possibly limited food production. Attempts to miniaturize these systems have failed thus far, but it has been demonstrated successfully that by using certain selected plants and domestic animals occupying areas smaller than a modest backyard, organic human wastes can be recycled, and vegetables and fruits can be produced as a by-product.

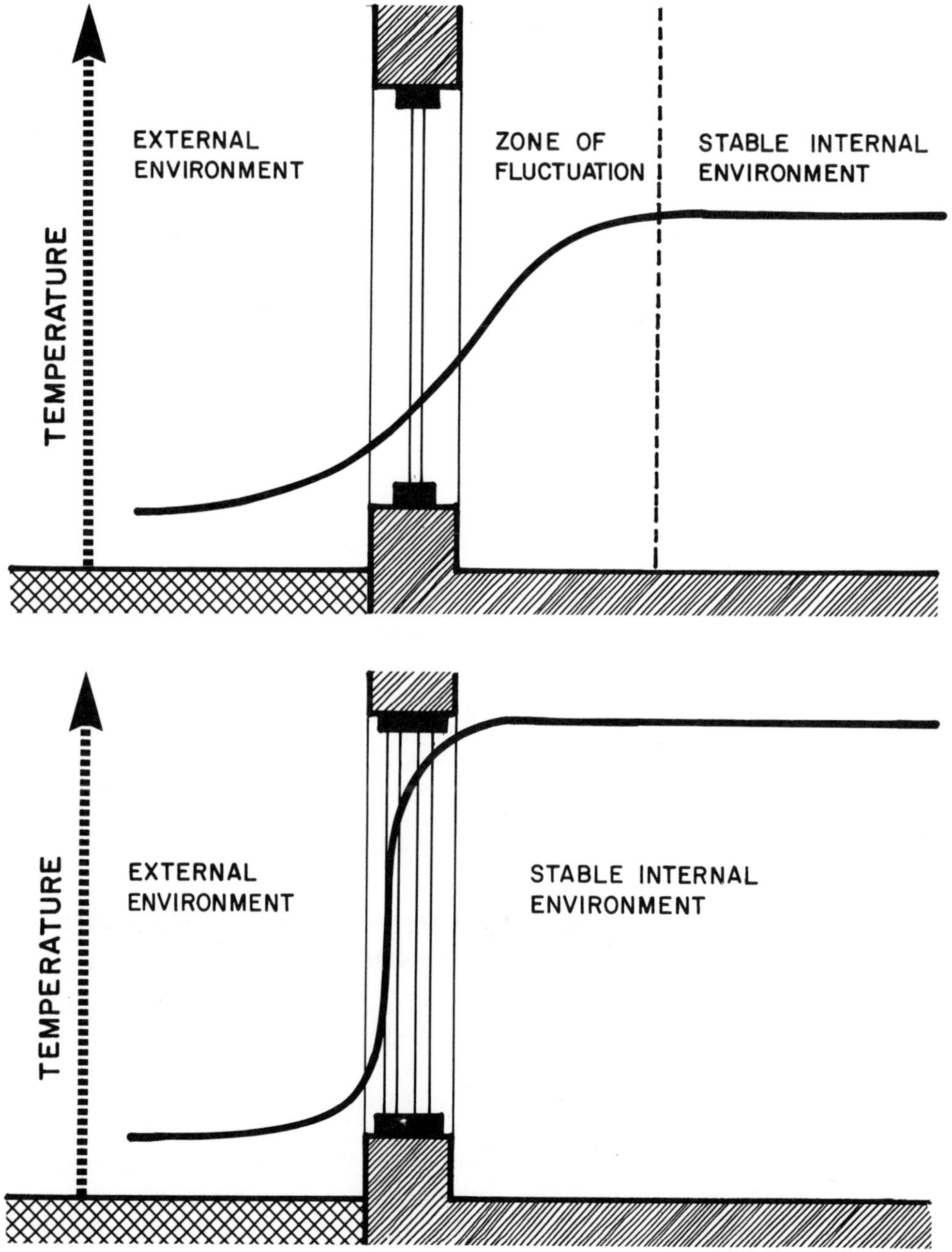

2-9. The Super Window eliminates a zone of fluctuation.

Placed in the metaclimatic zone, such a compressed ecological system would, in addition to its basic function, tangibly demonstrate the logic and flow of natural processes. This may improve, in yet another way, the quality of life in severe climatic zones.

In a final evaluation, an integrated, intelligent life-support system could reduce overall energy consumption to such a level that a small photovoltaic panel or wind generator may be sufficient for most energy needs. This would lessen dependence on a central energy distribution grid. Since the high costs and other difficulties associated with northern development inevitably lead to regimented conservation measures, an intelligent life-support system provides an opportunity to handle energy and resources with the precision of pharmaceutical scales. Waste can be practically eliminated and the quality of life enhanced.

A line of integrated systems could ideally perform "trick recycling," where the refuse of one system would become fuel for the next. Coupled with direct monitoring of internal metabolistic processes and the outside environment, it could permit smooth, efficient navigation through the external dynamics. Gradually, this may transform a human dwelling into a quasi-living system controlled by its own hypothalamus. External stress, particularly in harsh environments, may accelerate the evolution of this quasi-biological shelter system.

PERFORMANCE AND DESIGN CRITERIA FOR THE NORTHERN HOUSE

At present, the energy and resources required for a small medieval village must be pumped into a single home in order to have it meet contemporary standards of habitablity. To cope with such an abundance of power, the house must be plugged into a complex, interdependent system that inputs energy and discharges refuse. The opportunity cost (that of using scarce resources, making them unavailable for other purposes) of long-distance associations is particularly high in northern regions. The following is a summary of the challenges facing the contemporary northern house.

Most northern dwellers have either moved from the south or have been influenced by southern values through the media. Southern standards of comfort and well-being are therefore set and unyielding. Thus, the northern house must be as comfortable and as well equipped as its contemporary southern counterpart. But in addition, the internal environment of the northern house must also compensate for its unique environmental stress.

Although the northern house is an outpost of the southern energy/refuse—distribution/collection network, it should be designed to be as autonomous as possible. Insofar as current technology permits, it must harness the maximum amount of energy without polluting the environment.

The indoor environment of a house located in very harsh climates must compensate for the partial or total absence of those stimuli that are essential to human physical and mental health and that are an integral part of the southern natural environment; it must also compensate for the presence of life-threatening environmental forces. Such compensation can be divided into four major groups: hygiene, health and fitness, food preparation, and the general internal layout.

Personal Hygiene

Compensation as it relates to personal hygiene can include heated bathroom floors regulated by a bathroom thermostat, heat lamps, and full-body air dryers; shower stalls may be equipped with dispensers of heated body oils and moisturizers as well as a set of massage nozzles. Furthermore, shower stalls should also function as saunas and/or steam baths. Bidets and foot baths must become standard, and a traditional bathtub should be replaced by variants of a Jacuzzi, since these can compensate for the physical and mental stress related to northern environments.

Because skin care is very important in harsh environments, where dry, harsh winds and the reflected sun can damage the skin, the bathroom should be equipped with a facial steamer and hot-wax machine; in addition, vitamins (an absence of sun can result in vitamin deficiency), fragrances, conditioners, and face masks can be incorporated. The bathroom should be equipped with a built-in hair dryer and mirrors that do not fog.

Health and Fitness

Extremely cold weather makes it difficult for individuals to exercise as they do in southern climates. A fitness room should therefore include machines that measure heart rate and blood pressure, a computerized scale that records weights, and other self-diagnostic devices. This fitness room may be equipped with workout gear, a massage table, a videoscreen for interactive aerobics, or a hot tub equipped to accommodate aquatic exercises. Sun tan lamps that do not emit harmful rays are available and should be used under medical guidance. Light treatment lamps can be an integral part of a system coping with a harsh, sunless, environment.

The currently accepted strategy for staying healthy is based on prevention—a life-style and system of body care tailored to mitigate the effects of environmental stress. Because this strategy is gaining wide acceptance, the time spent in such bath–body-fitness rooms may generate a need for additional equipment and furnishings, such as refrigerators and drink dispensers, waterproof audio, video, and television systems.

Food Preparation

Technology permits the design of stunning, miniature kitchens that are highly functional—every inch of space is used and storage areas can be concealed or revealed at the touch of a button. The art of food preparation might ultimately be reduced to a purely mechanical process. However, the overpowering environmental stress present in harsh climates combined with the absence of a whole range of stimuli to which humans are biologically attuned, could result in a strong desire for the sensuous experience of preparing food from basic ingredients. Sterile, laboratory-style kitchens with concealed processes should not replace the seemingly unsophisticated kitchens that do not conceal processes of preparation and that integrate eating areas and preparation facilities.

The General Internal Layout

Houses in far nothern locations must offer space to compensate for the limitations in outdoor freedom and comfort. The health/fitness area should be adjacent to, or connected with, a greenhouse to create a natural setting for exercise. The greenhouse must be positioned to take advantage of any available sunlight. In the absence of sufficient natural light, fluorescent and incandescent lamps must provide the light for plants. The psychological importance of a greenhouse is particularly evident in isolated houses; the flowers, vegetables, fruits, or spices produced satisfy practical needs and require the kind of care that has great potential for reducing stress.

It must be remembereed that the design challenge lies in providing maximum internal space given economic limitations. Certain trade-offs are possible, for example, the traditional bedroom may be reduced to "walk-in" or "sleep-in" furniture, or developed into a high-tech capsule used for meditation, relaxation, entertainment, and sleeping. Convertible rooms and convertible furniture are other possibilities. Finally, there are a multitude of design illusions—tricks using space configurations, surface materials, and colors that extend the psychological space.

In summary, accommodations in cold and harsh climates require thermal comfort as well as the natural elements, processes, and variations under which we as a species evolved. In our exodus to less hospitable environments, it is necessary to carry along the warm air of our ancestral home, and every attempt must be made to recreate, at least symbolically, the basic elements of our original natural surroundings—the shallow warm sea in the bathtub, hot tub, or swimming pool; the lush forest in a garden, greenhouse, or plants in a window box; and the flame in the stove, barbecue, fireplace, or candle. In addition, we must establish a diurnal/nocturnal rhythm as well as a seasonal rhythm; diurnal and annual variation are part of our biological make-up, important needs that must be accommodated.

CONTEMPORARY LIVING IN EXTREMELY COLD ENVIRONMENTS

From a historical perspective, humans, in their slow exodus toward less hospitable regions, entered the cold environments relatively recently; even those northern areas with an apparently long history of human settlement, display, in many respects, an inadequate response to local climatic stressors. Archetypes of dwellings, buildings, and communities that are developed in civilizations with warm climates still dominate northern regions.

In spite of all past achievements, a desirable thermal milieu still poses a challenge to northerners. Thermal comfort is difficult to maintain. In certain houses or certain interior areas, usually in older housing stock, the indoor temperature traces too closely external temperature swings, and is intolerable by our contemporary standards.

The thermal milieu is achieved only by an extraordinary effort. Taking even the simplest scenario, such as putting a wood stove or fireplace in each room, it is difficult to maintain an indoor temperature that falls within the limits of the comfort zone throughout the day, night, and winter season. Although contemporary technological solutions offer manipulation of internal climates at the touch of a button, the costs of heat or energy generation and of long-distance delivery are usually hidden in a variety of subsidies.

In the north, slow natural processes and the slow healing capacity of the environment make it almost impossible to build a settlement without devastating the environment with pollution and the exploitation of natural resources. The bearing capacity of the northern environment is very low, so living off the land and in harmony with the natural surroundings is possible only if the population is thinly spread over very large areas. It is difficult to argue that, for the sake of environmental balance or harmony, the likelihood of killing or being killed, eating or being eaten should be roughly equal. This mode of existence is simply not acceptable, even to back-to-the-land enthusiasts.

In the process of mitigating the present deficiencies in human accommodation, it is apparent that there will not be, nor should there be, a single blueprint for a dwelling or for a community located in cold environments. Design solutions should be derived from, and depend on, the ratio of opportunities in the local environment, for example, what favorable elements must be harnessed, and what destructive forces must be filtered or deflected. The form of individual houses and settlements should also be determined by what is available, or what can be manufactured on the spot from locally available resources, and is thus possible to harness, and by what is necessary but must be transported from other locations.

A community or a single-family dwelling in any region of the world should provide an environment that serves both physiological and psychological needs. Within this environment, individuals should be able to realize their aspirations. Thus far, the deficiencies in southern solutions that have been imported to northern, cold, regions have had to be overcome by endurance and adaptation. For example, large double-hung windows that provide inefficient ventilation and are made draught-proof only at great expense are often included in northern dwellings. These dwellings often eliminate an air-lock entry, which protects living areas from direct contact with a harsh outdoor environment. Children's playgrounds tend to be typically southern—

they have only limited activities for winter. Some northerners actually believe that this is an inseparable part of northern culture and should not be altered; therefore, northerners should be forever condemned to live in southern-style houses and communities and to cope wih cold, darkness, fatigue, and misery unaided by the built environment. These environmental athletics may not be universally appealing, particularly to the very young and the very old. Neither does this attitude invite new settlers, nor does it discourage a steady exodus to warmer regions.

The fact is that in certain northern regions, a single-family house turns into a single-family prison for prolonged periods of time. Confinement amplifies the environmental stress, already sufficiently amplified by a geographic location that has a devastating impact on all but the strongest. The host of problems includes light and sensory deprivation, which, according to current medical journals, can cause headaches, irritability, depression, and poor physical performance. In a family setting, a higher level of regimentation is required to accommodate family members' diverse interests and territories within the physical limits of the dwelling. Again, the old and young are prime targets of any additional human-induced stress.

In regions with prolonged periods of darkness or prolonged periods of extremely low temperatures, a metaclimatic envelope can be incorporated into housing designs, providing a surrogate backyard that relieves the sense of confinement. All planning and design efforts might be coordinated to create clusters of individual dwellings within or adjacent to a common metaclimatic environment, filling the needs for outdoor space and an area in which to socialize.

Multiresidential clusters may grow or be incorporated into large neighborhoods. A group of integrated neighborhoods would constitute a community.

But in order to design human shelters and settlements more intelligently, environmental components specific to a particular area must be understood. This is the only reliable guide for planning, designing, and developing dwellings that ease the environmental stress in northern regions.

PART 2

THE NEIGHBORHOOD

CHAPTER 3

ENVIRONMENTAL FORCES AND LAND FORM

THE ECOCHART

Composite Force Action

A primary aim of architectural design is to achieve a stable internal environment to counteract the outdoor environment, which changes continuously. Changes in the environment result from the composite force action of the sun, wind, and water. In order to understand the cumulative effect of these forces, the individual components (for example, the speed, direction, and frequency of wind, the amount of sun energy received, and the ambient air temperature) must be isolated.

Once a component is isolated, it can be measured. Two types of measures are generally used:

1. the magnitude and direction of those components that are related to the static environment
2. the variation, interval, and range of those components that are related to dynamic environmental forces

These measurements of environmental components can subsequently be translated into useful guidelines for the construction of building forms and partial or complete neighborhoods. Although we do not yet have a complete understanding of the dynamic interaction between environmental forces and their cumulative effects, planning strategists can navigate with a reasonable degree of accuracy. The selection and synthesis of individual environmental components is based on professional intuition and the anticipated needs of practitioners rather than on the rigors of scientific inquiry.

Any understanding of local environmental forces must begin with the gathering and organization of local environmental data, particularly that which is related to the sun. Basic unanalyzed weather data can be of very limited value to an urban designer, architect, or building professional, who may not know how to use it fully. Thus far, climatology has rarely been applied to the planning of human settlements. Critiques of users' (planners, architects, builders) attitudes toward the usefulness of climatic data to their particular needs appear often in climatology publications and journals. Their message is clear: many planners, architects, and builders consider climate information, as it is being presented in conventional climatology publications, irrelevant, inapplicable, or incomprehensible. Climate is perceived as either too static or as something one can do very little about.

Part of the problem is that national agencies generally collect and release weather data for general forecasting or specifically for aviation, coastal and water resource management, forestry, marine navigation, or agriculture. Suppliers are not providing the type of data needed by planners and designers; urban climatology is primarily concerned with descriptive and comparative studies, highly academic or scientific research on the behavior of urban climate. In addition, communication between potential users and the suppliers of data (a variety of applied climatology branches) is often less than satisfactory.

It is not true that building professionals, architects, and urban planners ignore local climate completely. Climate is taken into account in building and design and always has been. Despite sophisticated technology, the construction process, for example, is still highly dependent upon weather. Precipitation and extremes in air temperature affect soil conditions and thus excavation, foundation construction, general traffic on the site, and the pouring and curing of concrete. Excessive wind may threaten the safety of cranes and lifting equipment, the erection of steel structures, and roof construction. Certain operations, like outside painting, require special weather conditions. Construction-site safety and general performance is, to a large degree, determined by the weather.

Taking climate into account, however, has primarily been a process of trial and error; there has been no direct, systematic approach. If a calculated and studied response were taken to local environmental stress, buildings and communities would be more cost efficient, since unnecessary features would be eliminated; they would be safer, since underdesigning may endanger property and human lives; they can also be more comfortable since microclimatic conditions (wind, temperature) would be improved, especially for pedestrians; finally, their socio-economic performance should improve as a result of increased use. To encourage building and urban design professionals to apply climatic data more consciously and systematically, it is necessary to make this data available in a more convenient, relevant, and digestible form.

Selection of Climatic Data

To meet architectural and urban-planning requirements, specific weather factors must be monitored and design criteria should be established by professional climatologists as well as building and planning professionals (architects, engineers, and urban planners). The list of climatic factors should reflect weather conditions unique to a given area as well as the presence or absence of individual weather elements. A generic list would likely include the following data:

1. air temperature (dry and wet bulb)
2. air movement (speed, direction, frequency)
3. hours of sun (solar radiation)
4. relative humidity
5. soil temperature (at different depths)
6. precipitation (snow, rain)
7. derivatives (cooling and heating degree-days)

The selection of data must be user specific. For example, engineering design criteria may require the inclusion of data related to the performance and operation of a building during the winter cycle. Thus, the following information would have to be incorporated:

1. snow loads on roofs
2. freeze-thaw cycles
3. air pressure (air leakage and precipitation penetration)
4. material corrosion
5. frequency of extreme events (blizzards, hail storms, hurricanes, tornadoes, tidal waves, earthquakes)

For purposes of building orientation, site planning, and the determination of height, bulk, and relationships among buildings, the architect and urban designer may also require more detailed annual and daily data related to sun geometries. The landscape architect may also need special information regarding the duration of certain conditions, for example, the average duration of cold and hot spells or wet and dry spells, the growth processes of plants and hibernal animals, frost and frost-free periods, and so on.

Tabulation of Climatic Data

A method of tabulating climatic data is shown in figure 3-1. The individual stressors (A, B, C, D, and so on) are tabulated horizontally and indicate the intensity level and frequency of a particular stressor (weather element) over time. The composite force is read vertically. Consequently, each column represents values for listed environmental stressors at a given time.

This method can be further expanded into a format developed by the Hamilton Planning Department (fig. 3-2). Called an Ecochart, it gives an ecological description of a given area. The top portion presents a conventional description of the topography, including human-made and natural features. The lower half of the Ecochart tabulates mesoclimatic data. The individual environmental stressors, such as air temperature, air movement, humidity, or their derivatives, such as heating/cooling degree-days, are tabulated horizontally. The composite environmental force is read vertically. Thus, each column represents monthly values for the listed environmental stressors. The indicated stress levels should represent thirty- to fifty-year averages wherever possible.

In constructing an Ecochart, the availability, origin, and nature of the data should be examined. Certain data may not have been recorded in a given region, for example, solar radiation. Certain data might be less representative, because records have been begun only very recently. It is possible that data has been recorded for other purposes, such as agriculture, or that it may not have been recorded systematically. The method of recording could have been changed midstream, as in the case of air movement characteristics where, quite often, measurements were initially made for four directions and later changed to eight. In addition, recordings could initially have been made in imperial measurements and later in metric.

For every environmental stressor, the location of the data-gathering station and the time interval represented by the data must be stipulated. If certain stressors—temperature, wind, or barometric pressure—tend to be erratic, this information must be included in the Ecochart tabulation as well. The Ecochart should be updated every five years in order to incorporate long-term changes and trends.

Interpretation and Application of the Ecochart

Even a superficial glance at the Ecochart of a northern region would reveal that inhabitants suffer two major environmental "blows" every year—the extremes of summer and winter. Studying the Ecochart results in a more thorough understanding and an improved ability to cope with stress characteristics associated with high and low temperatures that exceed the human comfort zone. It is necessary to generate heat in the winter and cool air in the summer; summer extremes can lead to variable degrees of discomfort, whereas coping with winter extremes is a matter of survival. Mortality statistics, compiled by Statistics Canada, indicate that over a recent fifteen-year period, a total of 1,621 deaths could be attributed to excessive cold, an average of 108 deaths per year. On the other hand, cataclysms, earthquakes, tidal waves, and other natural events have claimed seventeen Canadian lives per year.

These facts may not, of course, be a shocking discovery to a local designer or builder. However, since the conventional response to this annual environmental stress is central heating and air-conditioning and since, in northern areas, energy consumption peaks during the winter, an alternative to conventional space heating presents an opportunity for significant savings. The Ecochart can provide detailed data on the extent, magnitude, and specific behavioral characteristics of winter. This may initiate or reinforce an interest in winter strategies that employ new construction methods, new building materials, find different uses for existing building materials, heat pumps, active and passive solar systems, heat-recovery systems, and result in more conscientious site-planning and road-alignment design.

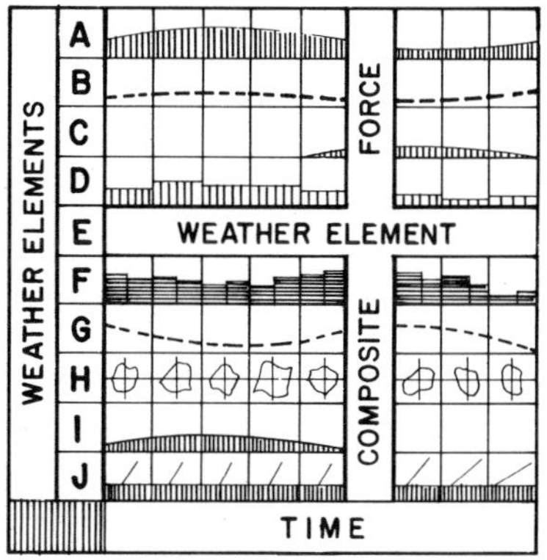

3-1. A method of tabulating climatic data.

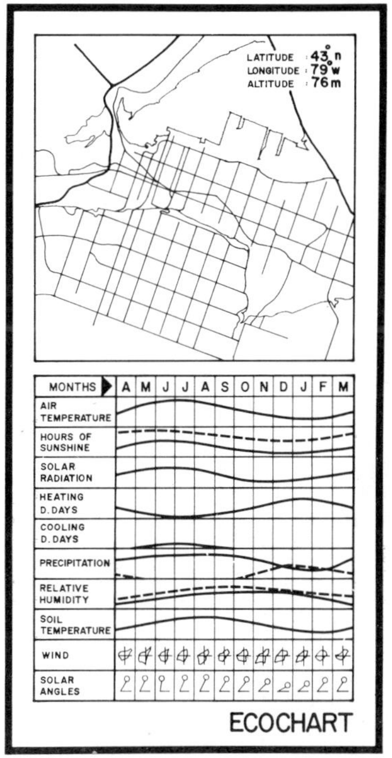

3-2. The Ecochart format.

Moreover, a more in-depth understanding of the characteristics of overall winter stress can have a significant impact on the strategic planning and development policies of state or local levels of government, and the large-scale development proposals of private industry. For example, in Ontario the cities of Hamilton and Kitchener, only sixty miles apart, have remarkably similar Ecocharts with two notable exceptions: the winter precipitation is much higher in Kitchener than in Hamilton, and they have different wind patterns. These facts alone have profound effects on engineering design, the turning radii of snow removal equipment, snow depository areas, municipal services, public transit, and the allocation of emergency funds and resources.

Finally, an understanding of the similiarities and differences in annual heat/cold stress characteristics allows for a better assessment of the project's complexity, construction techniques, construction schedules, sequence of trades, and construction costs without having to begin with an exhaustive on-site investigation (especially useful to developers, designers, or investors who are operating from another town or country).

Interpreting the Ecochart requires a scientific approach as well as professional intuition. By breaking down the composite weather force into environmental stressors that are measurable by conventional instruments, the Ecochart records the characteristics of each aspect of the composite force over a whole year. Each aspect carries specific directives to the developer, planner, municipal engineer, transportation engineer, architect, landscape architect, builder, interior designer, real estate broker, the mechanical, structural, and electrical engineers, and everyone else involved in the development process. For example, by understanding more fully the general direction, frequency, and speed of wind in a given area, wind can be used to cool areas opened or exposed to it and to warm areas shielded from or closed against it. Analyzing wind force makes it possible to calculate penetration and thus design features that prevent penetration. These might include air tightness, structural details (bracing, or heavy self-adhesive roof shingles), and a structural system (for instance, one that uses poured concrete instead of wood) to resist destruction. Finally, the Ecochart enables one to evaluate the potential costs and benefits of incorporating systems that harness wind energy.

Profiles of soil temperature help determine safe depths for burying municipal services, foundations, and all other underground constructions. Furthermore, the subsurface heat can be utilized for heat pumps, air intake, preheating, and precooling. Precipitation data helps establish performance specifications for all exposed material—roof construction, gutters, water leaders, systems for handling storm water, foundation design, as well as designs of all openings, including doors and windows.

It is the *combination* of environmental stressors found on the Ecochart that helps determine the most appropriate design for buildings and the best layout for a residential development in a given area. Soil reflection, for example, is a product of soil temperature, precipitation, and sun angle at a given time. Soil temperature has a direct bearing on thermal comfort inside a house, and therefore, should affect the orientation, landscape design, layout, interior design, and mechanical systems of the house. The urban-planning challenge lies in the development of a spatial framework within which individual buildings can be incorporated to maintain their thermal equilibrium in the midst of natural cyclic forces.

LAND FORM DESCRIPTION

Understanding Environmental Domains

The terrestrial environment is composed of domains, areas that are dominated by some physical characteristic. Domains vary over time and across space. The complexity of domains discourages planning and design professionals from using natural forces more fully; in comparison, manufactured energy from remote locations becomes a tempting and easy solution. However, planners and designers who feel responsible for the environment and for creating a physical setting that enhances individuals' well-being, and that lasts for several generations, must work with available natural forces.

The topography of land creates a variety of environments within a given area. In very simple terms, every domain has its own borders, which occasionally overlap with those of other domains and which continuously change with time. The center of a domain is therefore the most predictable area (fig. 3-3). Any system whose survival depends on maintaining an inner stability has a better chance if it is located in the center of an environmental domain. A system located toward the periphery is more difficult to stabilize and requires advanced engineering techniques to counteract a higher energy penalty.

Figure 3-4 shows boundaries between domains. A striking case is the air-water interface. In order for a boat to take advantage of the properties of both domains, it is composed of two fundamentally different shapes; one is permanently submerged in the water, the other is in the air. The air–earth's surface interface is another challenging boundary and requires two separate design criteria—one for the below-ground and one for the above-ground environment (a tree and its roots, a house and its foundation are two examples). Interface between the earth's surface, the air, and water (as is necessary in the construction of a seashore cottage) presents a most challenging engineering task, one that must respond to each of the three domains.

In order to arrive at some comprehensive design guidelines for the construction of human-made forms, the interaction between environmental forces and land form has to be better understood. The major task is to translate conventional topography into a description of land forms that would indicate the nature of its relationship to other dynamic environmental forces, particularly the sun. Topographical contours are an established medium for conveying information about land form and its height variations. Although contours are very convenient and useful for many purposes, a responsible ecological strategy requires a better tool.

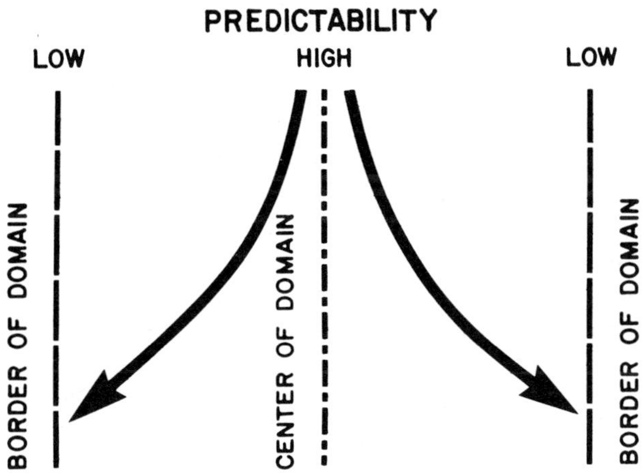

3-3. The level of predictability in the center and on the periphery of a domain.

3-4. Dramatization of boundaries between domains.

Two Methods of Describing Land Form

There are a number of alternative methods of describing topography. Two methods will be examined here. A description of the slope gradient and slope orientation is one quick and simple method particularly suitable for analyzing larger areas.

Conventional maps describe topographical variations by contours. If a contour represents a height differential of one unit, then a distance of 100 units between two contours would represent a slope of 1 percent (fig. 3-5).

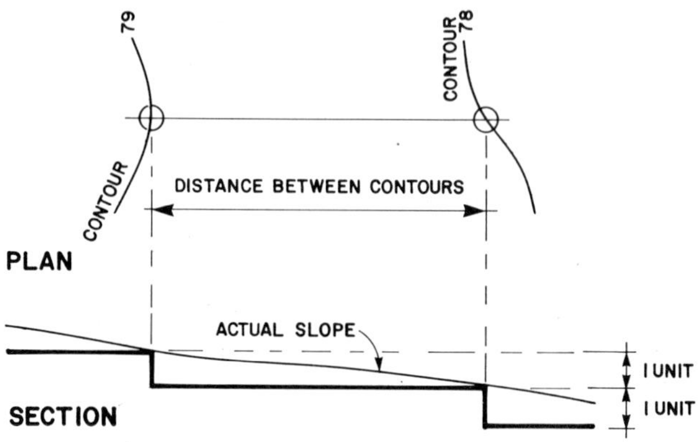

3-5. The geometry of conventional contours.

Thus, a table can be constructed:

A distance of 100 units	= 1%
50 units	= 2%
25 units	= 4%
20 units	= 5%

A given area, described by standard contours (fig. 3-6), is then divided into a square grid (fig. 3-7). The grid must be oriented toward true north/south and east/west. The square size of this basic grid is determined by the following formula: the height differential represented by a contour multiplied by 100.

Consequently, where a contour represents a height differential of one unit, a square of 100 by 100 units is the optimum size for the basic grid. Therefore, a contour representing a height of two units would require a basic grid of 200 by 200 units, five units would require 500 by 500 units, and so on.

This relationship between contours and the size of the basic square determines that a square with no intersecting contours represents a slope of less than 1 percent; for practical applications, it would be taken as a flat, horizontal surface (0° slope). A square inter-

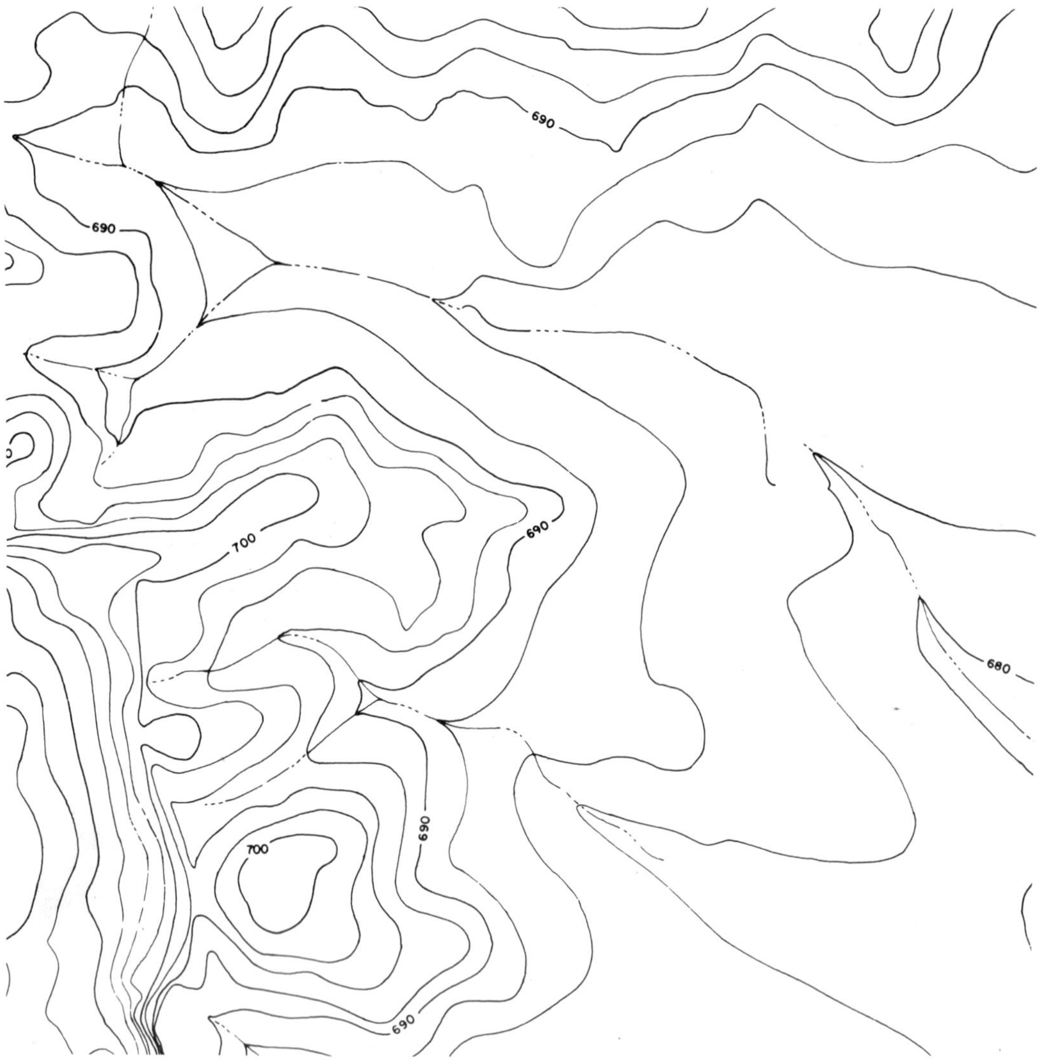

3-6. A given area described by standard contours.

sected by one contour would represent a slope of 1 percent.

If two contours happen to intersect a basic square, the basic square would be divided into four squares. The subsequent division of squares (fig. 3-8) would result in a map where no square is intersected by more than one contour, and the size of each square represents a certain degree of slope (fig. 3-9). Finally, by removing the contour lines, the conventional map is transformed to one that uses a different method to describe land form (fig. 3-10).

The graphic expression arrived at with the slope gradient description is a mosaic of squares. Each square carries information about the magnitude of slope variation. In the next step, land form is related to the dynamic environmental forces, the cyclic motion of the sun, by adding information concerning slope orientation. In simple terms, each square is taken in turn, and the nature of the intersecting contours is examined. The actual slope is then translated into either four or eight nominal slope directions (fig. 3-11).

Figure 3-12 illustrates the process of slope-orientation construction. First, the nature of any intersecting contours in a square is examined, and the actual direction of the slope is determined. Second, the actual slope is translated into a nominal slope in either a four- or eight-direction system.

Arrows are a convenient method of indicating the orientation of a slope. Alternatively, shade or color coding can be used. In a system involving four nominal directions, five nominal slope categories would result:

SOUTH FLAT (any slope less than 1%)
WEST EAST NORTH

Slope directions arranged in the above order also express the degree of sun energy that northern latitudes in the northern hemisphere receive:

S F W E N
MAXIMUM ——————— MINIMUM

It is important to note that although eastern and western sun paths mirror each other, the west receives energy from the sun under more favorable conditions. This is because the morning sun works against a dip in temperature at night and other associated obstacles (for

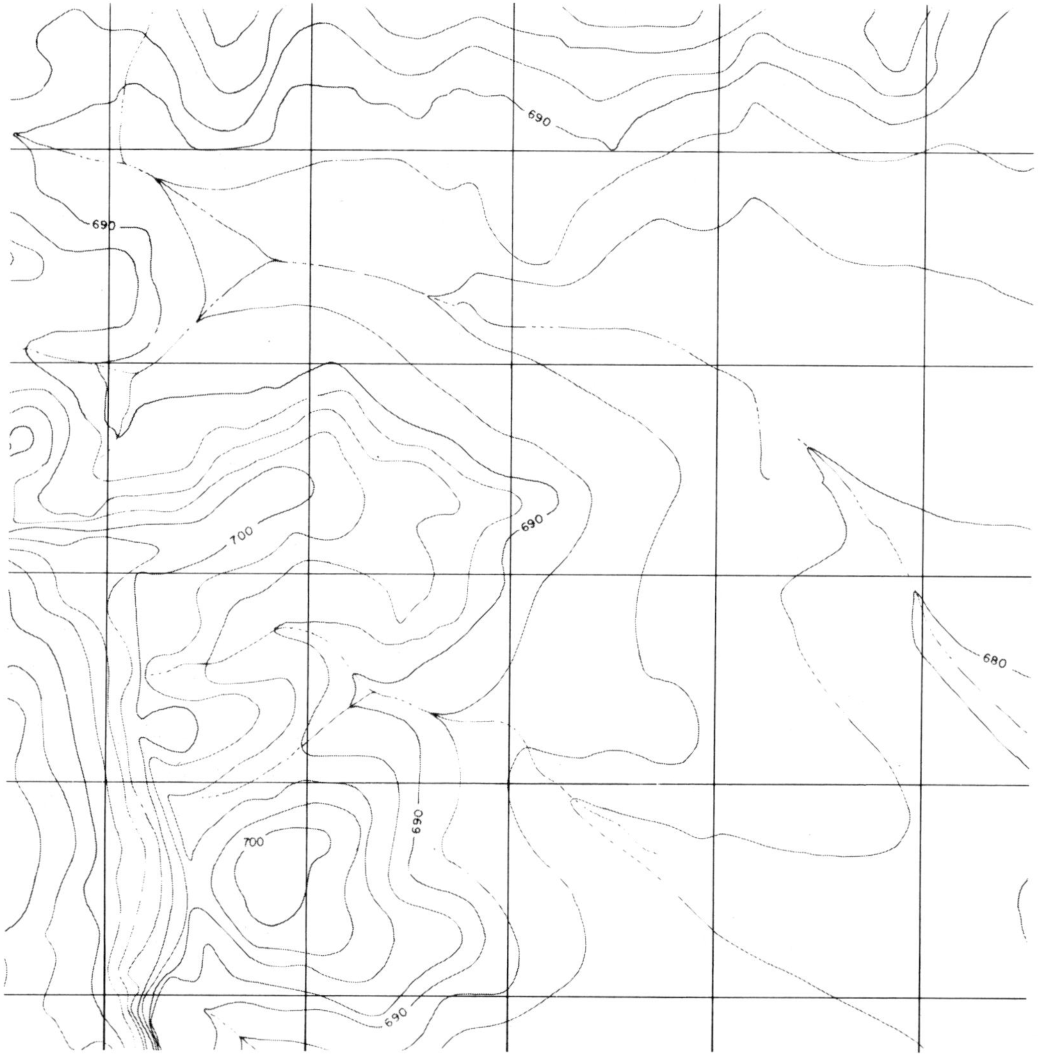

3-7. A basic square grid superimposed on standard contours.

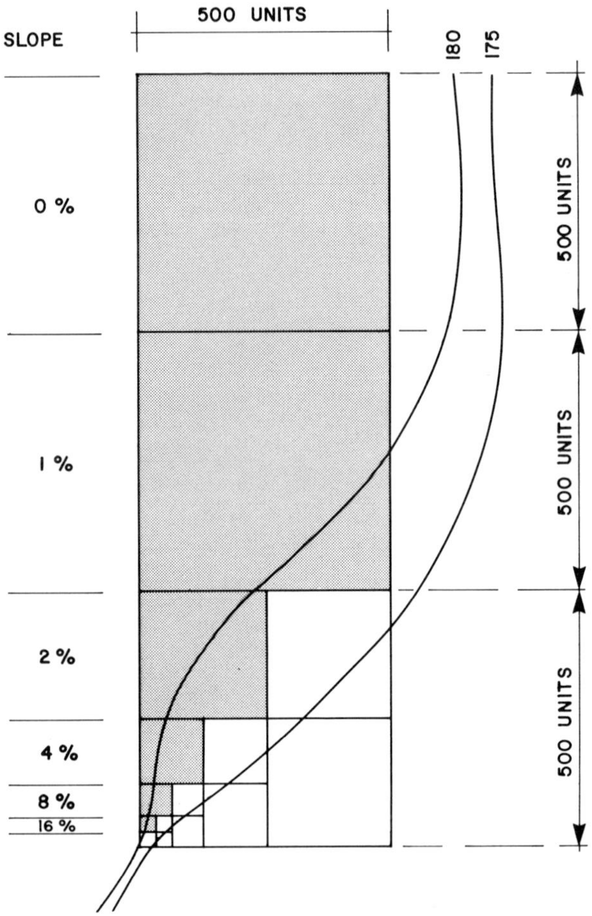

3-8. A method of dividing the basic square.

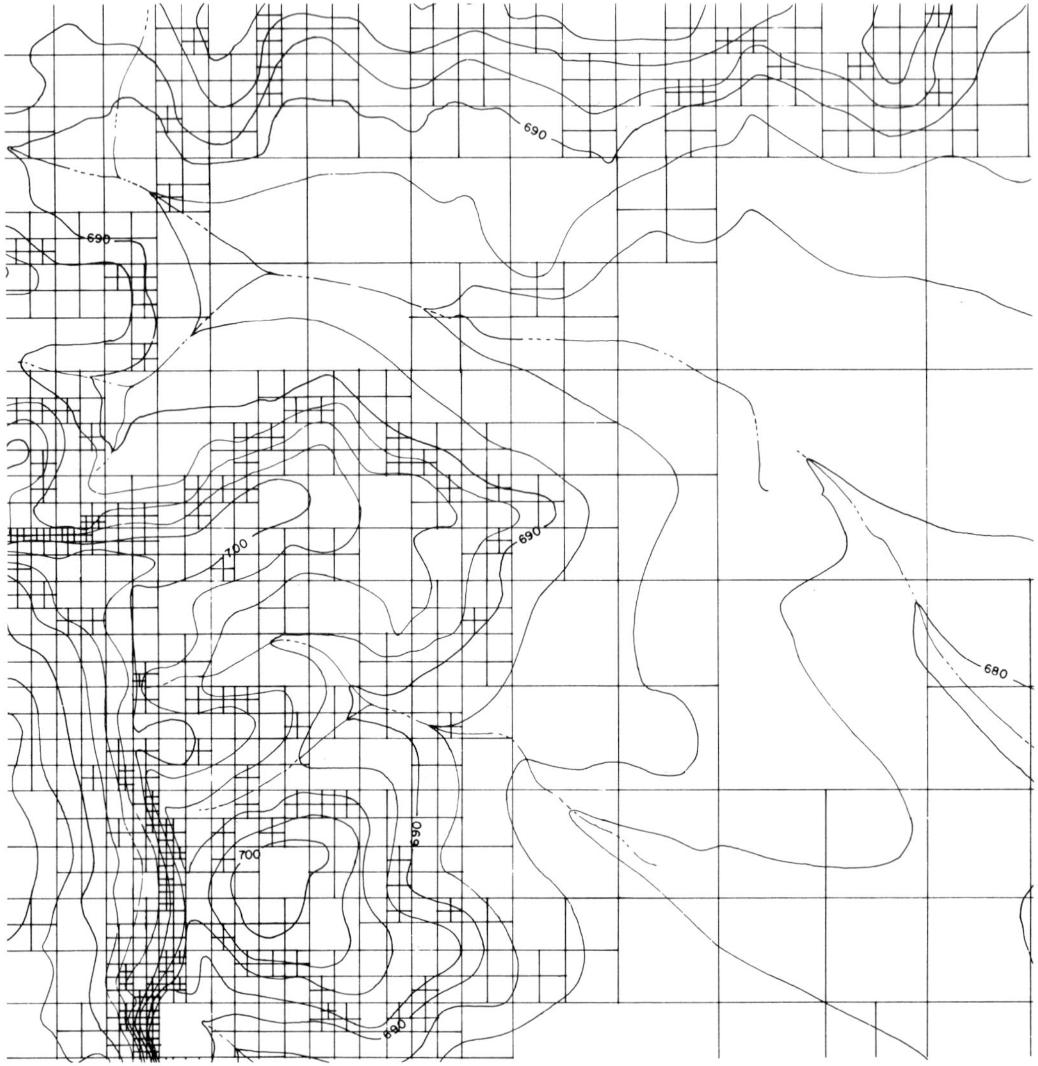

3-9. A map where each square, regardless of size, is intersected by only one contour.

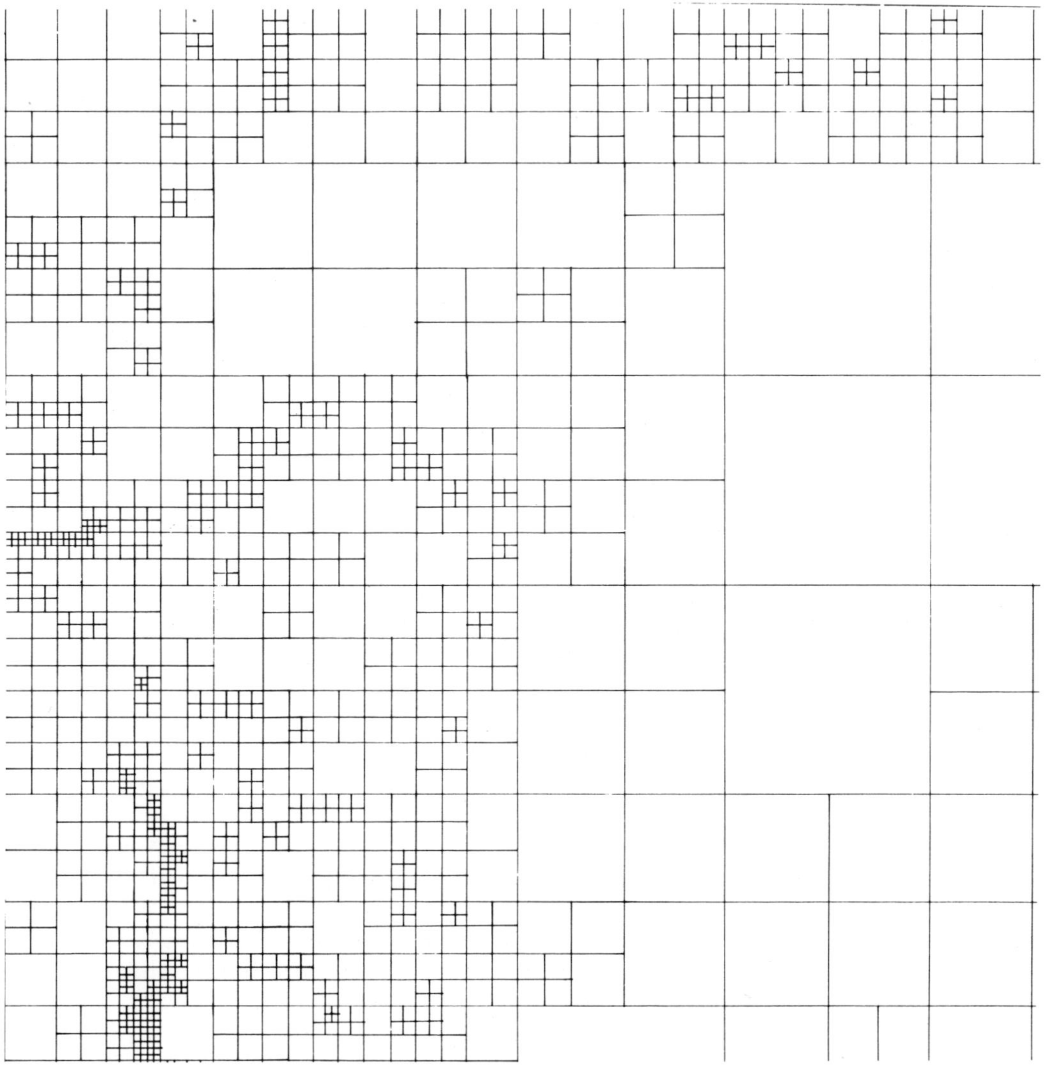

3-10. A map where slope gradient is described by the size of a square.

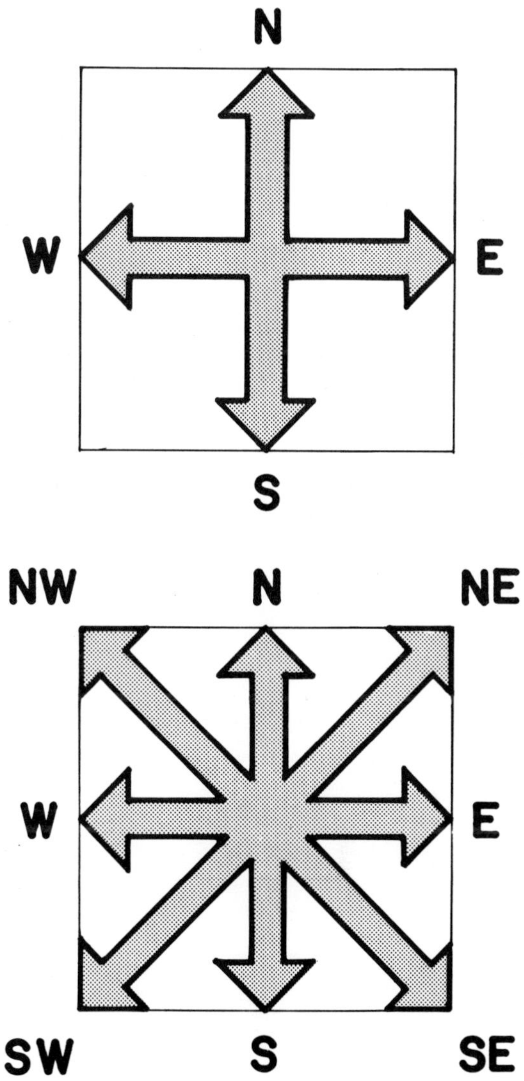

3-11. Four and eight nominal slope directions.

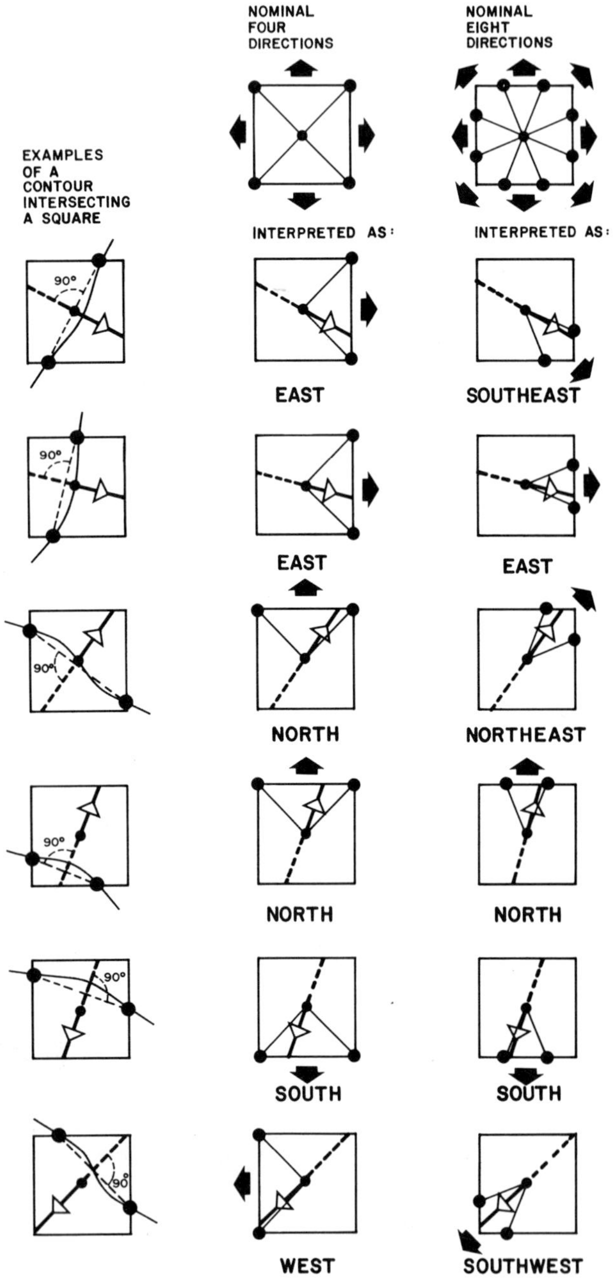

3-12. The orientation of contours in a system with four and eight nominal directions.

example, vapor in the air), whereas the afternoon sun shines on a preheated environment.

This method of describing land form can answer a whole range of questions important to a developer, environmental engineer, land-use planner, landscape architect, and those in related fields. For instance, it is easy to tell whether any areas are steeper than 25 percent, and how large areas are with a 0° slope. The location of all southern slopes up to 20 percent can also be determined. This information permits an analysis of development potential and a comparison with other areas. Figure 3-13 shows areas of 1° slope; figure 3-14 shows southern slopes; and figure 3-15 shows western slopes. A chart that shows both topographical variables (for example, a slope gradient description) and a description of slope orientation is called a *Slope Descriptive Synthesis* (fig. 3-16, see insert at back of book).

This first method, introduced by Knowles, is suitable for a rapid analysis of large areas of land. For individual subdivisions and other smaller areas, a more precise method has been developed. In this second method, the process begins with a basic grid that (as in the first method) must be oriented toward true north/south and east/west directions. Each square in the basic grid, however, would represent the smallest practical size. For example, the size of the square may be selected as 10 by 10 meters (30 by 30 feet) in areas where 10 meters (30 feet) is the typical frontage of a subdivision lot.

Next, the center of the square must be established according to the two nearest contours. If four nominal orientations had been chosen, there would be two cardinal directions, north/south and east/west (fig. 3-17). In the case of eight nominal orientations, there would be four cardinal directions, north/south, northeast/southwest, northwest/southeast, and east/west (fig. 3-18).

The shortest direction is then identified and scaled. A special, customized measurement tape or ruler may be constructed (fig. 3-19), or a conventional scale and a pocket calculator can suffice. Having measured the shortest distance between the two contours, the slope percentage can be computed. Consequently, it is possible to identify the slope gradient and slope orientation of each square in one step (see figs. 3-17, 3-18).

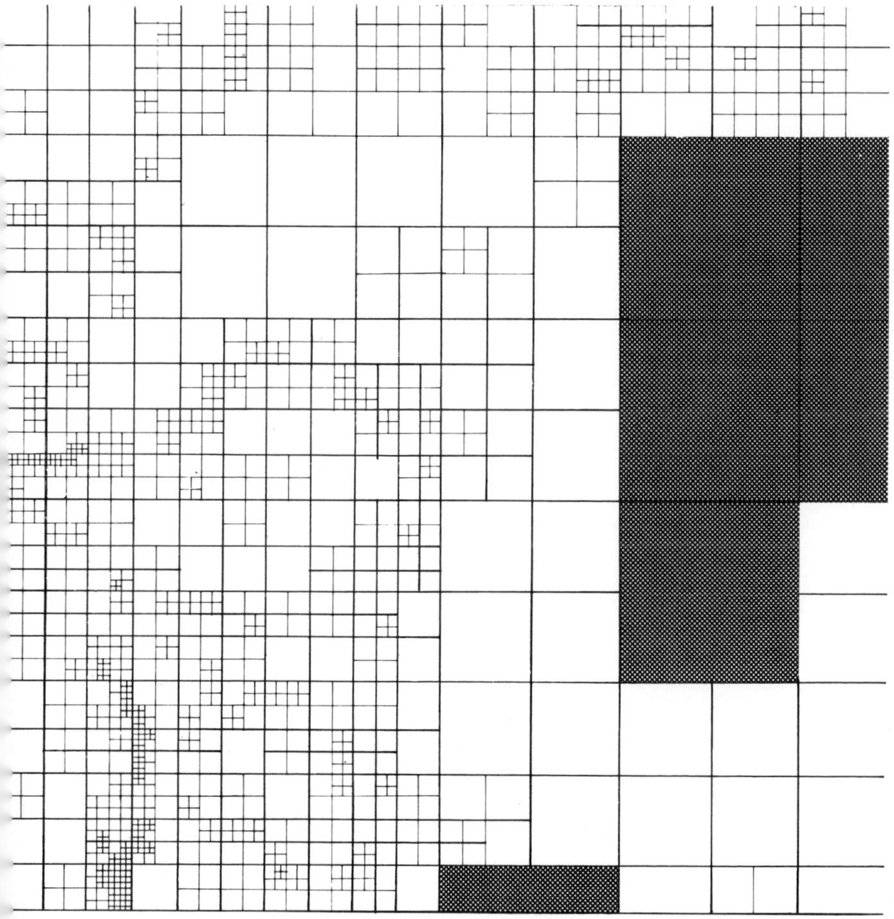

3-13. Areas of 1° slope.

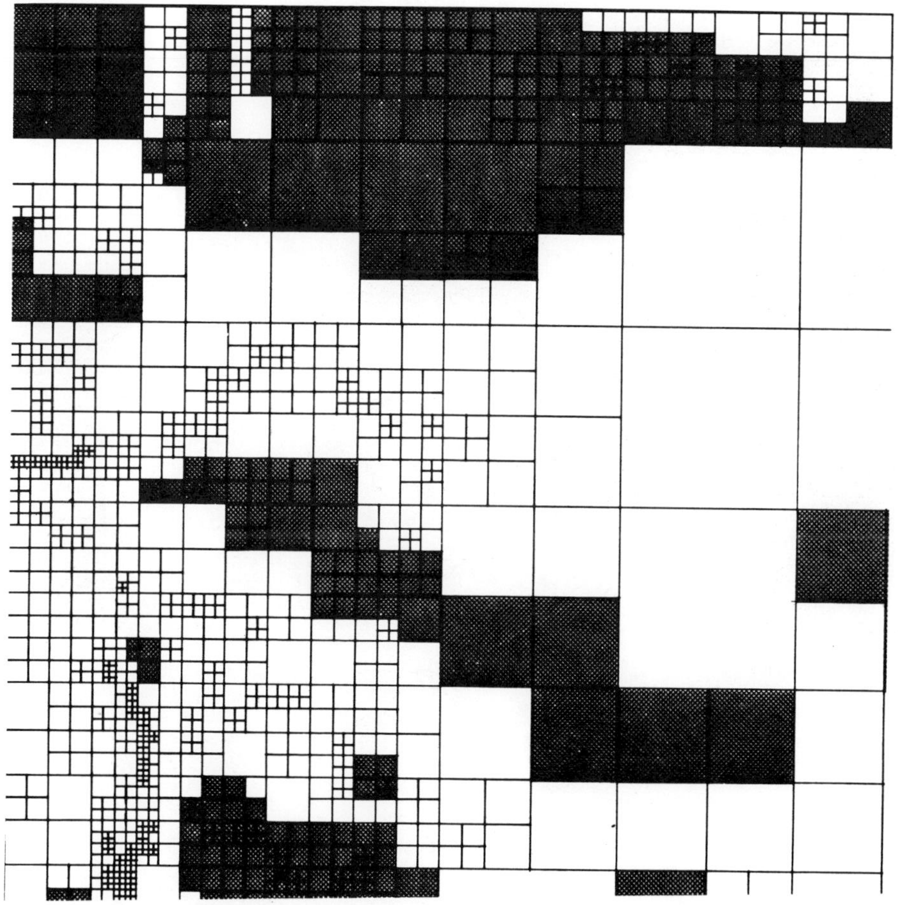

3-14. Southern slopes.

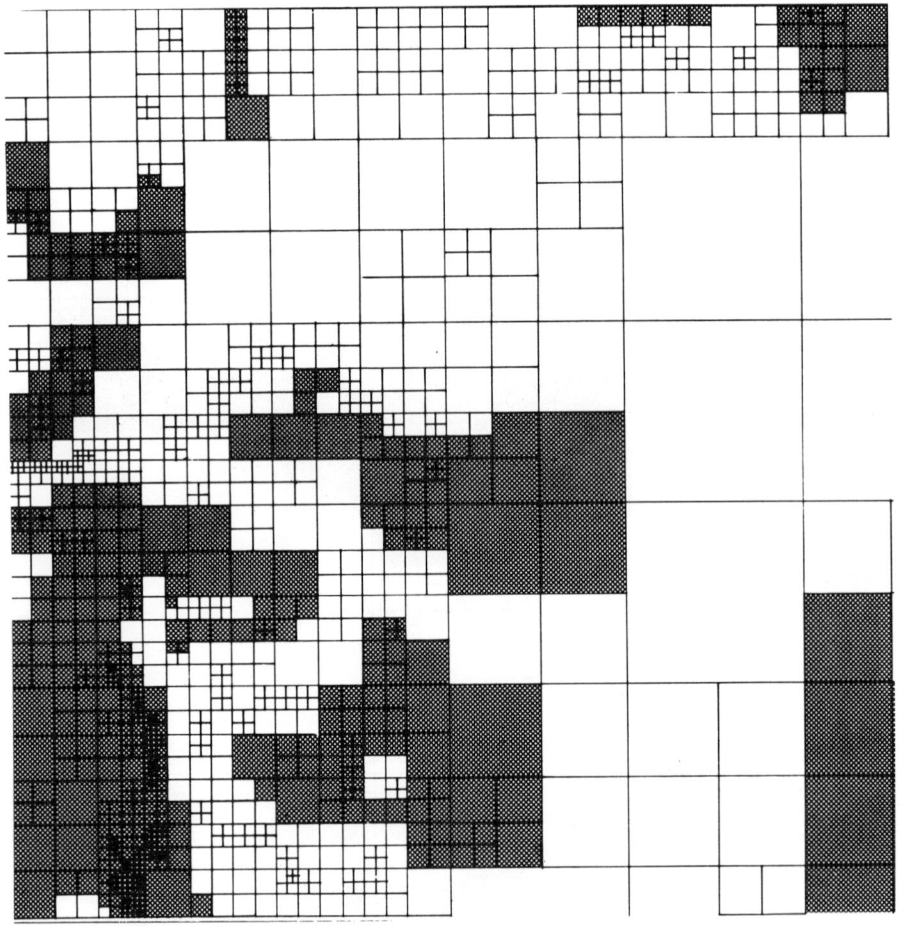

3-15. Western slopes.

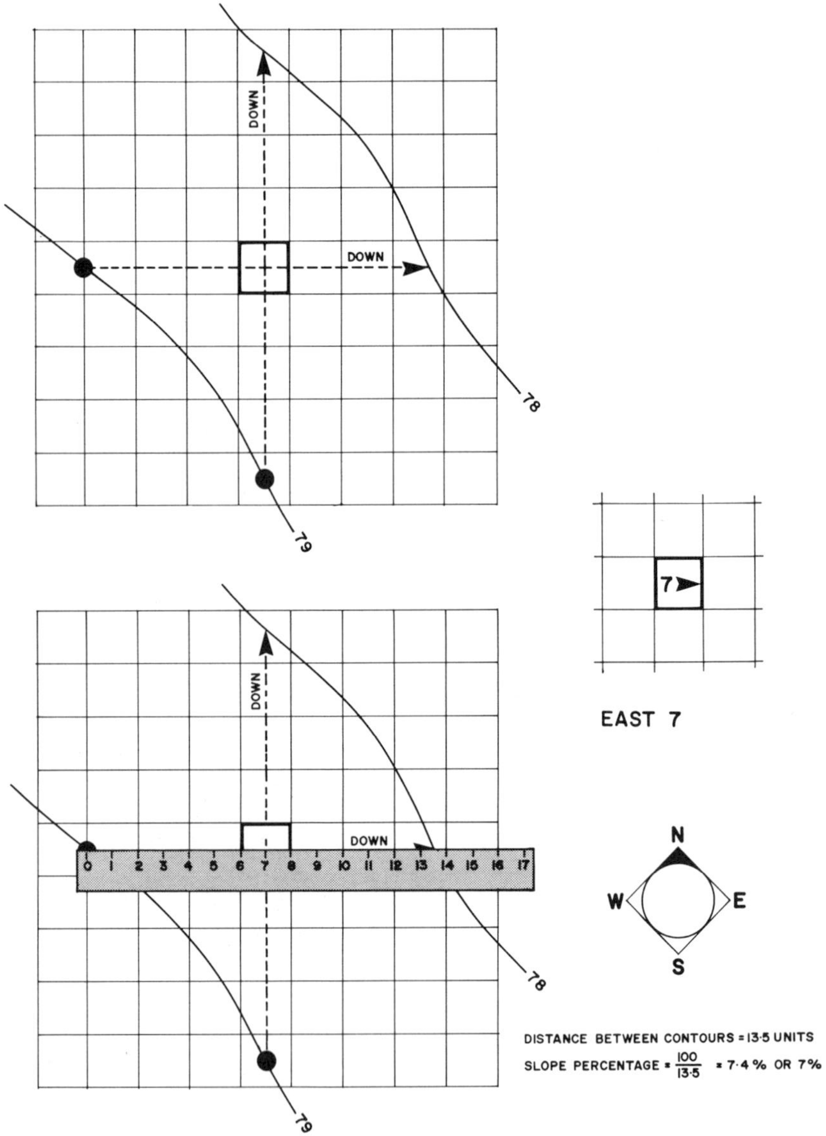

3-17. Slope gradient and slope orientation can be determined in one step using a method that considers four nominal orientations.

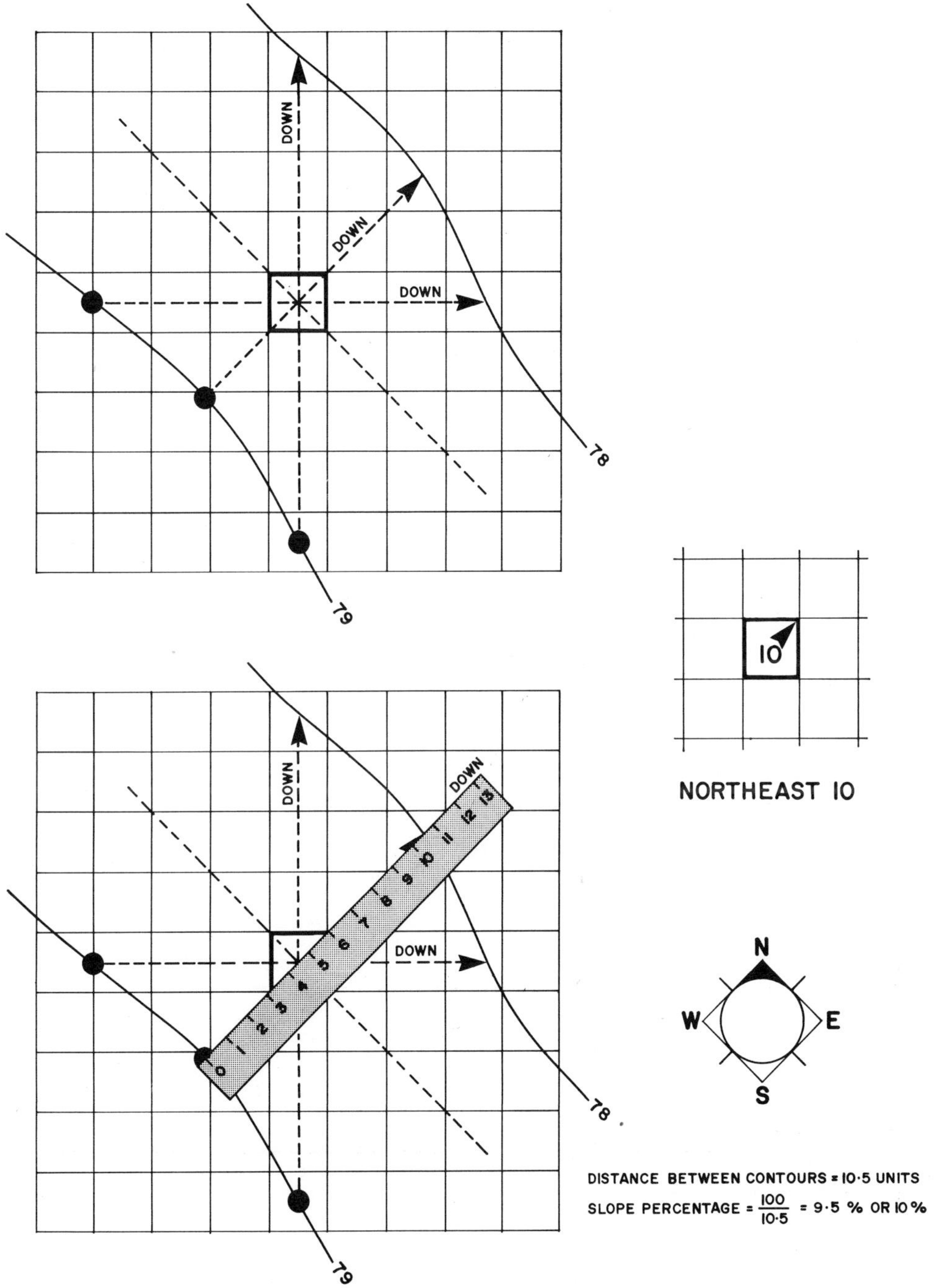

3-18. Eight nominal orientations can also be incorporated.

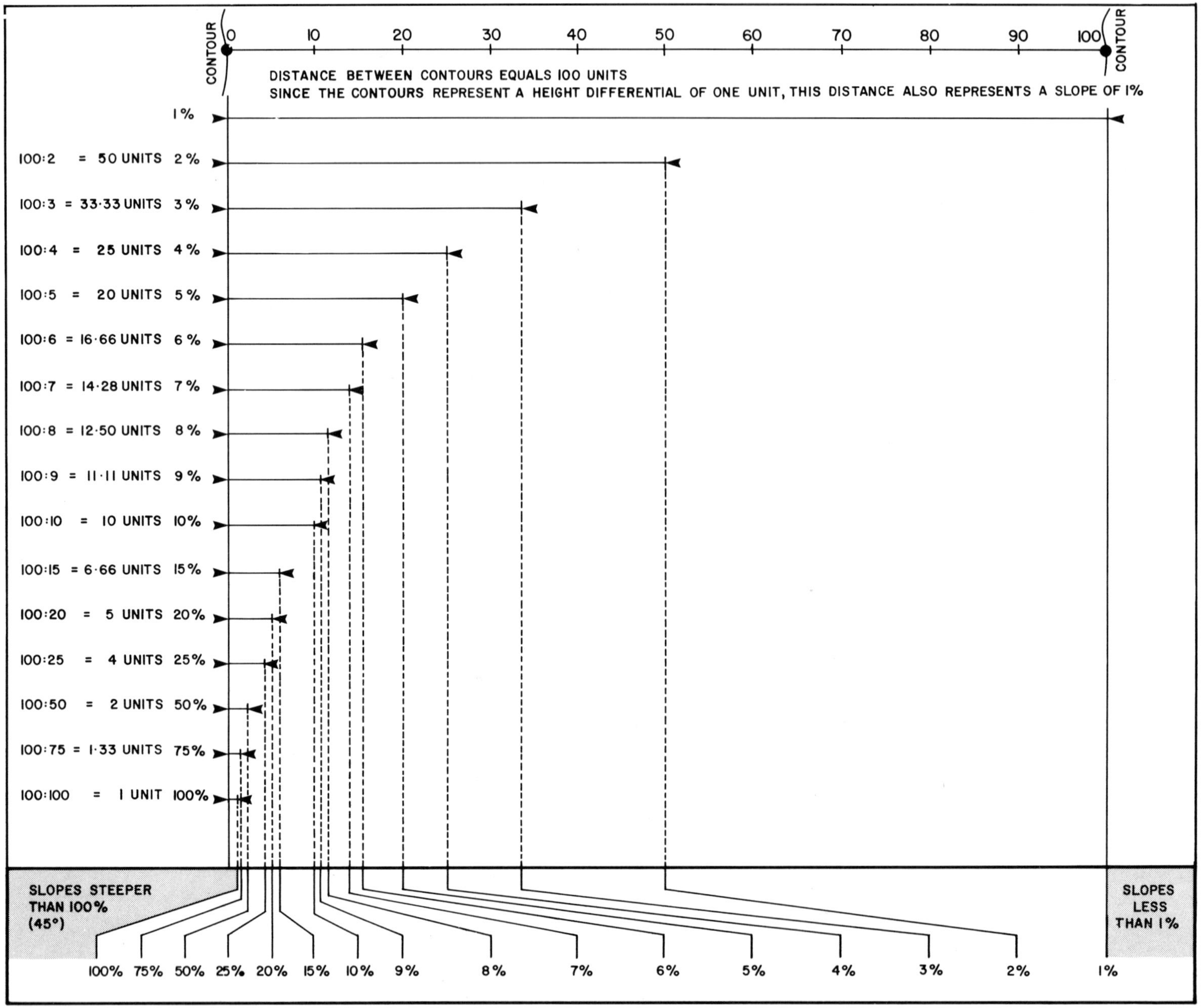

3-19. A measure tape can be constructed to read the direct slope percentage. All dimensions are expressed in units to allow application of both metric and imperial systems.

Land as a Natural Solar Collector

As a result of the exercise to describe land forms, or the Slope Descriptive Synthesis, a portrait of a given area as a natural solar collector emerges. Different parts of an area receive different amounts of solar energy. The size of a square indicates the severity of a topographical change and the arrow, shade, or color indicates the orientation. Furthermore, all the information can be expressed in numbers.

This general description of a given area can be very useful to a land economist, planner, developer, and architect. Since it consists of data related to static as well as to dynamic environmental forces, many strategic decisions can be defended with crystal-clear logic. Such strategic decisions could involve selecting areas for particular land uses, comparing two or more areas economically, or locating different land uses within a given area. Where a particular decision about land use is based on reasons outside the ecological context, the penalty, in terms of energy consumption and cost, can be calculated.

Interpretation and Application of the Land Form Description

In the first scenario, a single-family detached house is the only building form allowed in planning a new residential neighborhood. If the individual houses are to be energy efficient and take advantage of local environmental conditions, particularly insolation, then the most advantageous locations would be southern slopes and flat areas. Following the instructions of the Land Form Description, minimum densities should be proposed for northern slopes. But in certain cases, such as steep northern slopes, minimum densities might lead to other cost penalties, like longer roads and services. Thus, when the maximum permissible lot size limit is reached, a different strategy for developing northern slopes would have to be introduced.

Houses built on smaller northern lots would have a limited ability to use direct exposure to the sun. The superinsulated house might be one of several architectural/engineering solutions to steeper, northern slopes. It requires almost no energy input to offset diurnal and seasonal variations. Superinsulation uses heat produced by the human body and home appliances to secure thermal equilibrium without any need for a furnace, including a solar one. It is only necessary to retain heat.

In a second scenario, a planner is at liberty to create different residential densities within a neighborhood. A structure that can accommodate two or more families is inherently more energy efficient than a detached single-family house; the surface/volume ratio of a larger structure is more favorable, and the structure is therefore less susceptible to environmental stress caused by diurnal and annual variations. Furthermore, a multifamily building with its larger volume generates more energy from the human body and from appliances, which, in turn, reduces the expense of maintaining the necessary thermal equilibrium throughout the year. A planner might conclude that for maximum cost efficiency, dwellings with the highest permissible densities should be located on northern slopes.

In the absence of a holistic system of monitoring energy, the design process must combine art and intuition. One option that makes efficient use of energy and creates a logical layout is a northern slope as part of an open-space system designed for winter recreation.

Both the Ecochart and Land Form Description can show how to use land form to result in greater energy efficiency and land that is separated into a variety of lots that are thus aesthetically pleasing. In the past, lots with only a few standard widths and depths were used; a pie-shaped lot on a dead-end street provided the only visual excitement. From the point of view of aesthetics, function, and economy, airfield flatness was considered the ideal. The detailed topographical report provided by the Land Form Description about an area and its environment presents a picture of the area as a natural solar collector; the different levels of available sun energy suggest opportunities for energy-efficient land use. It also shows that this natural solar collector (natural topography) was not created with contemporary human needs in mind. The constraints, however, are usually more than offset by the opportunities, since, in practice, one will find that flat areas or some southern slopes exist in areas where the topography seemed at first glance to be unyielding. A closer scrutiny of the Ecochart and the Land Form Description might show that certain readjustments in the land form could result in greater energy efficiency for a developed area. In fact, these tools enable a designer and planner to design an ideal energy-efficient landscape. And if the cost of fuel justifies it, bulldozers should be kept busy creating this ideal.

CHAPTER 4

HELIOMORPHIC FORM GENERATION

GENERATION OF FORM

In the recent past, an energy-devouring mode of living all but eradicated the West's intimate relationship with natural forces. Mechanical systems made the existence, duration, and direction of sunlight irrelevant. Houses and communities were no longer tethered to the solar furnace. Recent energy shortages, however, make it necessary to renew the ties between the sun and the human-made environment.

Buildings constructed when energy was relatively inexpensive were neither planned nor located in response to their natural surroundings. The shape of an individual building was, and often is, the product of a variety of economic, political, and social pressures, such as its resale value, the commercial value of its land, the size of the land, building codes, and other related restrictions. The relationship between buildings is therefore often the product of an inadequate accounting system, the by-laws governing the site and, of course, pure accident. A planning system based on an understanding and the insightful manipulation of local environmental conditions makes it possible to plan, shape, and situate a building in a different manner. The proposed planning framework can mitigate seasonal and diurnal variations and can thus replace, limit, or reduce the need for mechanical, cooling/heating, energy-devouring systems.

This planning system is based on the premise that any building in a human settlement should be in harmony or have a complementary relationship with its natural and human-made surroundings. It is the planner's role to determine the physical limits of this specification by constructing an envelope that contains the maximum developable space. The size and shape of this volume is derived from the size and shape of the impact or influenced area. As already stipulated, one building should not negatively influence, that is distort, the thermal equilibrium of another building. This, of course, means that the envelope must also be derived from the geometry of the diurnal and annual rhythm of the sun. The following is a digest of the process involved in generating a Solar Envelope.

Skydome

An observer facing south sees the sun move from east to west. The sun's path has three important points:

sunrise—the lowest point
midday peak (noon)—the highest point
sunset—the lowest point

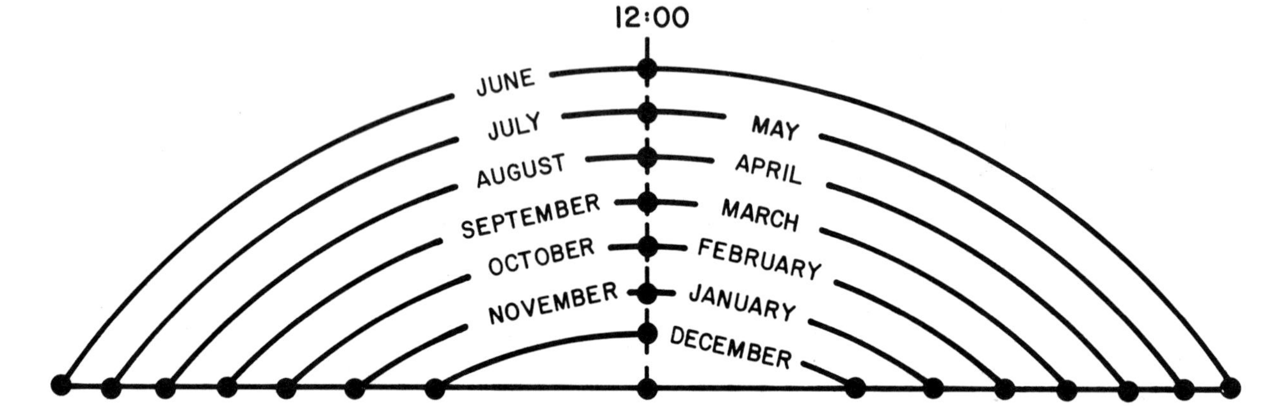

4-1. A diagram showing the annual sun path.

All three points change position continuously throughout the year. Standing in the center of a transparent dome would allow one to see and record the sun's paths year-round. Lines representing the movement of the sun would exist between two extreme situations, its path on June 21, the longest day of the year, and its path on December 21, the shortest day of the year (fig. 4-1).

All paths traveled by the sun can be drawn on the wall of the Skydome. In the northern hemisphere, this drawing stretches from the eastern to the western horizon; its width encompasses the two extreme sun paths. This model represents the most important part of the sky for a person or a system located in the center of the Skydome. In the absense of another energy source any obstruction of the sun is not just a minor nuisance but could be a matter of life and death for systems, either living or human-made.

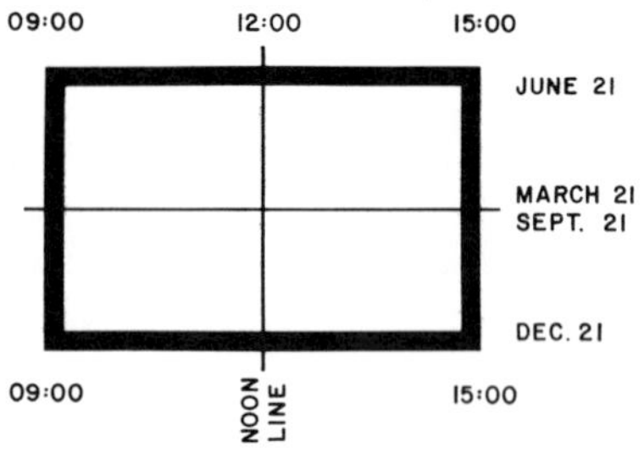

4-2. An example of Skywindow size.

Skywindow

The path taken by the sun is vitally important to living systems on the earth's surface. If obstructing this path were prohibited, then neighbors could not exist side by side, but would position their dwellings one behind the other.

The value of energy from the sun, however, changes throughout the day. The angle of incidence and the atmosphere make the most valuable sunlight for practical consumption available near the peak of the daily sun path. Years of research have led to a general consensus that the sunlight with the highest energy density is harnessed between 09:00 hours and 15:00 hours. If this limitation is drawn on the Skydome, it can be used to create a Skywindow (fig. 4-2).

In essence, a Skywindow is an opening in the Skydome (fig. 4-3). The geometry of the window guarantees direct access to the sun within a stipulated daily and annual time frame. The Skywindow is a physical expression of the right to sunlight and can easily be translated into the legal language of any municipal bylaws.

Solar Pyramid

A Skydome with a Skywindow can serve not only as a passive tool to demonstrate a legal right to sunlight, but also as an active, useful design instrument. As previ-

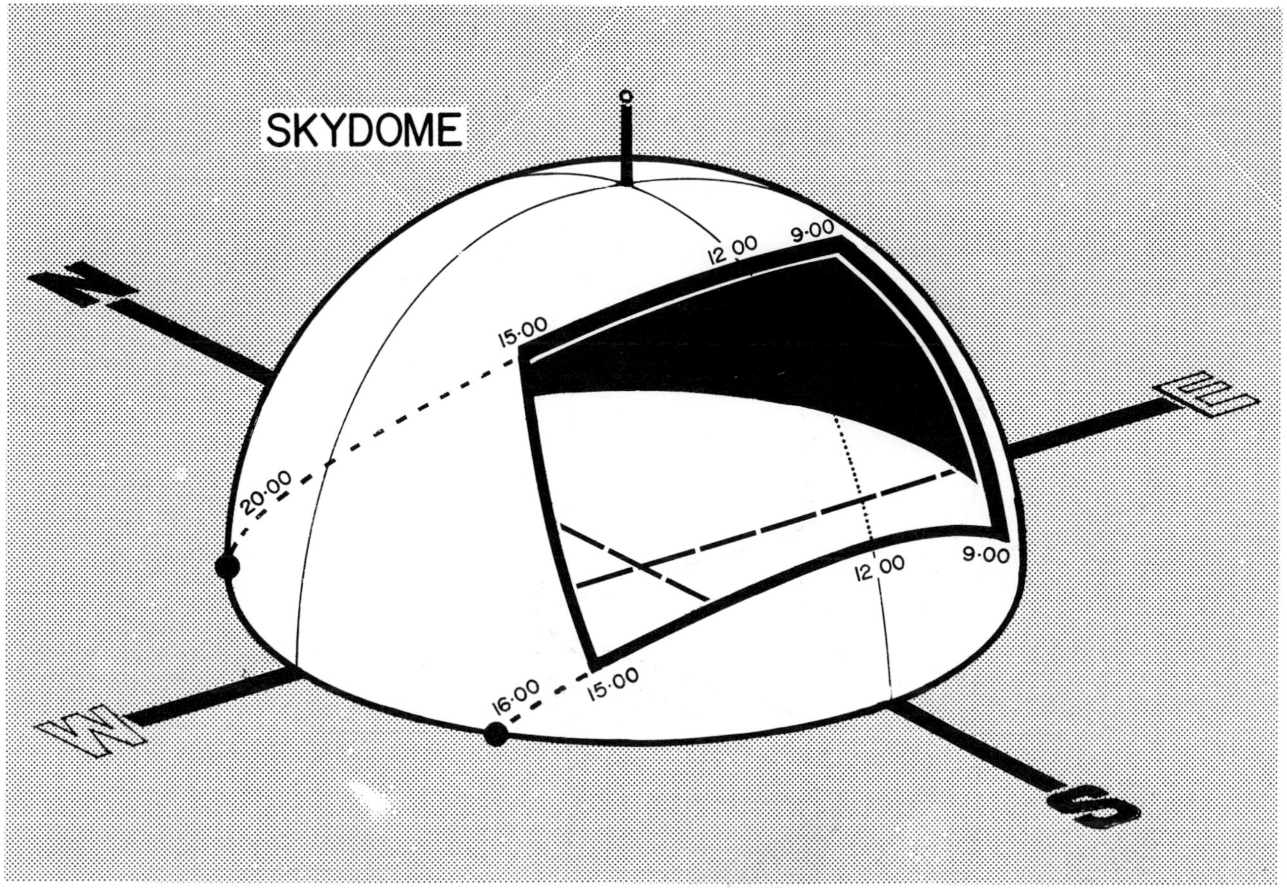

4-3. A Skydome and Skywindow.

ously described, only at the center of the Skydome can the full effect of the sun's rays that pass through the Skywindow be felt. In order to provide larger areas with the level of sun exposure that the dimensions of the Skywindow have determined are available, it is necessary to develop a Solar Pyramid.

First, the Skydome must be elevated from the horizontal reference plane (the ground). Consequently, the Skydome will function as a lens and will project the Skywindow onto the horizontal plane (fig. 4-4).

The size of the projected Skywindow is directly related to the height of the Skydome above the horizontal plane. The Solar Pyramid is created by connecting the four corners of the Skywindow projection with the center of the elevated Skydome (fig. 4-5). The size and volume of the Pyramid can be manipulated by a vertical movement of the Skydome.

Any object constructed within this volume will have a predictable impact on its surroundings. A building might completely fill a Solar Pyramid (the Solar Pyramid would represent the shape of the building). In this case, the base and the impact area of the building

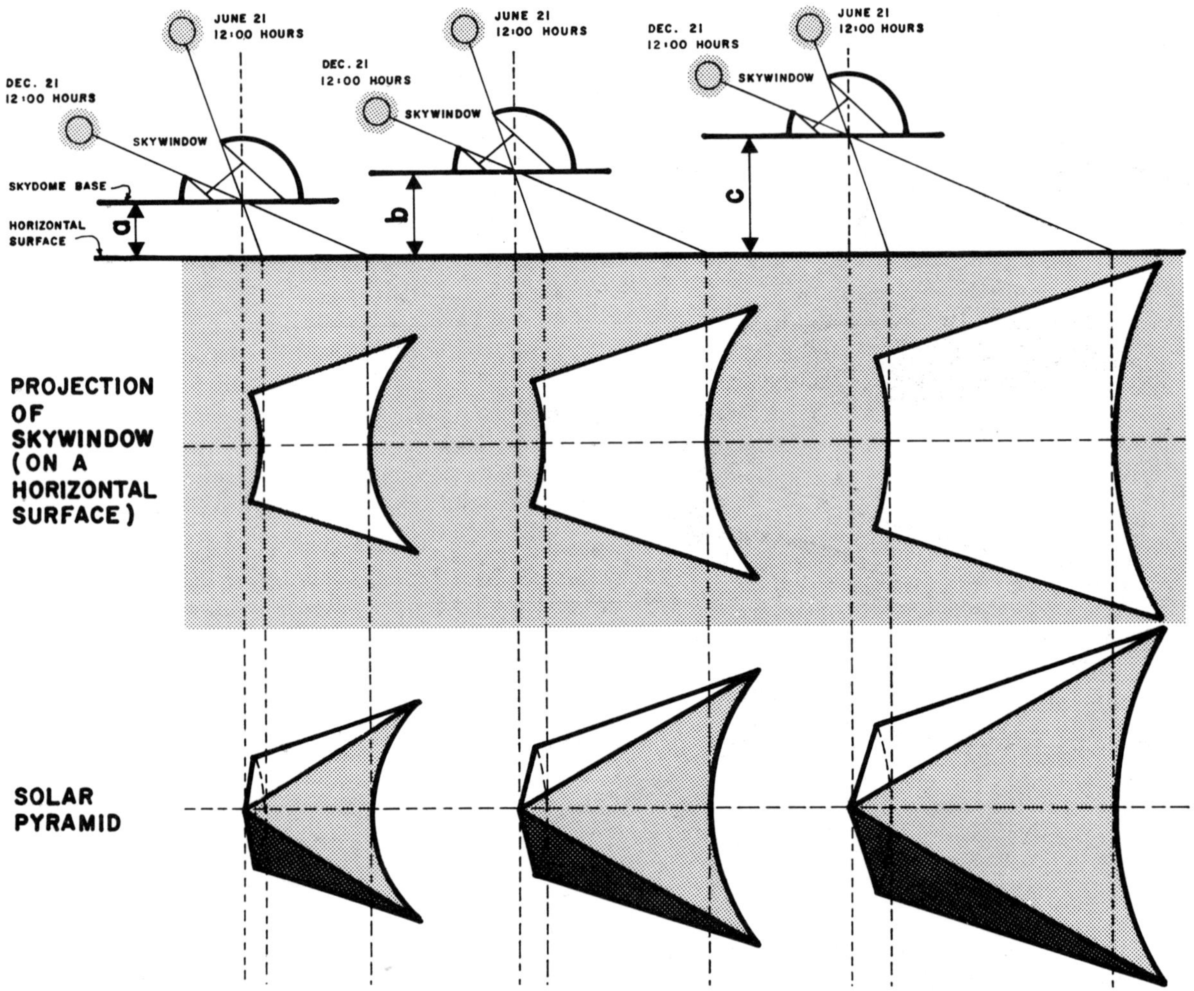

THE THREE PROJECTIONS OF THE SKY WINDOW ON A HORIZONTAL (FLAT) SURFACE ARE EQUIVALENT TO A COMPOSITE (ANNUAL, DIURNAL, 09:00 – 15:00) SHADOW PATTERN OF POLES OF THREE DIFFERENT HEIGHTS.

VOLUME OF THE SOLAR PYRAMID IS DETERMINED BY THE HEIGHT "a", "b", AND "c".

4-4. The generation of a Solar Pyramid.

would be equivalent. If the Solar Pyramid is derived from the geometries of the Skywindow, as shown in figures 4-2 and 4-3, then this building would not cast any shadows on its neighbors between 09:00 and 15:00 hours.

Because a building shaped like the Solar Pyramid may be cumbersome to construct, the actual building usually takes less space than that which is stipulated or allowed by the geometries of the Solar Pyramid; the Solar Pyramid thus represents the maximum developable space.

The construction of a Solar Pyramid is a relatively

simple drafting task with the help of Solar Templates (see chapter 8). The base of the Solar Pyramid is generated by a composite of the diurnal and annual shadow pattern of a pole. The height of the pole is equivalent to the height of the Pyramid (fig. 4-6).

4-5. The axonometry of Solar Pyramid generation.

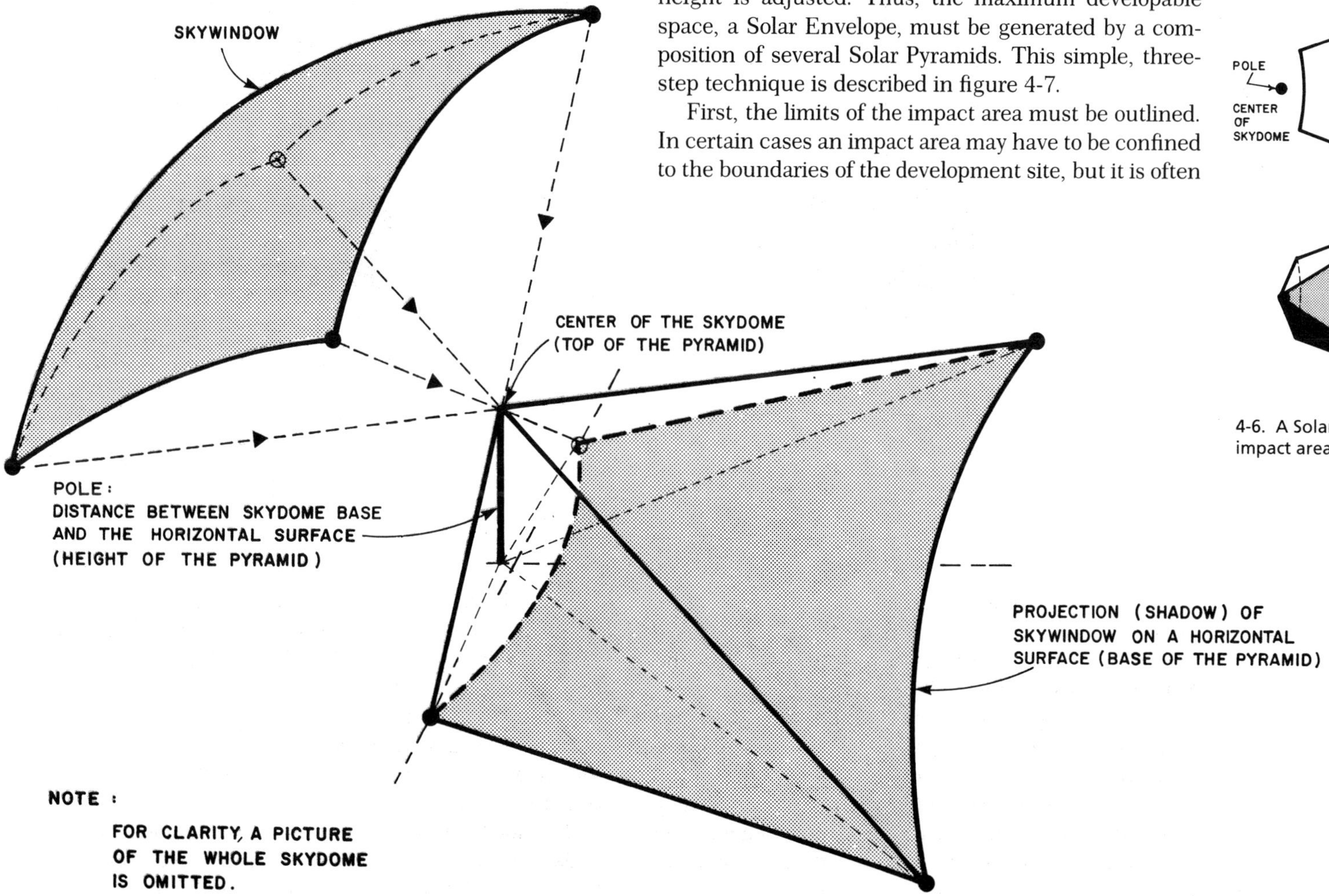

Solar Envelope

The Solar Pyramid is the simplest, most basic Solar Envelope. In planning and design practice, the maximum developable space is usually dictated by the size and shape of the maximum permissible impact area. However, such an impact area may be oddly shaped and impossible to cover with a single projection of a Solar Pyramid, regardless of how much the Pyramid's height is adjusted. Thus, the maximum developable space, a Solar Envelope, must be generated by a composition of several Solar Pyramids. This simple, three-step technique is described in figure 4-7.

First, the limits of the impact area must be outlined. In certain cases an impact area may have to be confined to the boundaries of the development site, but it is often

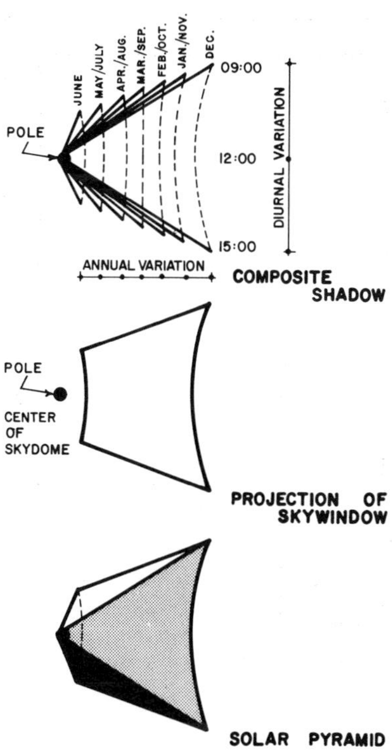

4-6. A Solar Pyramid generated by the impact area of a pole.

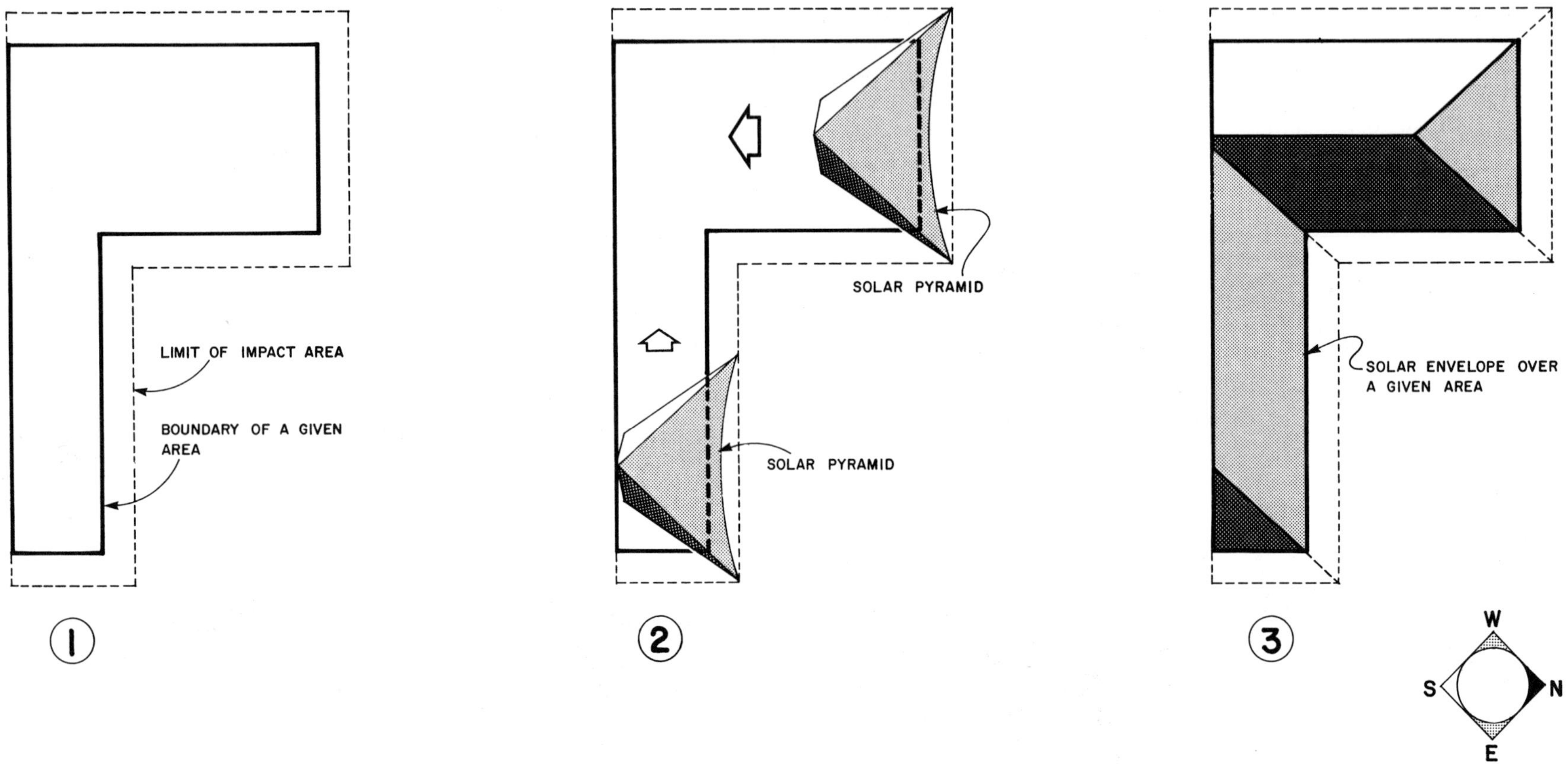

4-7. The generation of a Solar Envelope.

acceptable to extend it partially over public spaces, streets, back lanes, and even a neighbor's private property. As can be seen in figure 4-7, in order to shape the Solar Envelope, two Solar Pyramids of different heights (volumes) must be used. The heights of the Solar Pyramids are derived from the shape of the maximum impact area by raising or lowering the Skydome. In the second step, the Solar Pyramids are moved horizontally in the direction shown by the the arrows. Finally, the resultant Solar Envelope is trimmed along property lines.

Solar Envelopes for development along east/west streets are created by a simple east/west movement of the Solar Pyramid (fig. 4-8). Note how the volume of the pyramid is dictated by the size of the maximum permissible impact area. A Solar Pyramid may be moved in any direction, but it must be understood that the shape of the Solar Envelope will differ accordingly (fig. 4-9, elevations 1–5).

The following example illustrates how Solar Pyramids may be used to generate Solar Envelopes over large areas. A wide tract of land that is adjacent to a

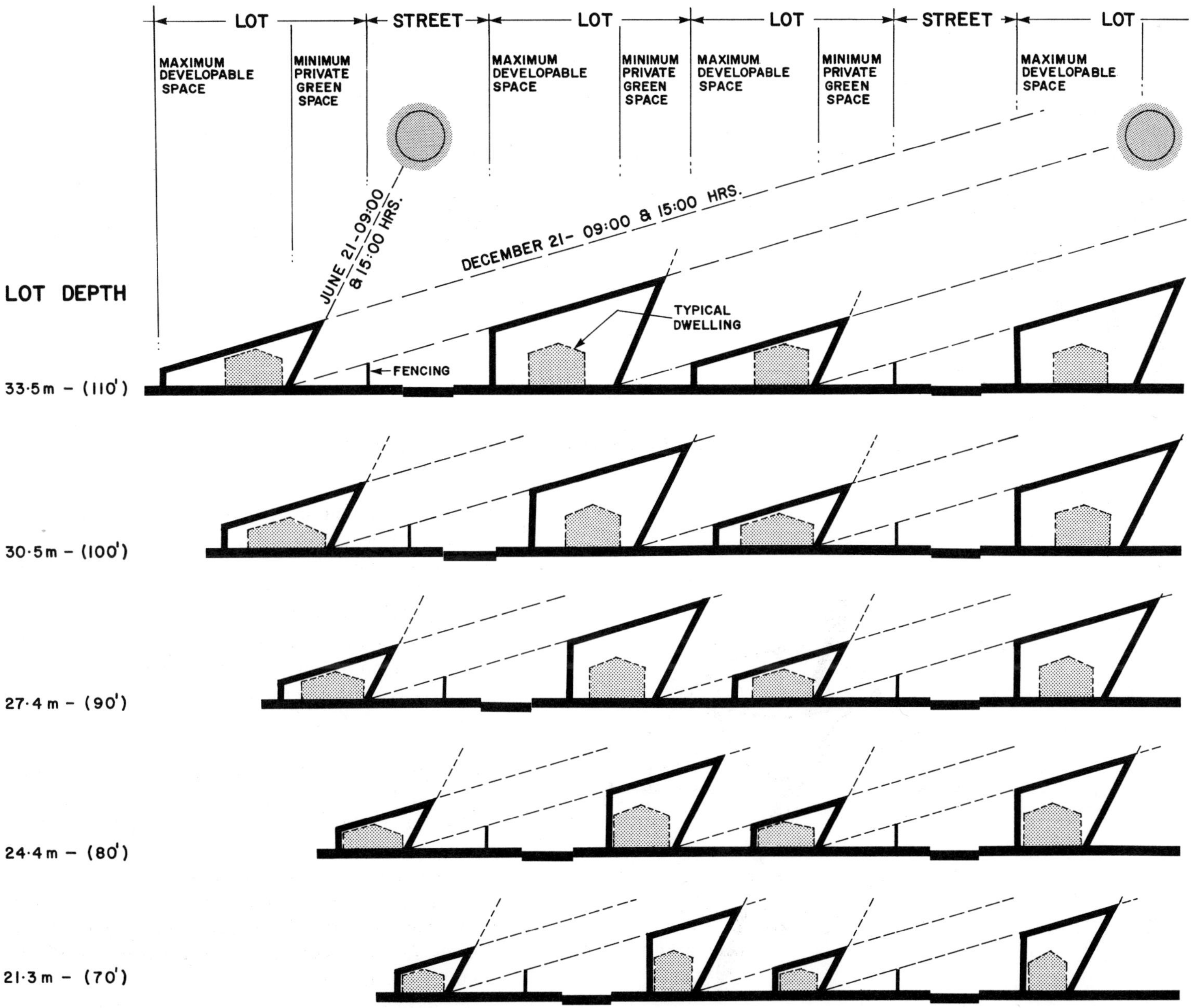

4-8. Solar Envelopes on an east/west street.

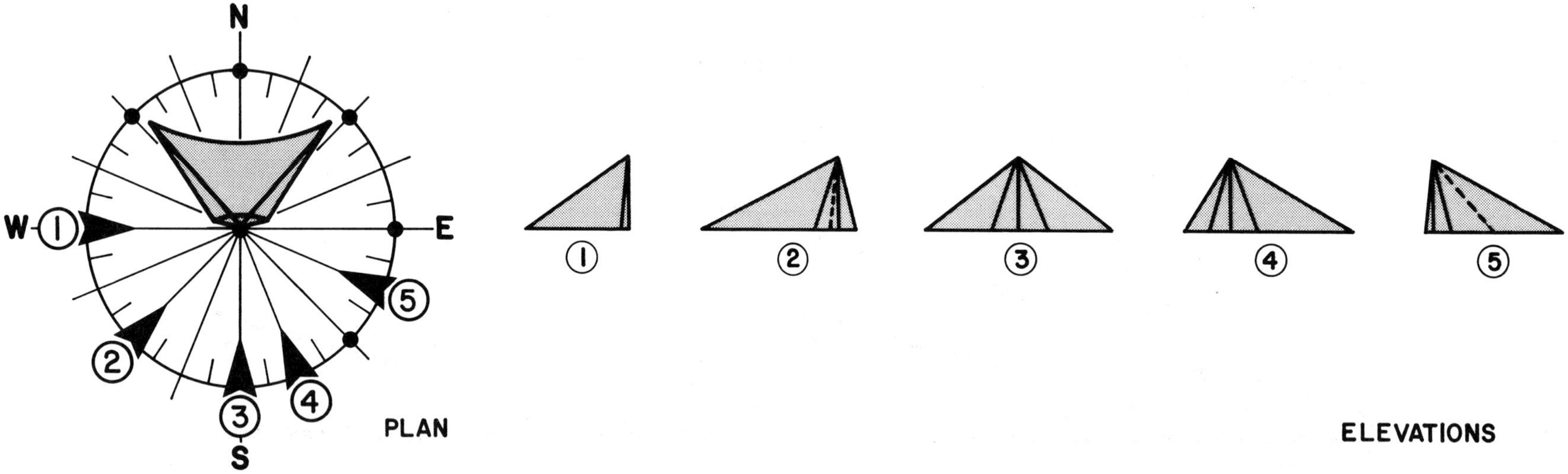

4-9. Solar Envelopes generated by the omnidirectional movement of the Solar Pyramid.

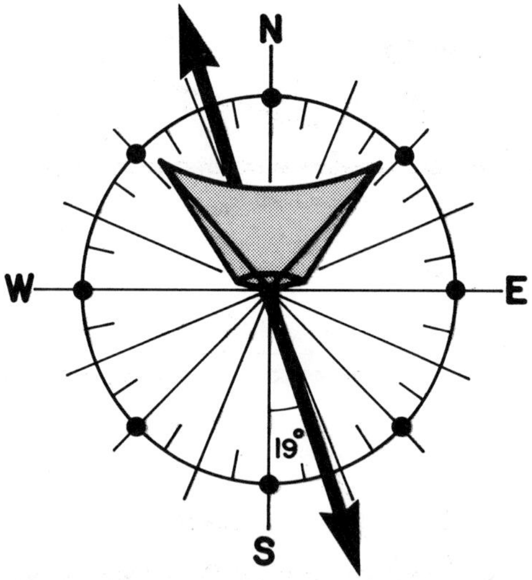

4-10. Direction of the street corridor.

major thoroughfare is being rezoned to accommodate commercial development. A justifiable concern is expressed regarding the potential overshadowing of existing residences in adjacent areas, mostly single-family housing. The axis of the street is 19° eastward off due south (fig. 4-10); thus, the Solar Pyramid must be moved along in the direction shown by the arrow. First, it is necessary to determine the extent of the impact area and to construct a projection of the Solar Pyramid. In figure 4-11(1), a projection of the Solar Pyramid, or "Superenvelope," is shown as seen from the middle of the street, facing southward. In figure 4-11(2), the actual street corridor is carved out. If setbacks are a legal requirement, additional volume must be taken out of the Superenvelope as shown in figure 4-11(3). Any development within the limits of the Solar Envelope would have a predictable impact on the adjacent residences; no shadows would be cast between 09:00 and 15:00 hours throughout the year.

Performance criteria may be added. For example, if pedestrians are anticipated, and pedestrian amenities and comforts required, the street would require direct access to sun. The same predictable impact may therefore be required for the pedestrian part of the street. Figure 4-11(4) shows how this may be accomplished.

Since height limits are often stipulated by law, a fifth step is included (fig. 4-11[5]). In this case, the additional volume must be taken from the Solar Envelope. On the other hand, when the Solar Envelope is constructed, such arbitrary height ceilings are rarely either justifiable or sensible; the Solar Envelope precisely determines the height of a planned structure and will lead to a site-specific height determination.

In summary, a Solar Envelope can be constructed over any permissible impact area shape. In most cases, this is done by changing the height and horizontal movement of the Solar Pyramid. A method of constructing tailor-made Solar Envelopes over very complex impact areas is described in chapter 8.

Solar Envelopes are carved out of invisible urban space. They can be either partially or fully filled by a building. In reality, however, an architect would use the Solar Envelope as a general description of a built form, specifying building height and area of influence. To comply with the current professional apartheid, the construction of a Solar Envelope, the maximum developable space, should be the prime responsibility of the urban planner.

Solar Envelopes and Land Form

Thus far concepts for generating forms have been presented as if an abandoned airfield were being developed. Zero-slope situations, however, are actually quite rare. Even a seemingly flat plateau is usually composed of rather definite topographical variations, as was shown in the Land Form Description exercise in chapter 3. Thus, the maximum developable space has to be constructed in conjunction with the topography, be-

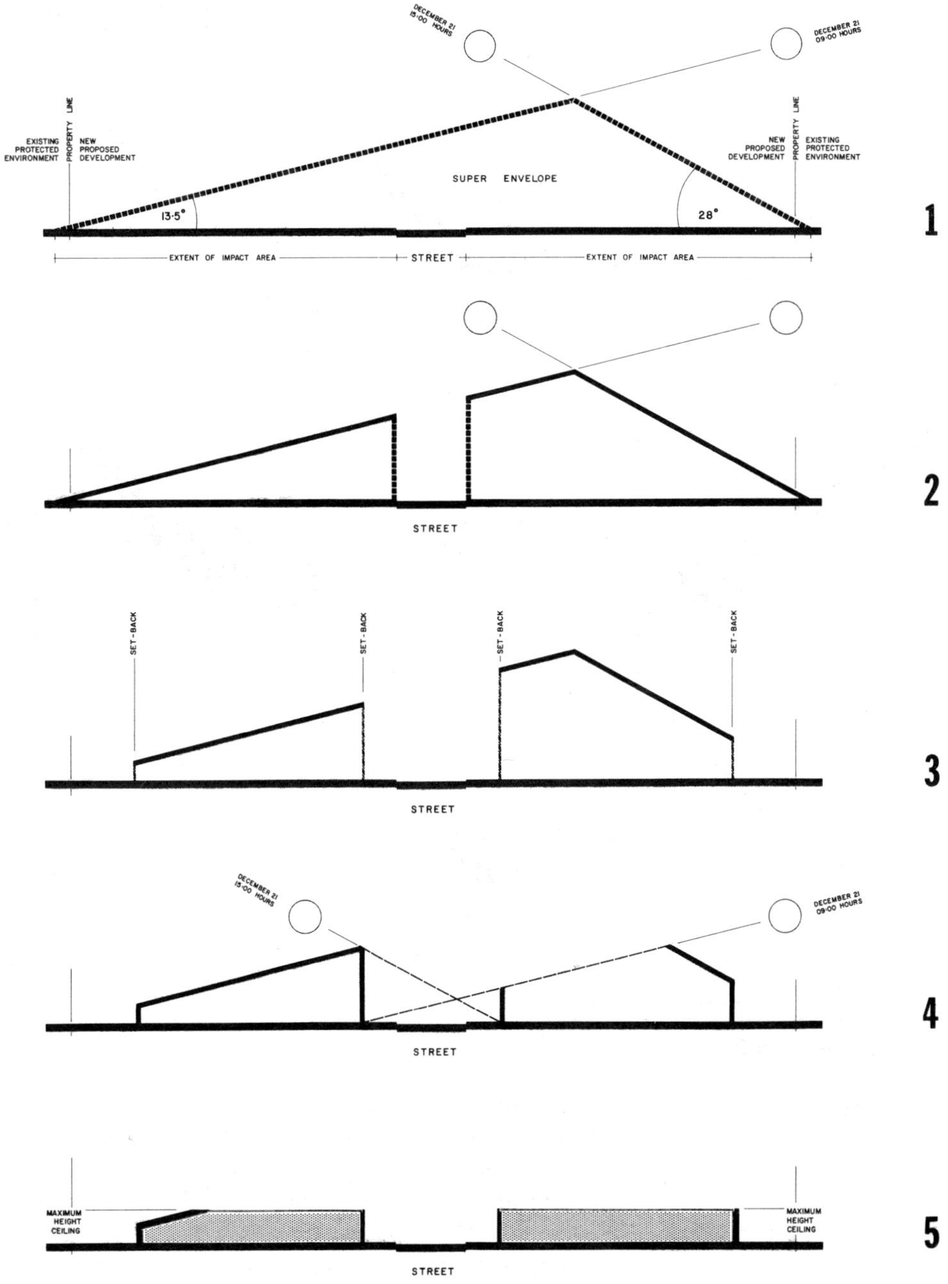

4-11. The construction of a Solar Envelope over a specific corridor.

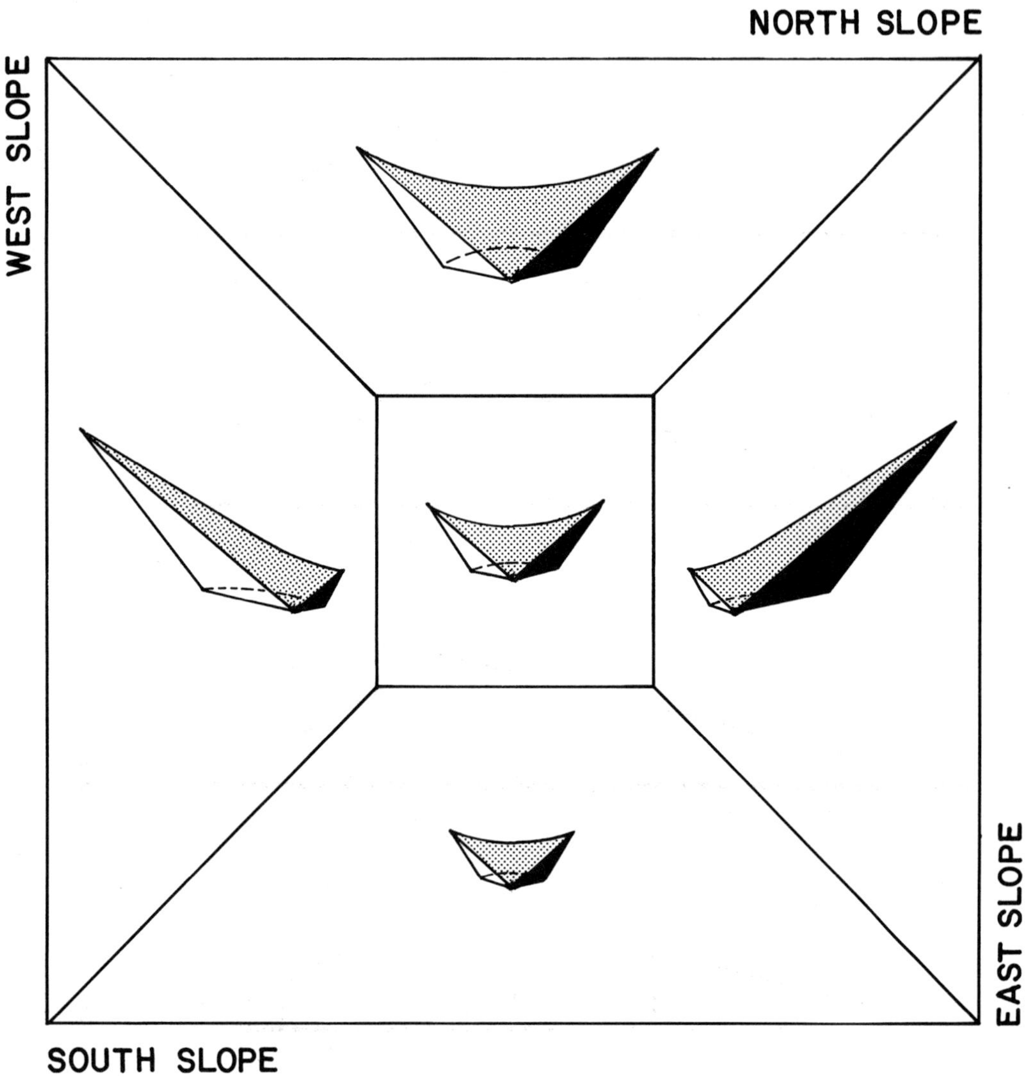

4-12. The Solar Pyramid and land form.

cause Solar Envelopes are affected by the sun's rays and the land form in planning and design practice.

In figure 4-12, five Solar Pyramids of the same height are placed over a nominal hill with four sides and a flat top. The shapes of the Solar Pyramids are distorted by slope orientation and slope gradient. At first glance, the task of constructing these distorted Solar Pyramids may appear to be a formidable exercise in descriptive geometry, but it is actually quite manageable. Land Form Description recognizes either four or eight nominal slope orientations; this is also the range of directional variation used by the planner. Solar Templates can be used to construct Solar Pyramids for these nominal slope orientations. Once the concepts of the Skywindow and Solar Pyramid are understood, the construction of a maximum developable space or envelope for any nominal slope direction becomes a typical drafting task (see chapter 8).

GENERATION OF BUILT FORM

The format of this book does not allow for details on the generation of built forms. Concepts associated with the design of built forms, however, should be of interest to the urban planner. If a Solar Envelope were to be filled with two-by-fours, concrete, bricks, and mortar, it would result in a building with a predictable impact on its neighbors within a stipulated time frame. Such a building, however, might be costly to build and have internal divisions that are inefficient for housing human activities. A building designer traditionally reaches a compromise between two considerations, that of providing a comfortable space for human activities and of protecting inhabitants from external environmental forces. In many cases, the shape of a building is the result of a compromise between these conflicting requirements. Two additional concepts that any urban planner and architectural designer must consider are surface/volume ratio and the way sun energy is received by the built form.

Surface/Volume Ratio

Every building is subject to pounding by environmental forces. One way of measuring its susceptibility to the influence of the composite environmental force is by its surface/volume ratio. The less surface needed to en-

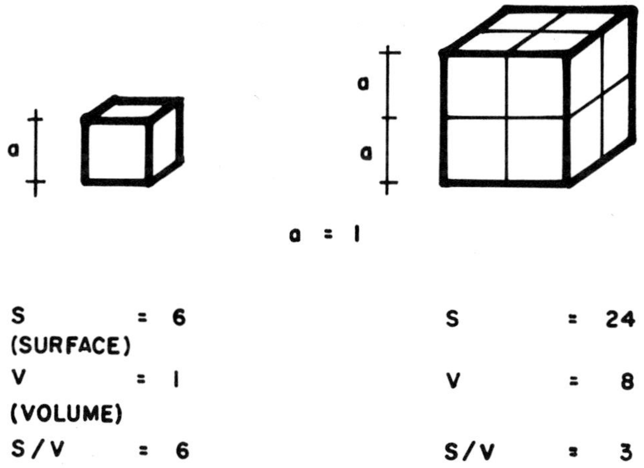

4-13. The surface/volume ratio of a cube.

close a given volume, the less the system is affected by variations in external stress. In other words, it becomes easier to maintain a steady state in unsteady surroundings.

The classic illustration of surface/volume variables is an expanding cube (fig. 4-13). A sphere possesses the best surface/volume ratio. All other shapes require a larger surface to enclose the same volume. It therefore becomes easy to evaluate the relative efficiency of other forms by selecting a number of the most familiar or most frequently encountered geometric forms and comparing their surface/volume efficiency with that of the sphere.

This exercise is presented in figure 4-14. Nine different solids were chosen, and the surface/volume ratio of the sphere was established as being 100 percent. It must be understood, however, that this chart has certain limitations.

1. Surface/volume ratio is not the only indicator of susceptibility to external stress. Other properties of a building envelope, such as the material of which it is made, may be more significant than its size.
2. Certain shapes are suitable only for certain purposes. For example, a sphere may be an excellent shape for containing liquids or gases, but it is very tricky to divide and use it for human occupancy.
3. Materials and methods of construction may indicate which geometric form should be used. For example, a sphere can be constructed from identical panels; thus, there is no distinction between the wall and the roof. On the other hand, there is a need for more precision during the manufacturing and construction processes.

Even with these limitations, the conclusion to be drawn by a building designer and an urban planner is clear: buildings with greater volume not only provide an opportunity to increase internal metabolism, but their surface is a relatively smaller target for external forces. There is a physical limit to the improvements possible with a favorable surface/volume ratio. In a holistic energy account, one would have to consider the energy penalty in terms of the cost of manufacturing special building materials that cope with gravity and the cost of manufacturing and maintaining the complex mechanical devices (for example, elevators) needed for human occupancy.

Sun Energy Received by a Built Form

A building's relative efficiency as a solar collector depends on its shape (assuming it has been constructed within the limits of a Solar Envelope). Under a continuously moving sun, different shapes receive different amounts of sun energy at different times of the day and year. Figures 4-15 and 4-16 illustrate these variations in the sun energy received.

Information about the relationship between the shape of a form and its ability to receive energy from the sun is gained by viewing forms from the sun's position. The area that can be seen from this position is

	SHAPE	VOLUME	SURFACE	POSSIBLE FLOOR SPACE	VOLUME SURFACE EFFICIENCY
1	PARALLELOPIPED no. 1 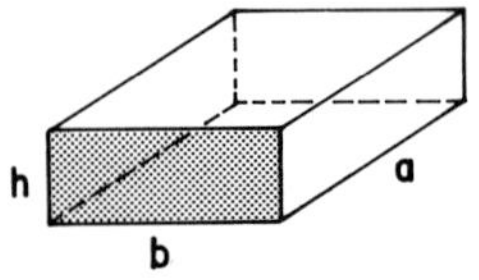 h = 10' a = 40' b = 25'	10,000 cu.ft. 283 m^3	3300 sq. ft. 306·6 m^2	1000 sq.ft. 93 m^2 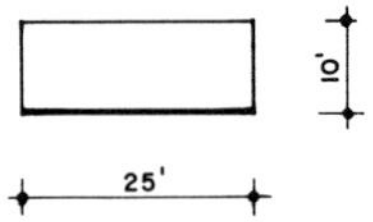	68·01 %
2	PYRAMID 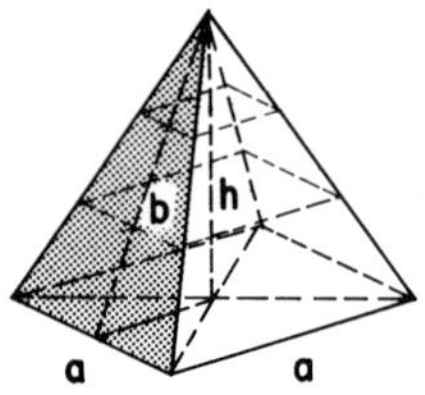 h = 30' a = 31.6' b = 33.9'	10,000 cu.ft.	3144·7 sq. ft. 292·2 m^2	1555·5 sq. ft. 144·5 m^2 a1 = 21·1' a2 = 10·5' 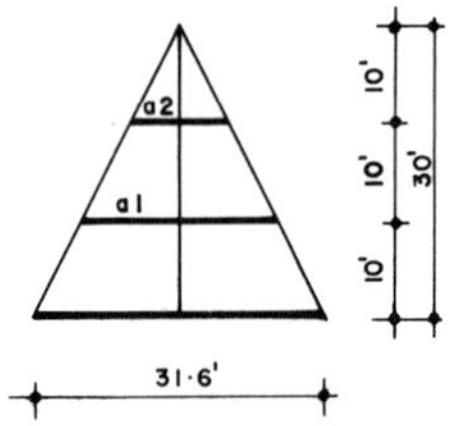	71·37 %
3	CONE 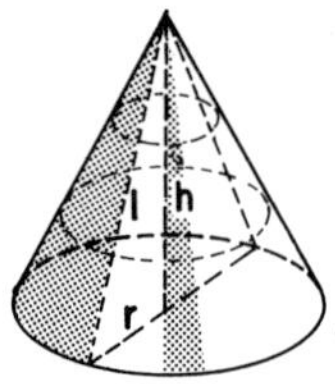 h = 30' r = 17·8' l = 34·9'	10,000 cu. ft.	2956·4 sq.ft. 274·6 m^2	1555·5 sq.ft. 144·5 m^2 r1 = 11·9' r2 = 5·9' 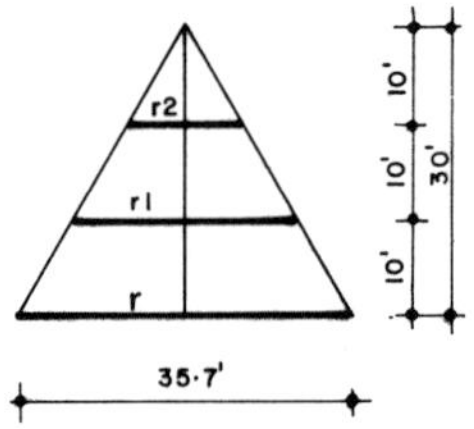	75·92 %

4-14. A chart comparing the volumes and surfaces of nine selected solids.

	SHAPE	VOLUME	SURFACE	POSSIBLE FLOOR SPACE	VOLUME SURFACE EFFICIENCY
4	PARALLELOPIPED no. 2 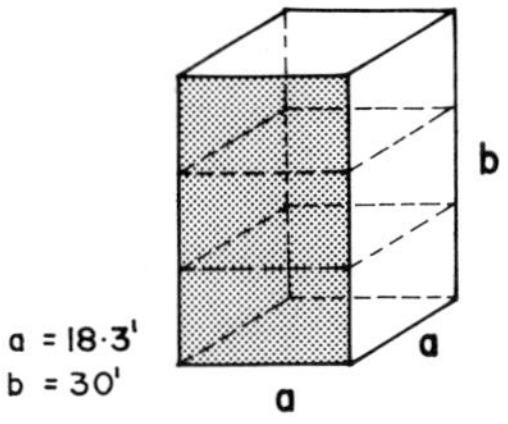	10,000 cu.ft.	2857·5 sq. ft. 265·5 m^2	1000 sq. ft. 93 m^2 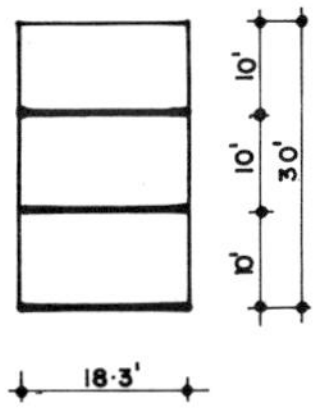	78·55%
5	CUBE 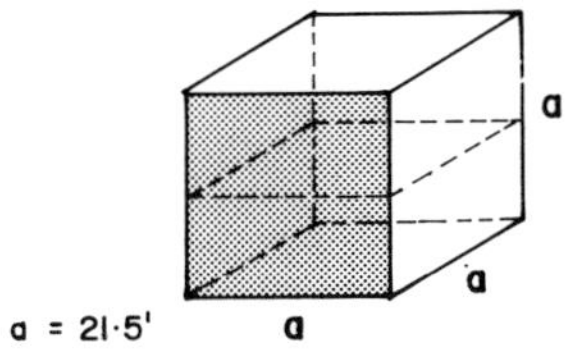	10,000 cu.ft.	2784·9 sq.ft. 258·7 m^2	928·3 sq.ft. 86·2 m^2 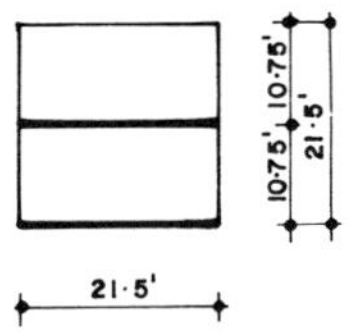	80·60 %
6	HEMISPHERE 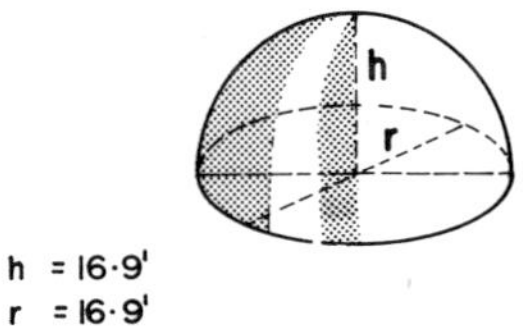	10,000 cu. ft.	2676·1 sq.ft. 248·6 m^2	892·0 sq. ft. 82·9 m^2 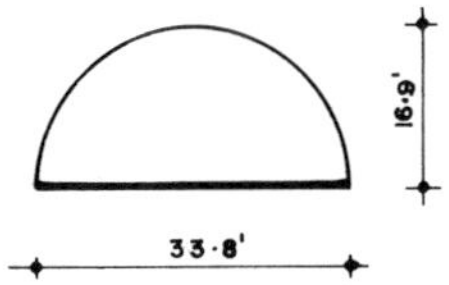	83·87 %

4-14.

	SHAPE	VOLUME	SURFACE	POSSIBLE FLOOR SPACE	VOLUME SURFACE EFFICIENCY
7	CYLINDER 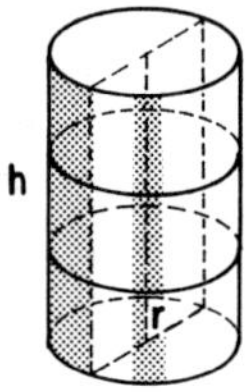 h = 30' r = 10·3'	10,000 cu.ft.	2608·1 sq. ft. 242·3 m^2	1000 sq.ft. 93 m^2 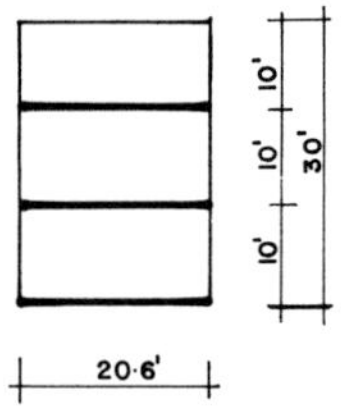	86·06 %
8	RIGHT CYLINDER 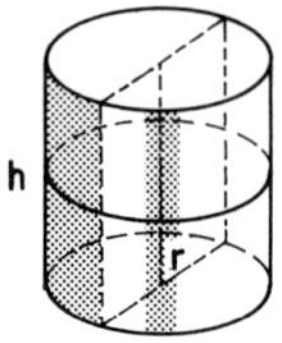 h = 23·4' r = 11·7'	10,000 cu.ft.	2569·5 sq.ft. 238·7 m^2	856·5 sq.ft. 79·6 m^2 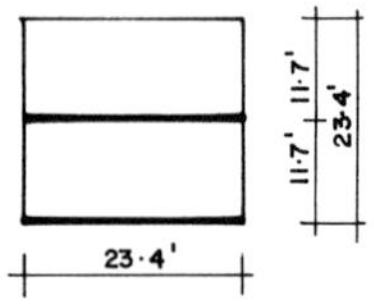	87·35 %
9	SPHERE 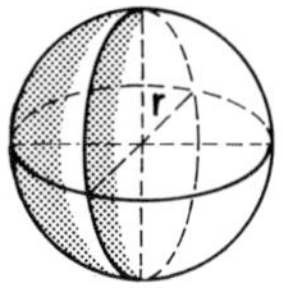 r = 13·4'	10,000 cu.ft.	2244·6 sq.ft. 208·5 m^2	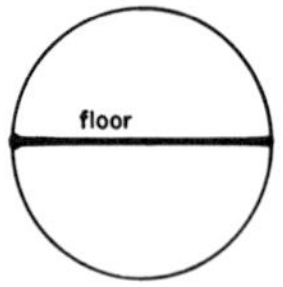 561·2 sq. ft. 52·1 m^2 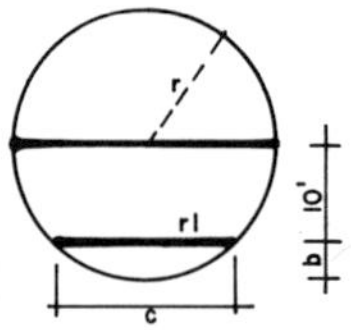 808·16 sq.ft. 75·1 m^2 b = r - 10 = 3·4' c = 17·7' $r1 = \frac{c}{2}$ = 8·9' 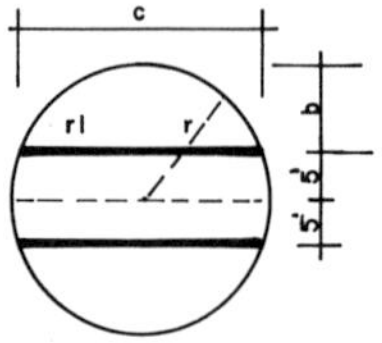 965·24 sq. ft. c = 24·8' $r1 = \frac{c}{2}$ = 12·4' b = 8·4' 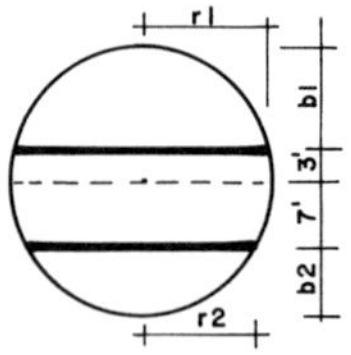 940·11 sq.ft. b1 = r - 3 = 10·4' b2 = r - 7 = 6·4' r1 = 13·0' r2 = 11·4'	100 %

4-14.

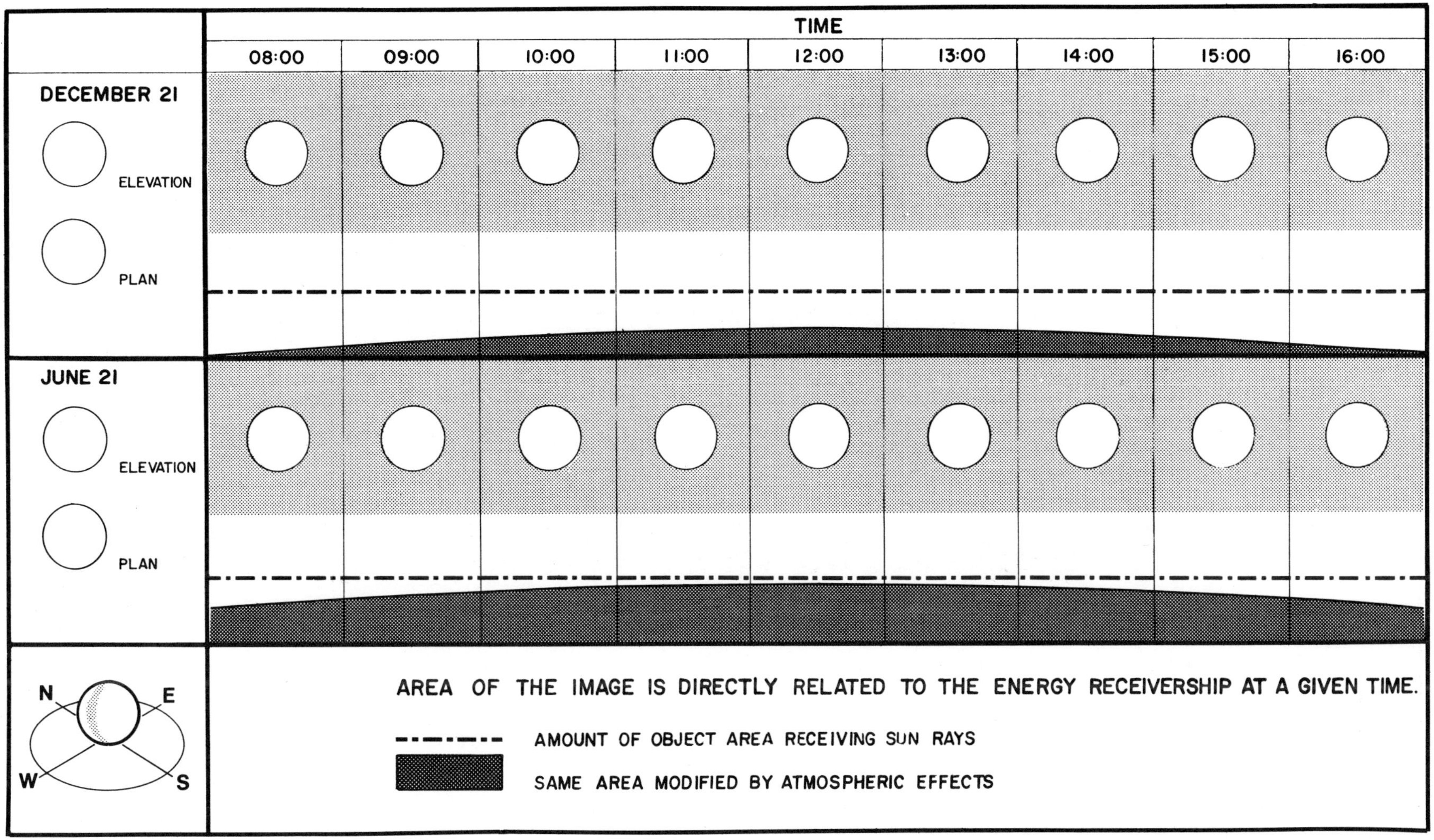

4-15. An image of a sphere generated by the sun's rays.

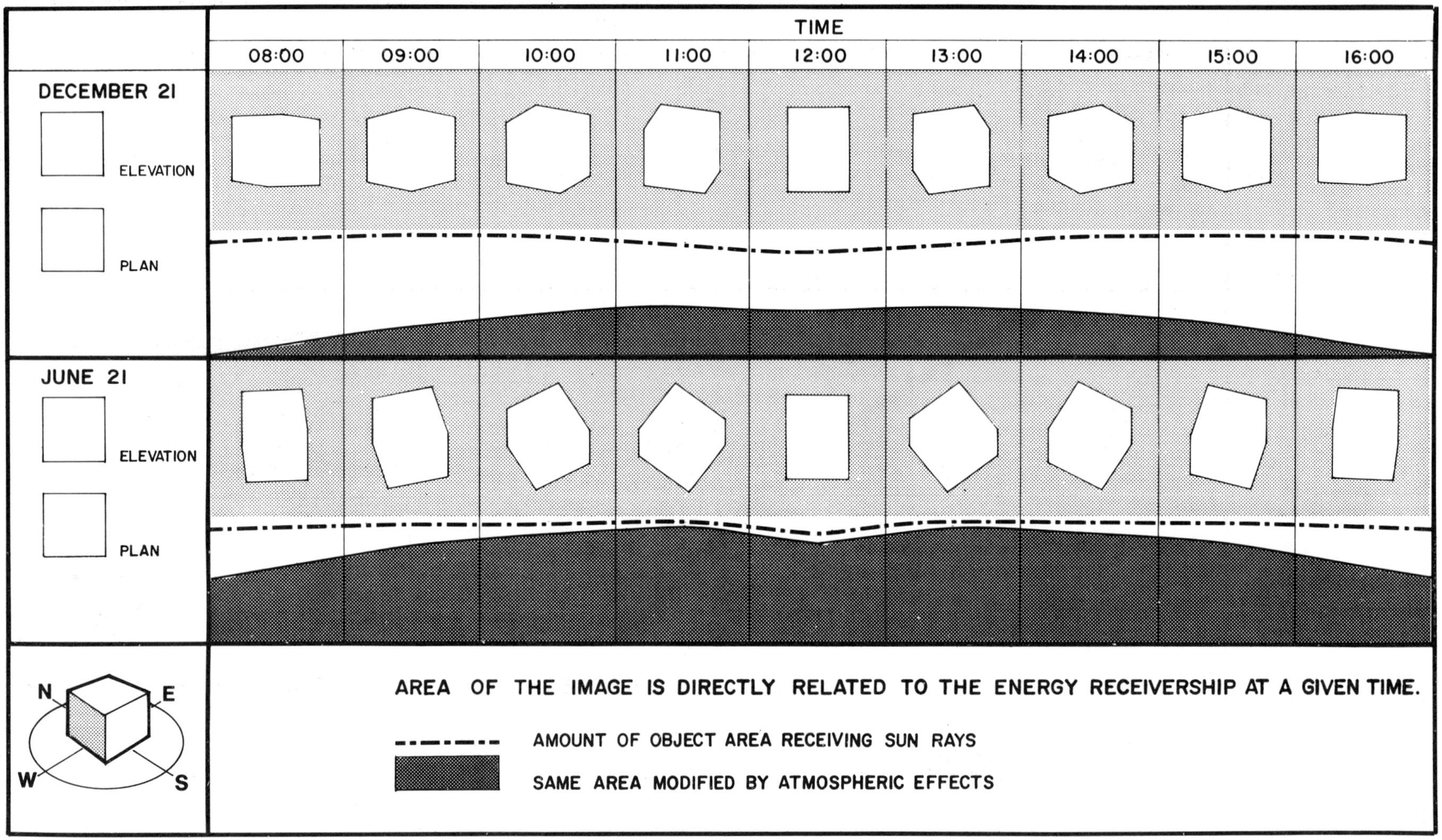

4-16. An image of a cube generated by the sun's rays.

directly related to the amount of sun energy an area receives.

Another way of describing this sun/shape relationship is to say that the degree to which sun energy is received is in direct proportion to the area of obstruction of the sun's rays caused by a given shape at a given time. Both definitions demonstrate the need to construct a projection (a shadow) of a given shape on a screen that is perpendicular to the direction of the sun's rays.

Figure 4-15 shows the image of a sphere as seen from the position of the sun on December 21 and on June 21 between 08:00 and 16:00 hours. This chart was constructed using sun path geometries at 43°N latitude. The sphere is a solid that, when viewed from the position of the sun, has the same shape from sunrise to sunset at any latitude. Therefore, the projected area of the sphere—the disk—does not change. Energy received would be constant, like that of a circular tracking collector. But, it would be constant only in space, because the presence of a gaseous atmosphere, vapor, and pollutants changes the energy received from the sun over time. The sun's rays have to travel through an atmosphere of different thicknesses, and thus energy received from a low sun cannot equal the energy received from a midday sun.

The same applies to the amount of energy received during different seasons. The annual variation is marginal between the tropic lines, but it increases dramatically at higher latitudes (see fig. 1-1), so the altitude of the sun must be taken into consideration. One way of doing this is to assume that when the sun is located immediately overhead (altitude 90°), it provides 100 percent energy, and at the moment of sunrise or sunset (altitude 0°), it provides 0 percent energy. This assumption would permit the multiplication of all values by the sine of sun altitude. The resultant sun energy receivership of a sphere then differs during the day, during the different seasons, and at different latitudes. As figure 4-15 shows, a sphere located in northern latitudes cannot continuously receive its theoretical maximum amount of energy.

Every object has its own unique capacity to receive energy from the sun. The cube in figure 4-16 shows a striking difference between potential and actual values of energy received on December 21, the time of the lowest sun path.

The first and obvious step in using this methodology is to evaluate commonly used built forms and, wherever possible, improve their shape and orientation. With the advent of powerful computer graphics software, however, it may be possible to use this tool for evaluation as an active design tool. The process would be reversed: the designer would specify the need for sun energy received (annual and diurnal) and then search for acceptable and suitable forms. Deriving a built form from optimum use of the sun's rays can be a design tool equally important to an architect and an urban planner in the creation of large urban areas.

INTERACTION BETWEEN TWO OR MORE BUILT FORMS

Optimal distances between buildings are latitude specific, since they are determined by the extent of the impact area. A rule that would govern ideal spatial relationships among buildings in cold environments would require that no building intended for human occupancy be constructed within the impact area of another structure.

Because impact areas are time specific throughout the year, for example, from 09:00 to 15:00 hours, such a rule is equivalent to requiring that a building intended for human occupancy always be exposed to direct sunlight between 09:00 and 15:00 hours. As a result, spatial relationships might be governed by access to the sun (the size of the Skywindow) or by shadow impact (the size of the impact area).

In some regions, the low northern sun, and thus the extended impact area, dictates distances between buildings that are too great to be economically justifiable. The penalties of long-distance associations may outweigh the benefits derived from direct sun exposure, so a planner should compromise. For example, he or she may insist that only the southern wall be exposed to direct sunlight within a stipulated time frame, and other walls could be overshadowed at certain times. Much higher land development densities would thus be possible. On the other hand, since the southern wall has become a "sun gate," this method would require an increased level of cooperation between the architect and planner, another argument in favor of eradicating the artificial boundaries that often exist between these two professions.

The following analysis of spatial relationships is based only on requirements to expose a southern wall. On east/west streets, in a low-rise residential neighborhood, the density of built forms is dictated by the depth of the site (see fig. 4-8). The Solar Envelope is derived from the maximum permissible impact area. The shallower the site, the lower the volume of the Solar Envelope.

A graphic representation of an impact area can be of great assistance to a designer of low-rise residential neighborhoods. On a north/south street (fig. 4-17), conventional land divisions can lead to the overshadowing of individual houses. The size of the impact area would determine the extent to which the size of the lot should be modified in order to allow total insolation of a southern wall throughout the year (figs. 4-17, 4-18).

The illustration of the spatial relationships between houses on a winding street is the most dramatically different (fig. 4-19). A true east/west street tolerates houses shoulder to shoulder without affecting solar access. Any deviation from the east/west direction, however, requires a widening of the space between buildings.

The design of a cul-de-sac is another challenge for a neighborhood planner. Unobstructed exposure of the southern wall to direct sunlight is rare (fig. 4-20), yet a relatively minor adjustment in lot design would rectify the problem (fig. 4-21).

DECEMBER 21 AS A METER OF SOLAR ACCESS

The cost of municipal services and other harsh economic realities may require such densities that, even in a so-called low-density residential neighborhood, it may be virtually impossible to guarantee that the entire southern wall is exposed to direct sunlight. In this case, only part of the southern wall would serve as a sun gate, with unobstructed exposure to direct rays of the sun at a stipulated daily interval throughout the year. Part of the wall would occasionally be overshadowed. In other words, such a building would be constructed within the impact area of another structure or natural object. The sun gate is therefore smaller than the southern wall, which may profoundly affect the fenestration, allocation of space for human activities, division of internal space, and the functions of a variety of mechanical systems.

In northern environments, where internal space is so important for the physical and mental well-being of its occupants, it is appropriate to require that no building be constructed within the impact area of another building. In figure 4-22, distance A between two houses ensures that no overshadowing takes place between 09:00 and 15:00 hours. Distance A must be calculated for December 21, the shortest day in the northern hemisphere, the day with the lowest sun path and the longest shadows.

If it is decided that the minimum distance for A is not economically justifiable (for example, because December may be too cloudy statistically to insist on no overshadowing, or because of simple density/land use economics), then planners might decide that no overshadowing will take place between 09:00 and 15:00 hours in January and November (fig. 4-23). This would result in the minimum distance between houses being

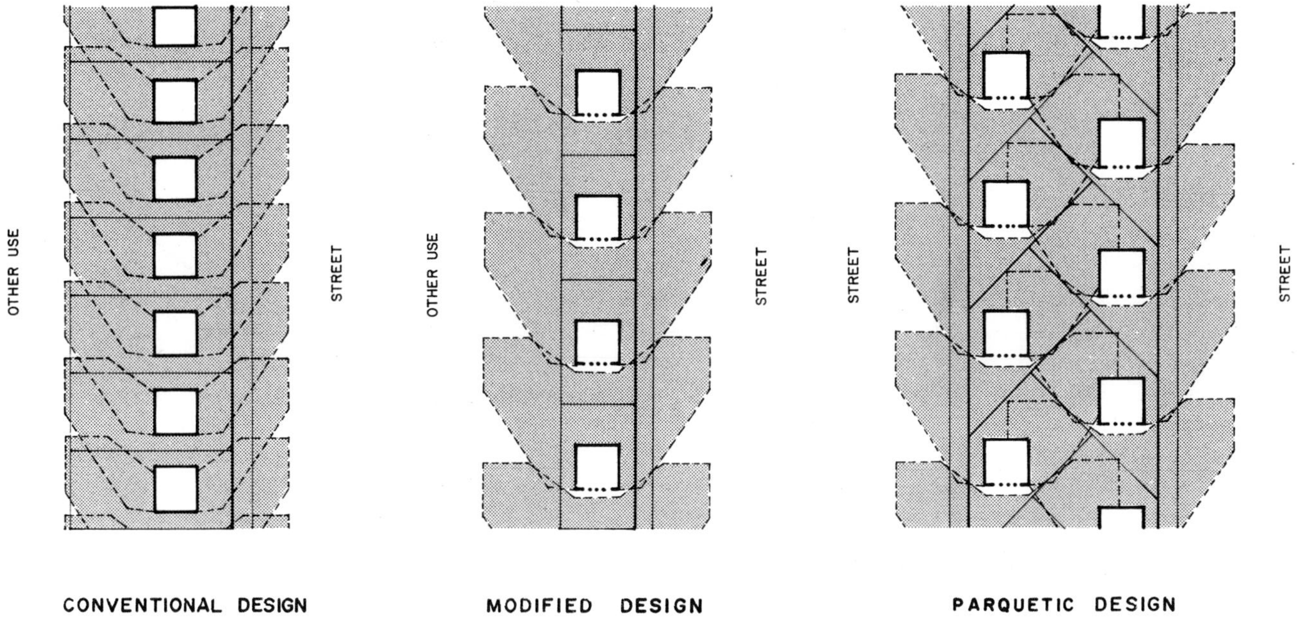

4-17. A shadow-casting analysis of a north/south street.

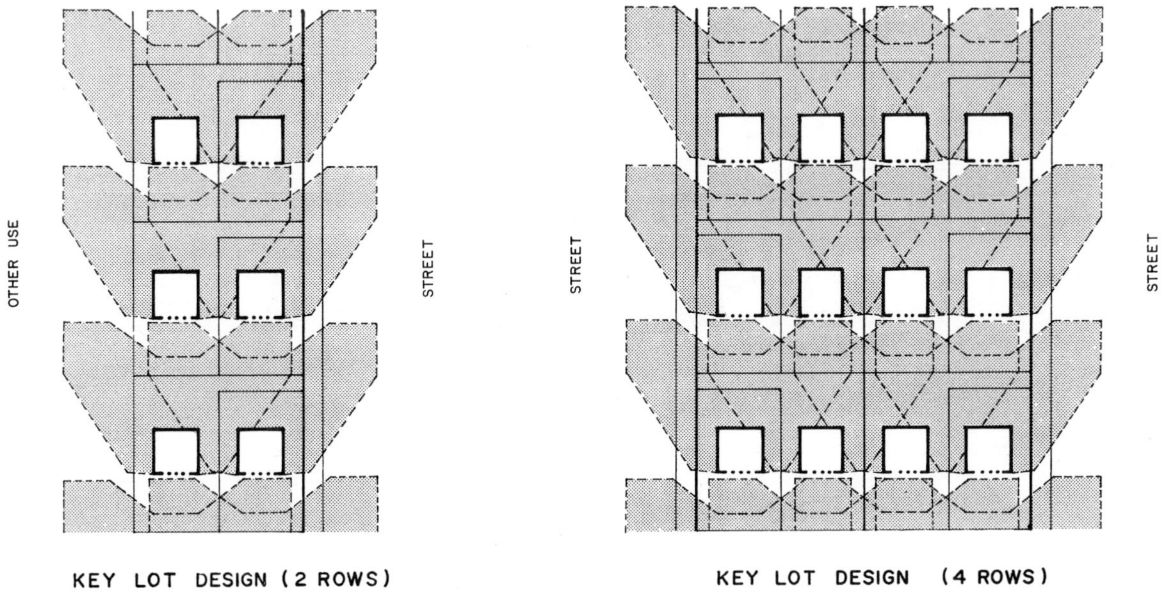

4-18. A shadow-casting analysis of a north/south street.

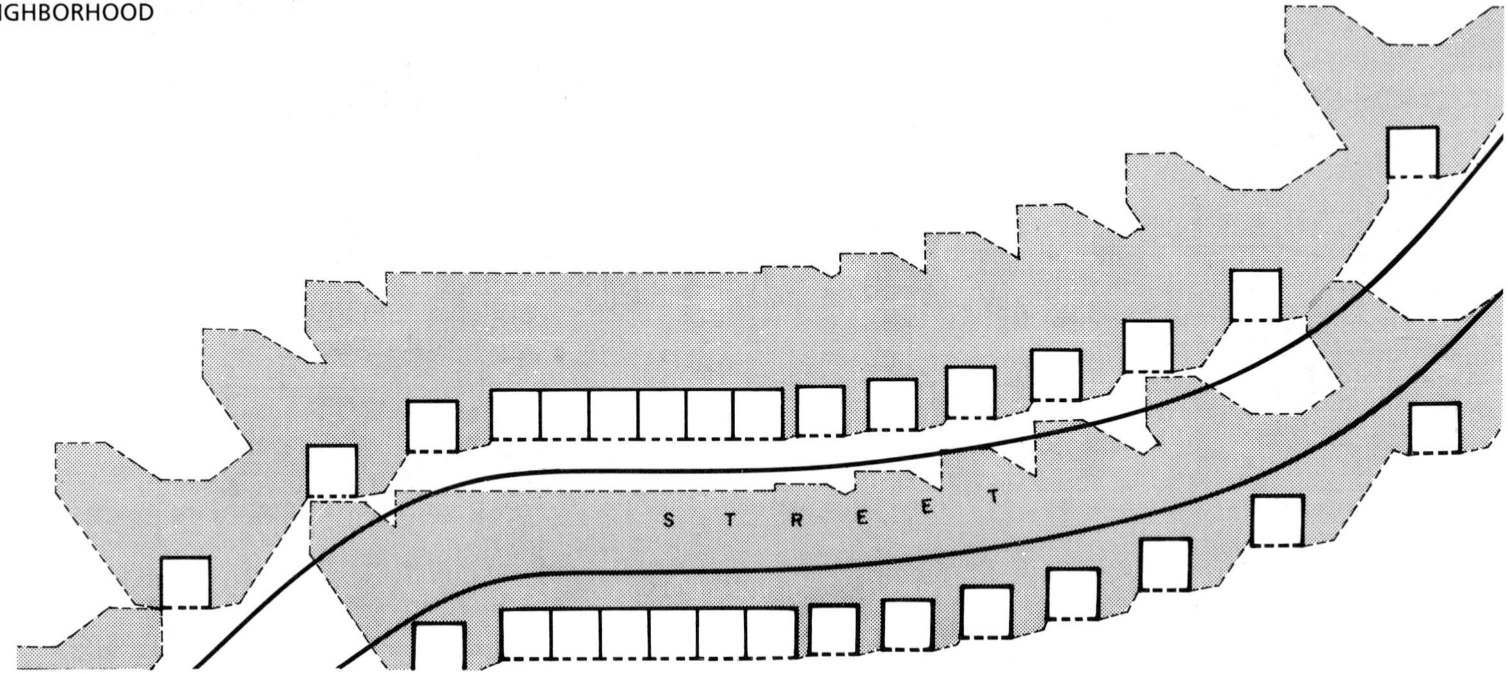

4-19. A shadow-casting analysis of a winding street.

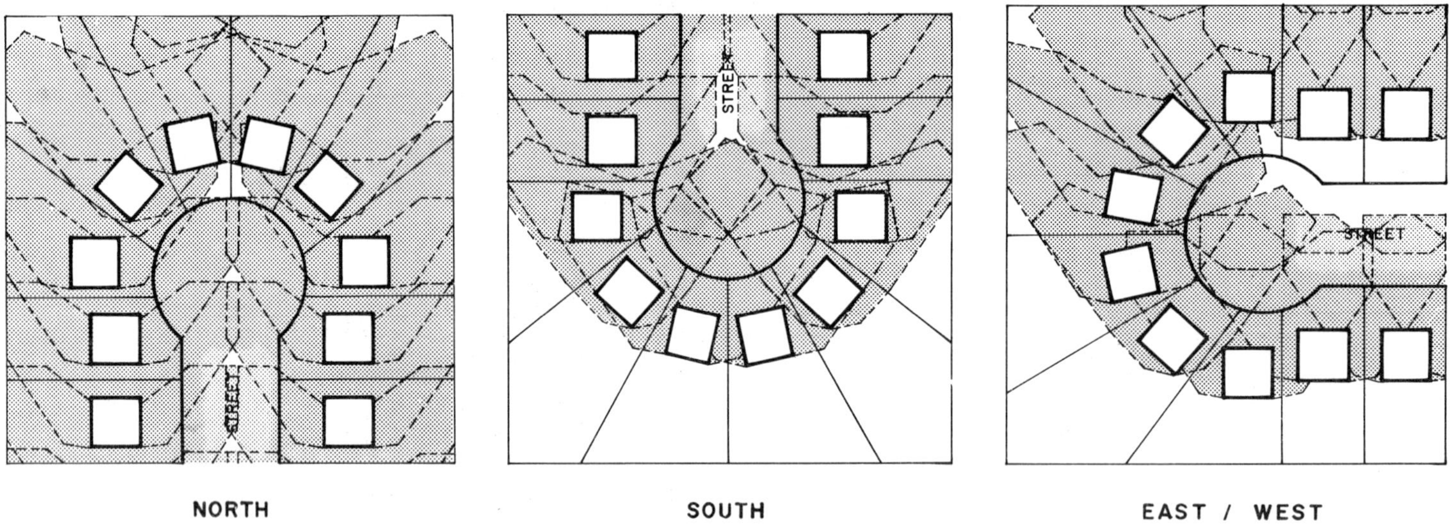

4-20. A shadow-casting analysis of a cul-de-sac (conventional design).

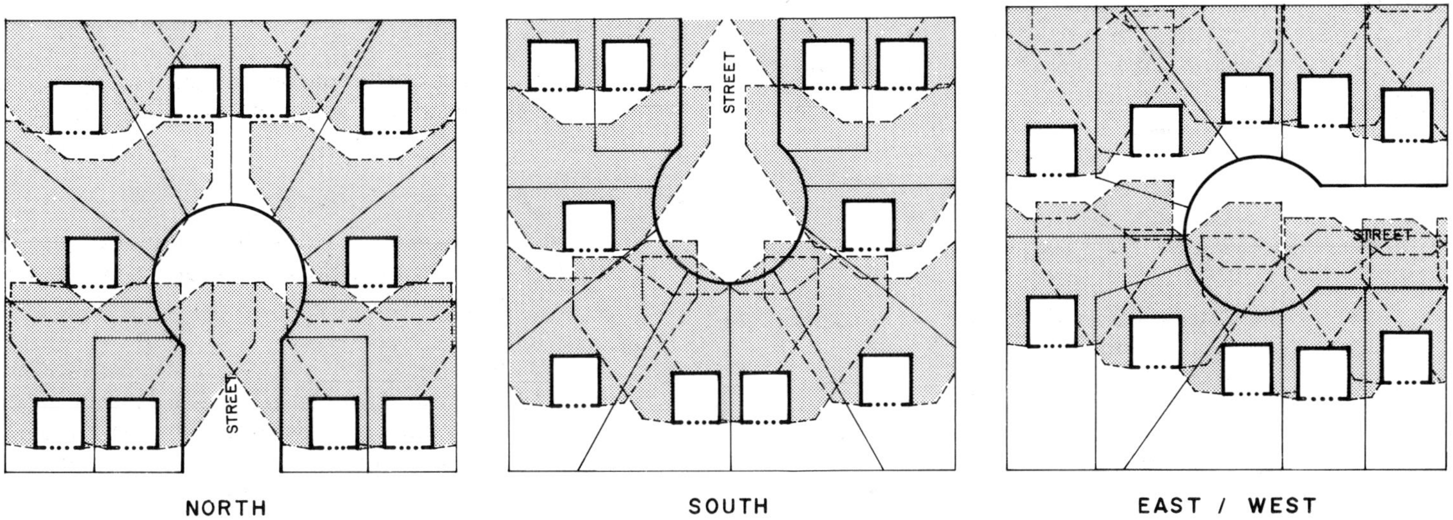

4-21. A shadow-casting analysis of a cul-de-sac (modified for solar access).

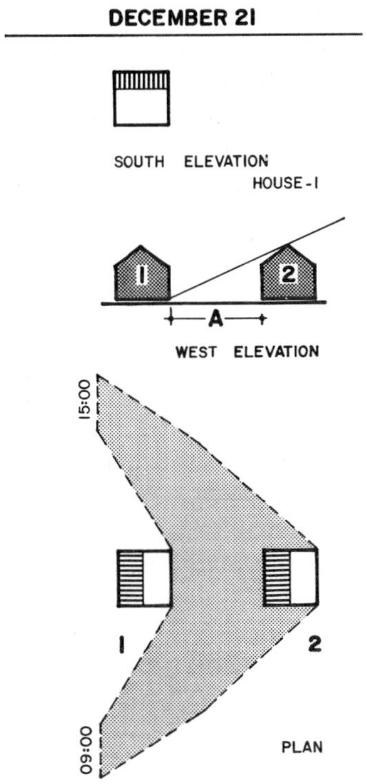

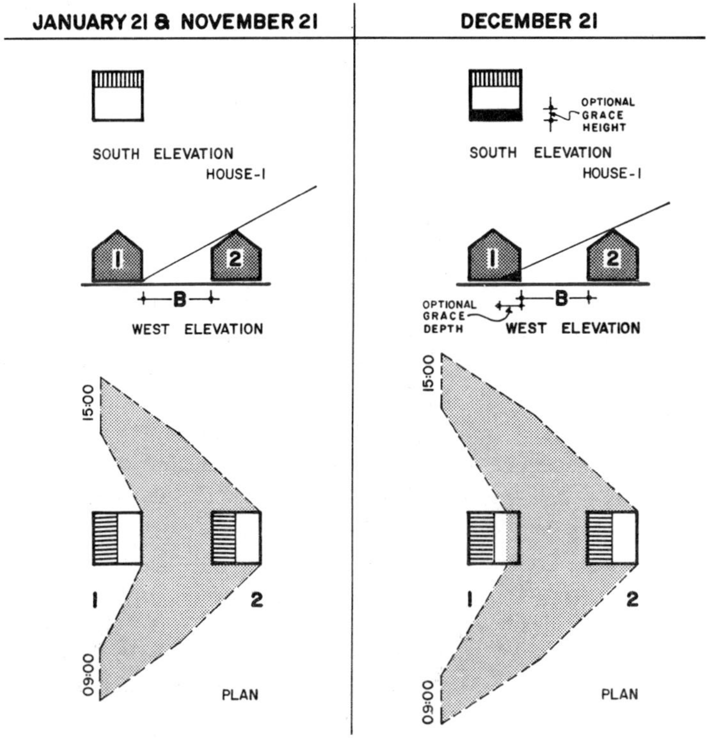

4-22. *(Left)* The minimum distance between two houses for December 21 equals A.

4-23. *(Right)* The minimum distance between two houses for January 21 and November 21 equals B; B plus optional grace depth equal A.

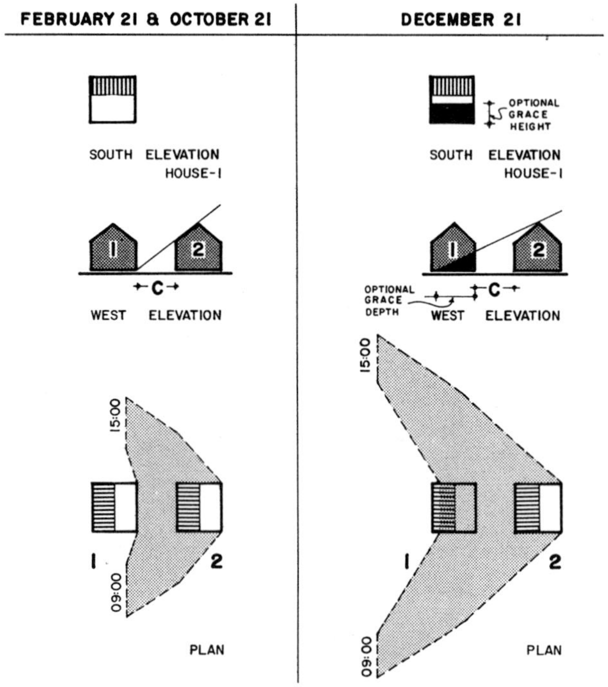

4-24. The minimum distance between two houses for February 21 and October 21 equals C; C plus optional grace depth equal A.

equal to B. A request that shadows are not cast during 09:00 and 15:00 hours in February and October would result in an even smaller minimum distance, C, between two houses (fig. 4-24).

It is absolutely essential that the minimum distances B and C should be reached by using the December Template to measure the degree of shading and by introducing the *optional grace depth* or *optional grace height* concept. Optional grace height permits a shadow to climb the southern wall. Alternatively, optional grace depth permits a shadow to penetrate the interior of a house. In this way, the month of December becomes a meter of solar access. Figures 4-23 and 4-24 show how a variety of spatial relationships can be translated in terms of December and optional grace height/depth.

There are several important reasons for stipulating spatial relationships in terms of December, where one building must be built within the impact area of another building.

1. The landscape architect is shown the exact size of the impact area and thus, to a large extent, the microclimatic characteristics of the immediate surroundings of a building.
2. The architect is given concrete instructions for the design of fenestration, the allocation of private green space, the outdoor extension (expansion) of the living area, the allocation of greenhouses and sunspaces. In addition, instructions are provided for designing internal spaces of the appropriate size, proportion, and with functional links for the anticipated type and rhythm of human activities.
3. The interior designer is provided with useful information about interior surfaces, materials, color coordination, and design for light and visual comfort.
4. The engineer is furnished with directions relating to the design or selection of mechanical systems for thermal comfort and the allocation of solar-harnessing devices.

Measuring distances between buildings using the December yardstick also provides for the future. If new technologies are developed that will allow for the harnessing of solar energy in less than ideal conditions, the winter sky, even when partially obstructed by haze or clouds and with limited hours of direct solar radiation, may be an economically viable source of energy.

Consequently, setting standards for spatial relationships in the northern hemisphere that are based on any time other than December 21 leads to uncertainty about the overall annual impact and practically eliminates any intelligent design based on the rhythm of natural forces. The magnitude of this uncertainty is illustrated in figure 4-25.

In this figure, the planner has stipulated a January and November 21 impact area to govern minimum distances between buildings. The solid line indicates the ceiling or maximum permissible height of buildings and structures south of a single-family house (shown in a darker tone). The given ceiling allows the construction of a variety of buildings of different heights and at different distances (shown in gray tones). The buildings all possess a common outer limit governing their impact areas during January 21 and November 21 at 12:00 hours.

The impact area of each building, however, would be very different on any other date. Between January 21 and November 21, it would grow smaller, and would

4-25. The shadow effect on December 21 and January 21.

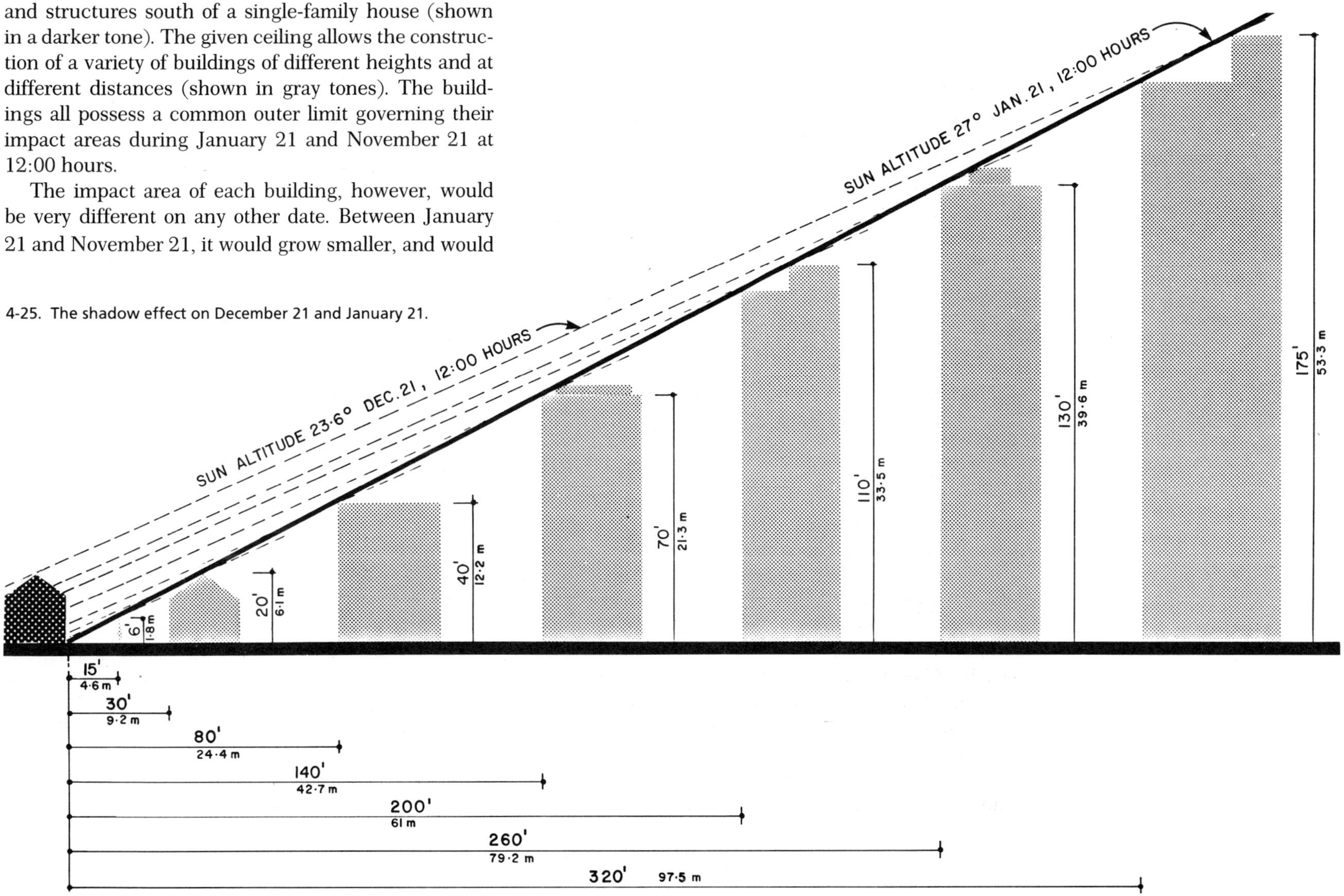

be smallest on June 21. Conversely, between November 21 and January 21, it would grow larger, peaking on December 21.

The problem is that on December 21, the maximum shadows cast by different buildings will differ in size, and thus, their negative impact will vary. Figure 4-25 shows that buildings closer to the single-family house would overshadow the southern wall only partially; taller buildings located farther away would overshadow it completely. As a result, where a building must be constructed within the impact area of another building, it is mandatory to use the December 21 data in conjunction with the provisions of optional grace height/depth.

Planners today are often caught between laws that do not recognize the right to sunlight and citizens' groups that demand it. The negative impact of any proposed development must be determined by planners and an acceptable compromise must be reached.

The following is an example of such a case. Figure 4-26 shows the plan and elevation of a proposed high-rise building. A rezoning application was submitted to accommodate a higher density than permitted by the existing zoning of the site, and a number of citizens living in the vicinity objected to the height of the proposed building. The Planning Department prepared a shadow-casting analysis and met with the developer. Figure 4-27 shows the annual composite shadow impact area. The composite shadow impact area for June 21 (fig. 4-28) and the composite shadow impact area for December 21 (fig. 4-29) were briefly analyzed. An axonometric drawing was constructed to illustrate the three-dimensional relationships between building and shadows (fig. 4-30).

It was determined that buildings 49, 55, and 59 would not be directly affected by the new development, because they were already located within the impact area of building 315. The first four floors of building 315 would be overshadowed in the afternoon. Certain fourth-floor windows would have their Skywindow reduced to 09:00 until 12:00 hours. Buildings 35, 37, 41, 43, 225, and 227 would be affected to varying degrees but for no longer than two hours between 09:00 and 12:00 hours. The eastern facades of buildings 35 and 37 were leaning marginally toward south, so that they would receive about one hour of direct sunlight from 11:00 until 13:00 hours. Eighty percent of the southern facade of building 41 would have a reduced Skywindow between 12:00 and 15:00 hours in November, December, and January.

After being presented with this analysis, the architect for the developer redesigned the project. Two stories were removed and the south-eastern corner was altered. This practically eliminated the overshadowing of buildings 35 and 37. In addition, the building's position on the site was rearranged. The improvements were so impressive and so substantial that objections to the development were withdrawn. Even after eliminating two floors, the developer still received a higher density than permitted by the original zoning of the site.

Another example is of an owner of a greenhouse who vigorously objected to the proposed development of a twin-tower high-rise in a low-rise residential area. Again, the Planning Department prepared a shadow-casting analysis. Figure 4-31 indicates that one of the twin towers of the proposed development would cast a shadow over the objector's property during nine months of the year (a more detailed analysis for December 21 is shown in fig. 4-32).

The architect for the developer rearranged the site plan and, without changing its height, was able to eliminate all shadows cast over the greenhouse. The Skywindow for the single-family dwelling was narrowed from 09:00 to 12:00 hours for December and from 09:00 to 13:00 hours for November and January. The objection was withdrawn, and the development was approved.

These two examples show that even within the framework of existing planning practice, the negative impact of one development over another can be less-

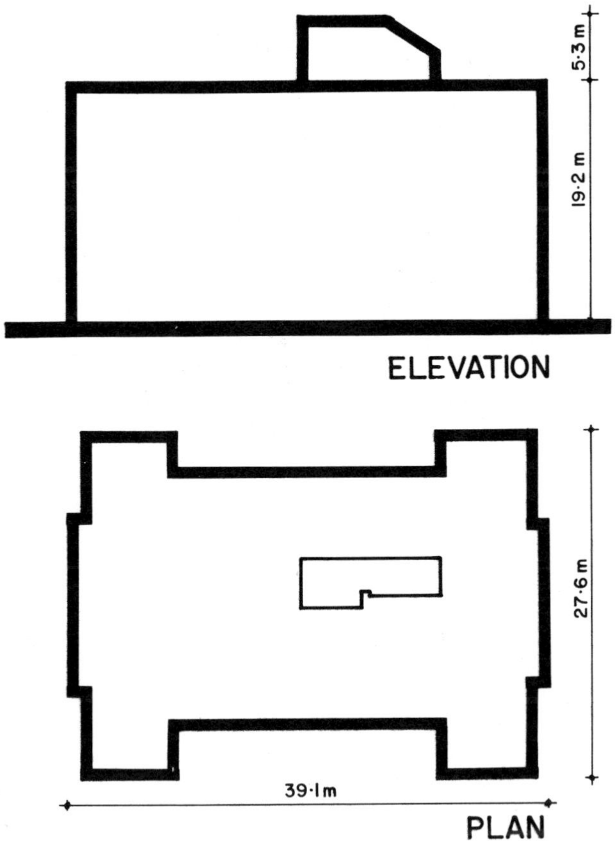

4-26. A proposed high-rise building.

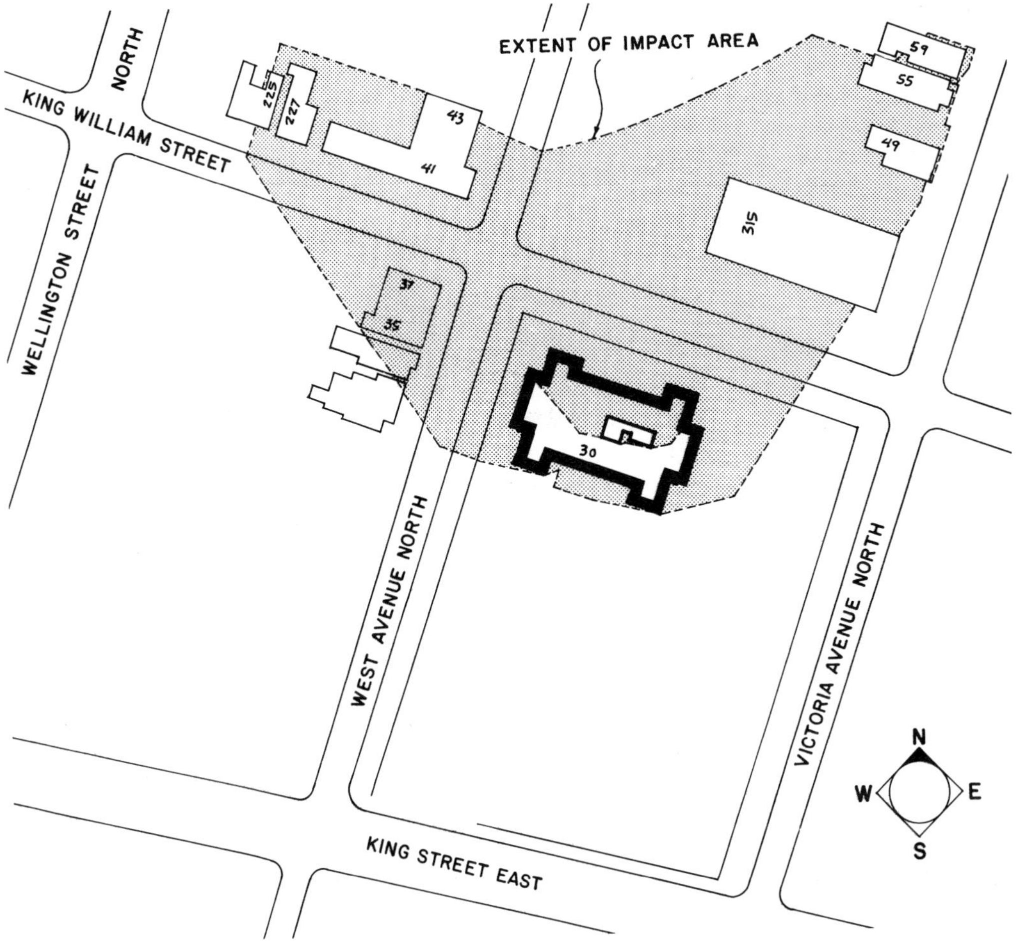

4-27. The annual composite shadow pattern, from 09:00 to 15:00 hours.

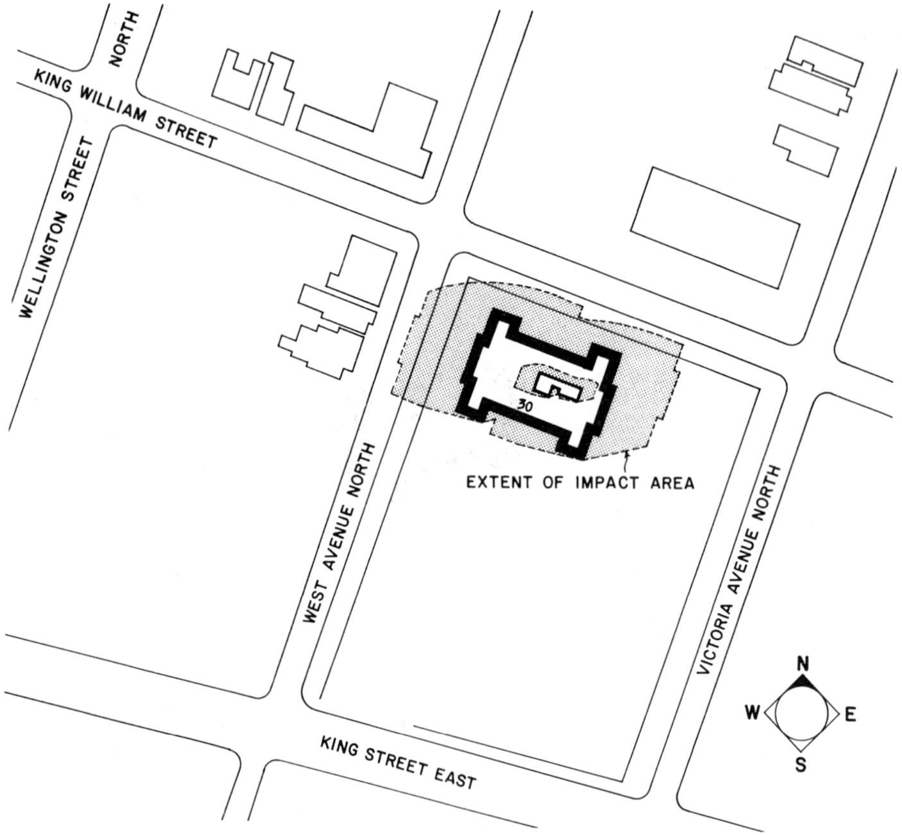

4-28. The composite shadow pattern for June 21, from 09:00 to 15:00 hours.

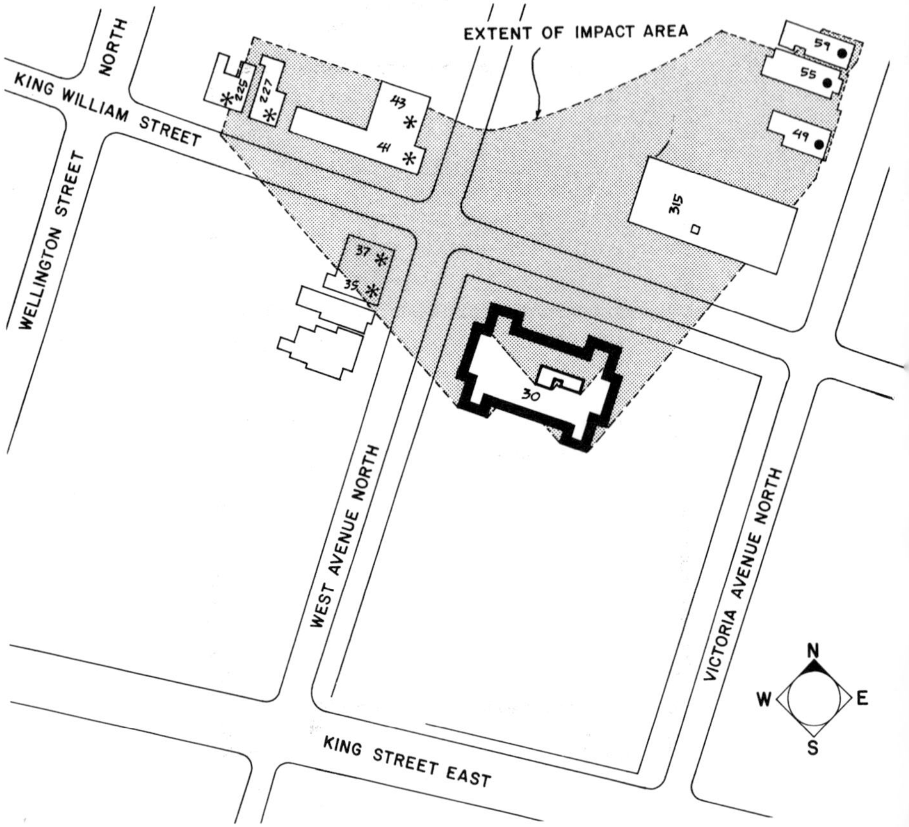

4-29. The composite shadow pattern for December 21, from 09:00 to 15:00 hours.

Impact area analysis:

□ The first four floors of this building are overshadowed for approximately two hours in the morning

● These buildings are overshadowed by the building at #315 King William Street

* These buildings are overshadowed in the morning, from 10:00 to 12:00 hours

ened on occasion. These examples also demonstrate the urgent need to review all regulations governing limits on physical form. The minimum developable volumes as stipulated by most conventional zoning by-laws must be adjusted to permit maximum enjoyment of life in harsh northern climates. This does not necesarily mean a crusade against high-rises. In reality, Solar Envelopes often permit developments of higher densities than is allowed under existing zoning.

The foregoing examples of shadow-casting analyses taken from municipal planning practice clearly illustrate that the acceptable compromise turned out to be the acceptable optional grace height/depth (see figs. 4-23, 4-24). However, an intelligent assessment of the consequences related to optional grace height/depth is impossible without understanding their impact on the internal environment of affected buildings.

SUN PENETRATION

Developing Links between Internal and External Environments

A prime goal when generating any structure for human occupancy is to ensure that internal spaces promote physical and mental health. The construction of a hollow mass makes survival possible by providing a shield against hostile environmental forces and by providing a storage area for nourishing natural energies.

Particularly in cold environments, buildings must negotiate a complex relationship with their surroundings, providing a steady thermal state at minimum cost and within the limits of human comfort. Light must also be admitted that corresponds with the human circadian rhythm. Exposure to direct and diffused sunlight is a necessity for any environment designated for prolonged human occupancy, particularly in regions characterized by extended light-deprivation periods.

Figures 4-33, 4-34, and 4-35 show the relationship between the orientation of a solar gate and the direct sunlight available at different times of the year at 43°N

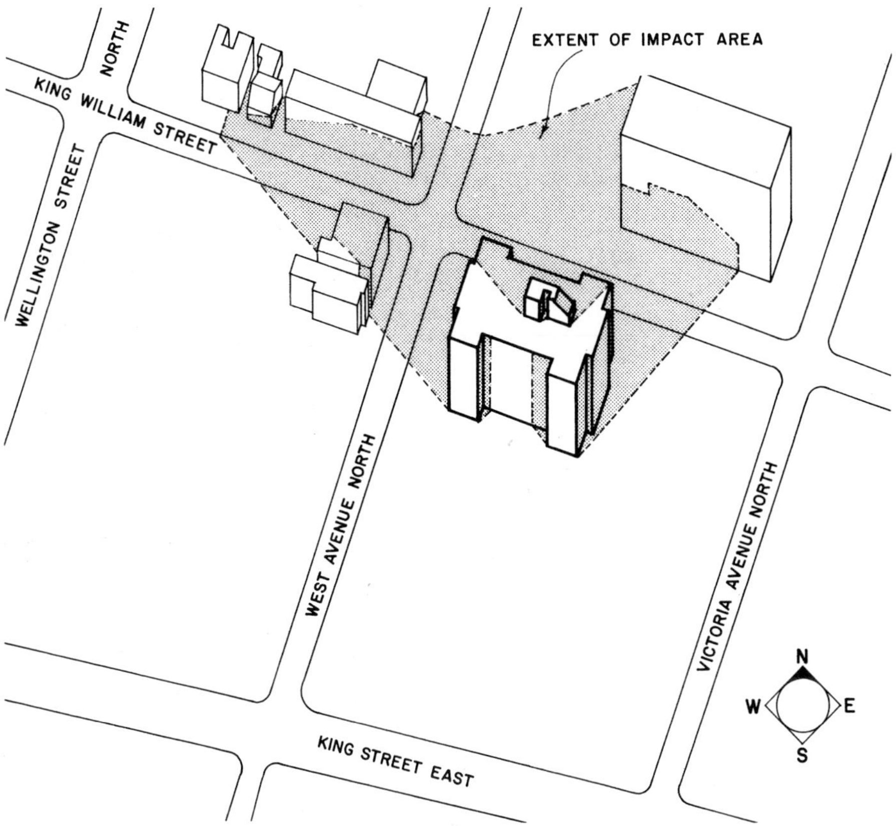

4-30. An axonometric drawing of the composite shadow pattern for December 21, from 09:00 to 15:00 hours.

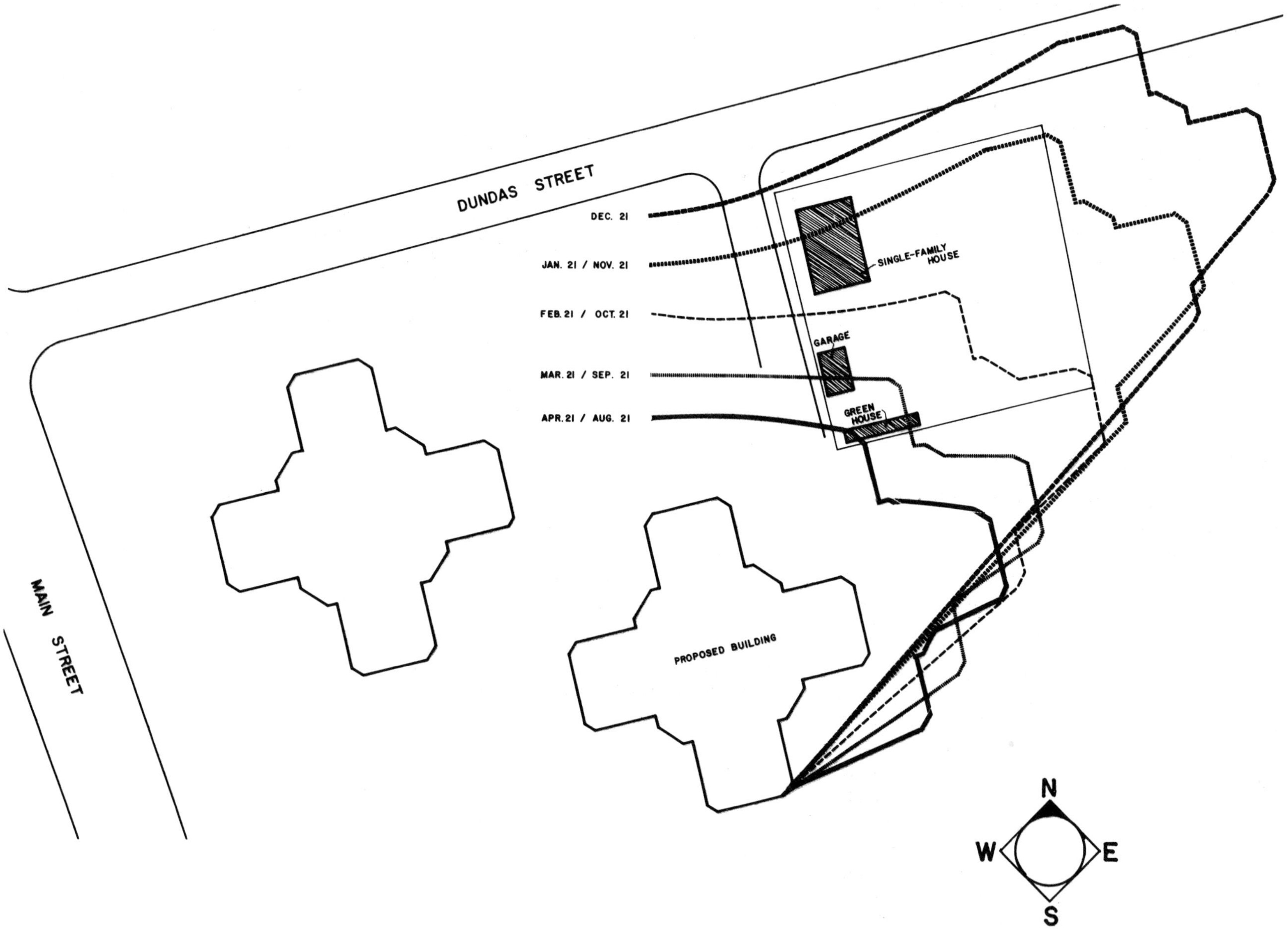

4-31. The composite shadow pattern showing annual impact for the proposed development of two high-rise residential towers.

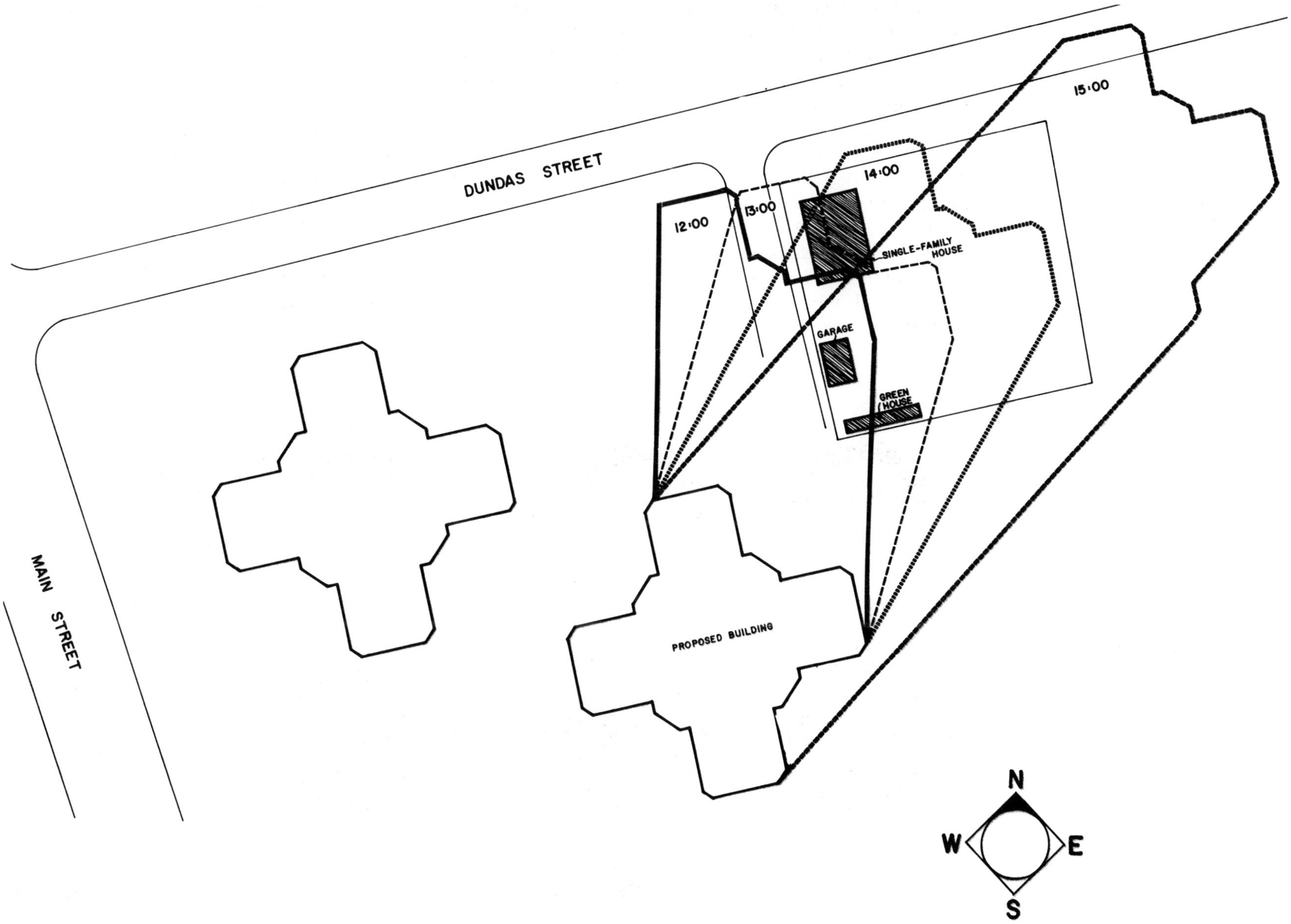

4-32. Shadow patterns showing diurnal impact for December 21 at 12:00, 13:00, 14:00, and 15:00 hours.

latitude. Conditions during equinox (March 21 and September 21) clearly illustrate the relationship between the orientation of the solar gate and the available sunlight. In figure 4-33, the day is divided into equal periods of day and night, and there are theoretically twelve hours of available sunshine. The most energy from the sun (twelve hours from sunrise to sunset) would be harnessed by a chamber with a sun gate oriented toward true south. The eastern and western gates would receive half the total available sunshine (sunrise to noon and noon to sunset respectively). The northern gate would receive no direct sunlight at this time.

On June 21, the peak of astronomical summer (fig. 4-34), all gates would receive direct sunlight, and the northern gate would receive direct sunlight twice a day—after sunrise and before sunset. In the winter, on December 21 (fig. 4-35), the maximum amount of direct sunlight is received by the southern sun gate. The eastern and western gates receive half the total of direct sunlight that is theoretically available at the time.

A further test using a chamber of unit size demonstrates the relationship between the insolation of internal spaces and the orientation of the chamber. The following figures show only the eastern orientation of the chamber, because true eastern and western orientations are exactly opposite in terms of sun penetration. It should be noted that in many locations, particularly during the summer, the sun is welcomed inside in the

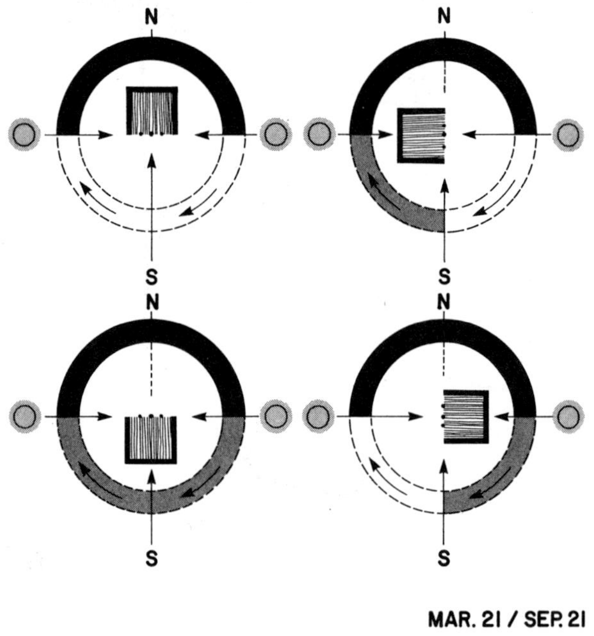

4-33. The amount of sunlight available and the amount that actually penetrates the chamber are shown for four chambers with different orientations at equinox.

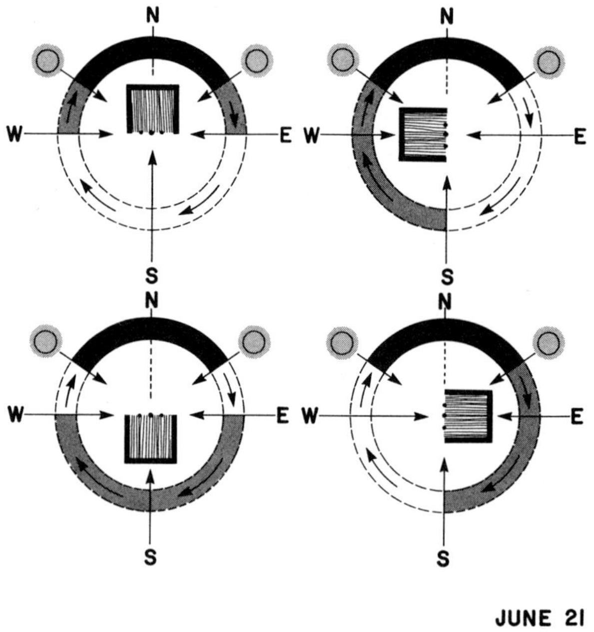

4-34. The amount of sunlight available and the amount that actually penetrates the chamber are shown for four chambers with different orientations at summer solstice.

morning when its rays are not so intense rather than in the afternoon.

In figure 4-36, the chamber gate is oriented toward true east. During equinox (March 21 and September 21), the floor of the chamber is fully lit at sunrise (06:00 hours). By 10:00 hours, the amount of floor area covered by direct sunlight is reduced to approximately 40 percent. At noon, all direct sunlight disappears. The reverse would apply to a true western orientation. The sun would appear shortly after noon and the amount of floor insolation would gradually increase. At 14:00 hours, approximately 40 percent of the floor would be lit, reaching 100 percent at 18:00 hours.

Figure 4-37 shows equinox, with a chamber gate

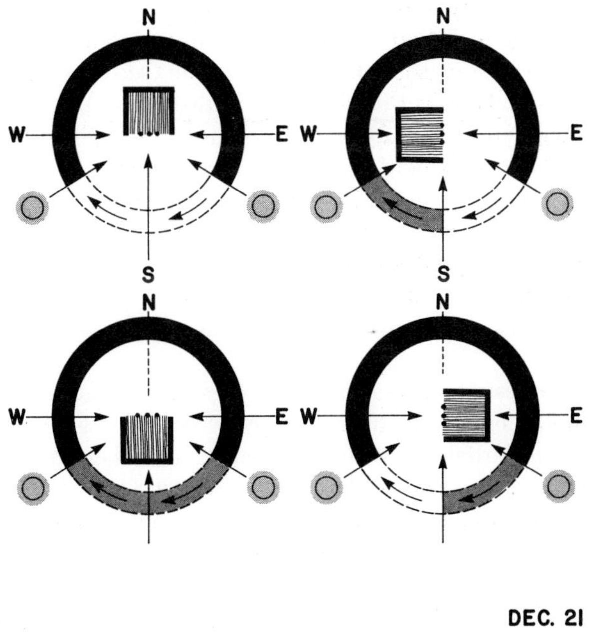

4-35. The amount of sunlight available and the amount that actually penetrates the chamber are shown for four chambers with different orientations at winter solstice.

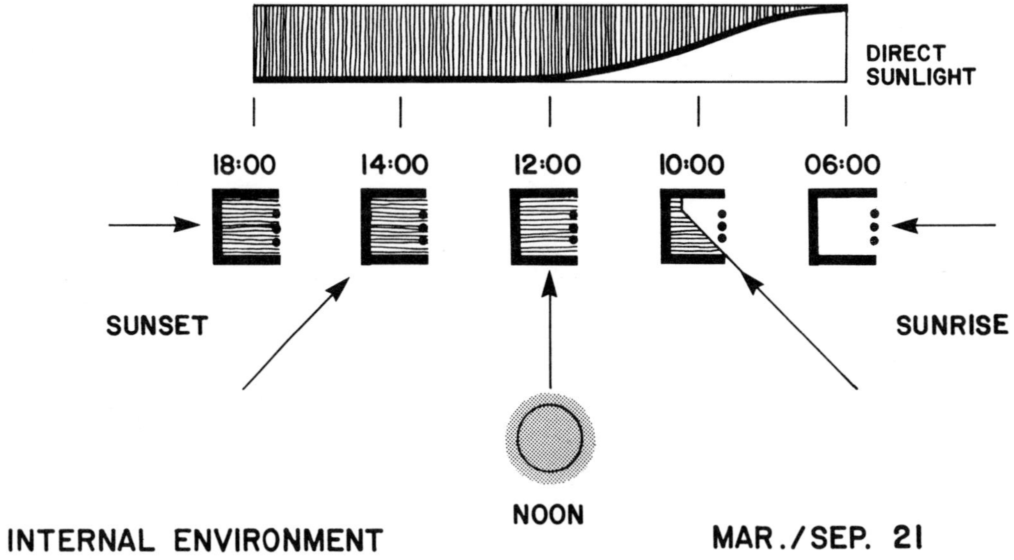

4-36. Sun penetration during equinox through an east-oriented chamber.

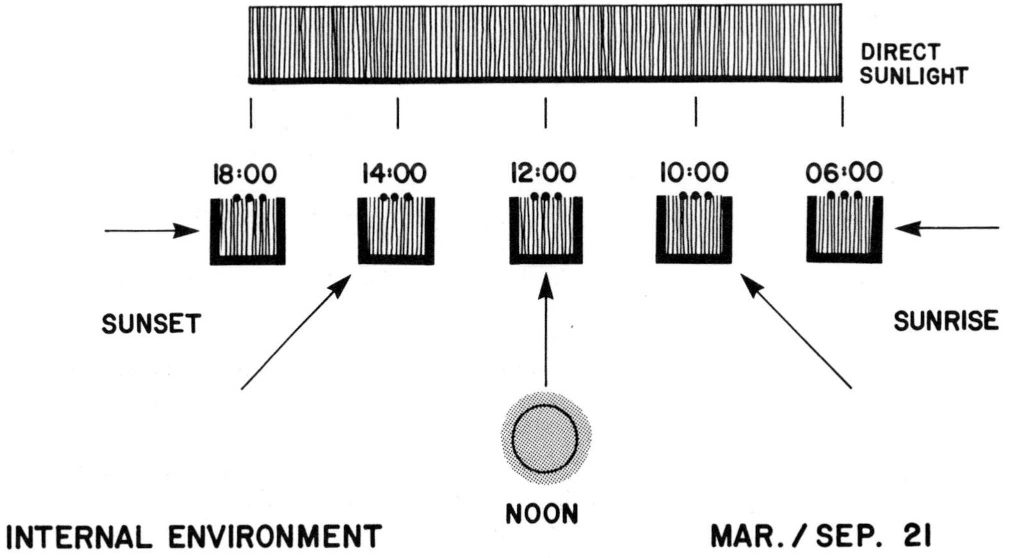

4-37. There is no sun penetration during equinox through a north-oriented chamber.

oriented toward true north. There is no insolation of the internal space. Figure 4-38 shows equinox, with the chamber gate oriented toward true south. The sun appears shortly after sunrise and the insolation gradually increases until it reaches about 70 percent of the floor space at 12:00 hours. In the afternoon, insolation decreases to 0 percent at 18:00 hours.

Figure 4-39 shows the situation on June 21, the longest day of the year. The chamber gate is oriented toward true east. Maximum floor insolation is 100 percent and takes place at 08:00 hours, gradually decreasing to 0 percent at 12:00 hours. The process would be reversed with a western orientation.

The chamber is oriented toward true north on June 21 in figure 4-40. Twice daily, the chamber gate receives about three-and-a-half hours of direct sunlight between sunrise and 08:00 hours and between 16:00 hours and sunset. The amount of sunlight received never reaches the 50 percent level. In figure 4-41, on June 21, the chamber gate is oriented toward true south. The sun would penetrate the chamber between 08:00 hours and 16:00 hours with two peaks at 11:00 and 13:00 hours.

In figure 4-42, it is December 21, the shortest day of the year. The chamber gate is oriented due east. Approximately 70 percent of the floor is sunlit shortly after sunrise, gradually diminishing to 0 percent at 12:00 hours. As previously stated, the reverse would take place if the chamber were oriented toward true west. In figure 4-43, the chamber gate is oriented toward true north on December 21. No penetration would take place during this time. In figure 4-44, the chamber gate is oriented due south on December 21, and there is generous insolation of the internal space from sunrise to sunset, peaking at 12:00 hours.

Table 4-1 summarizes the direct sun received at each gate at different times of the year.

The diagrams in the figures are, of course, latitude specific. Moving north during the summer season, the overall availability of sun would gradually increase until the 66°33′ parallel (Arctic Circle), where points of sunrise and sunset coincide. Beyond the Arctic Circle, the sun is present in the sky for twenty-four hours in the summer, and the day/night cycle ceases to exist. All the gates, regardless of orientation, come to have nearly the same capacity to receive direct sunlight. It is a unique challenge to design structures for human occupancy in areas with a seasonal light/dark rhythm (see chapter 1, fig. 1-1). However, in latitudes between the tropic of Cancer (tropic of Capricorn) and the Arctic Circle (Antarctic Circle), where a twenty-four hour day/night cycle exists, orientation of the sun gate is an important part of the design process involving the allocation of internal spaces. In latitudes between the tropic circles, seasonal variations are minimal, sun altitudes are high, and structures are oriented to avoid rather than receive the sun.

For design and planning purposes, the degree of sun penetration can be measured by determining whether there is partial or full illumination of a chamber floor. It should also be noted that noon radiation (sun at high altitude) and late-afternoon radiation (sun at low altitude) are not of the same density because of the angle of incidence and the varying thicknesses of the earth's atmosphere. The obvious advantage of the southern gate in terms of depth of penetration is clearly demonstrated:

TABLE 4-1
Summary of Distribution of Direct Sun Received

	NORTHERN GATE	EASTERN AND WESTERN GATES	SOUTHERN GATE
Summer	less than 50%	50%	100%
Equinox	0%	50%	100%
Winter	0%	50%	100%

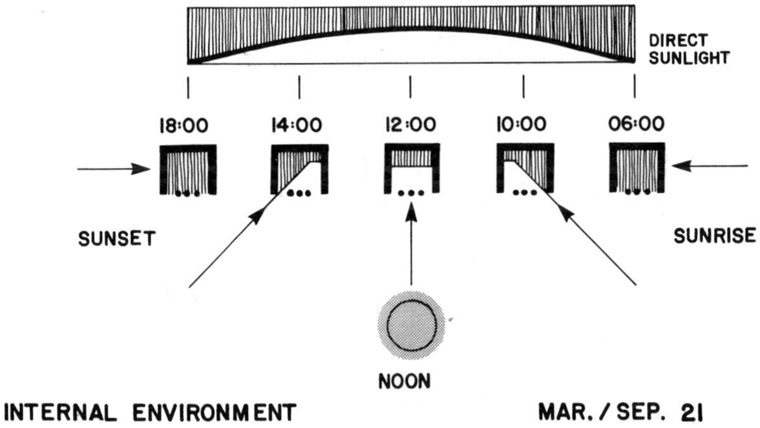

4-38. Sun penetration during equinox through a south-oriented chamber.

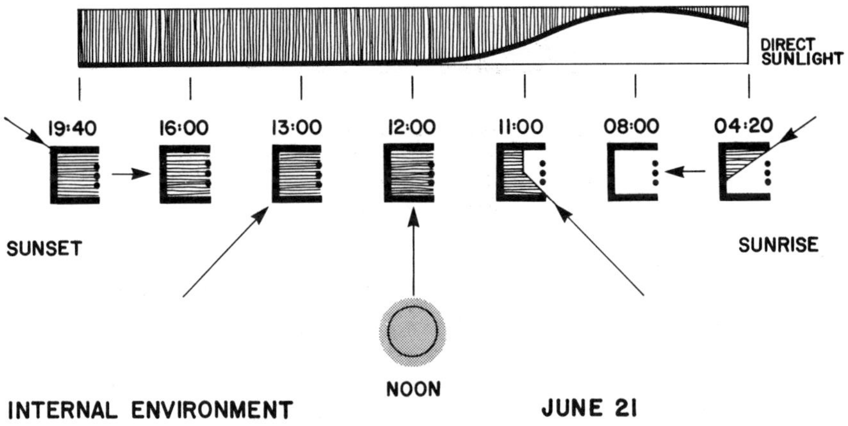

4-39. Sun penetration during summer solstice through an east-oriented chamber.

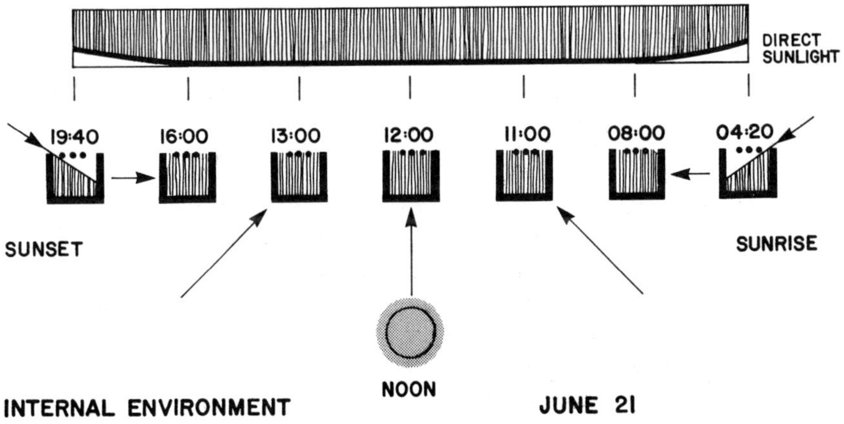

4-40. Sun penetration during summer solstice through a north-oriented chamber.

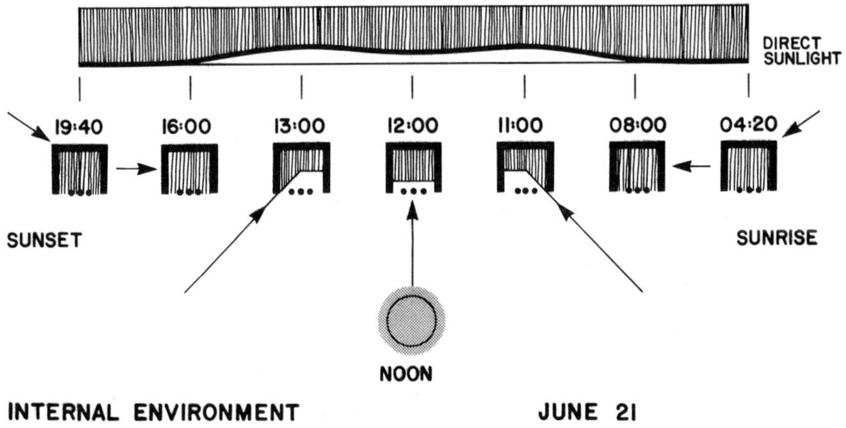

4-41. Sun penetration during summer solstice through a south-oriented chamber.

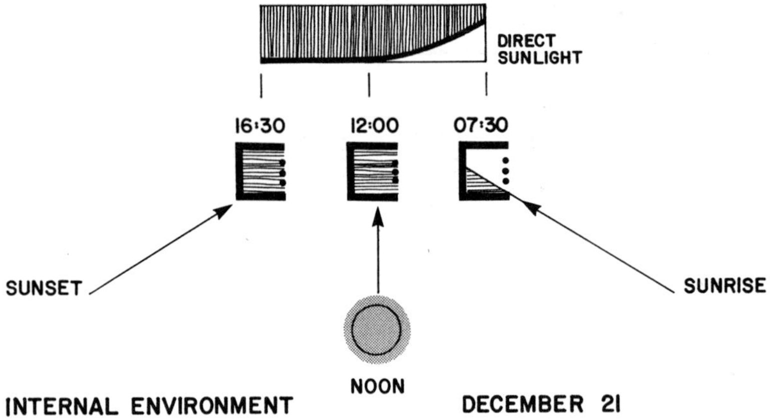

4-42. Sun penetration during winter solstice through an east-oriented chamber.

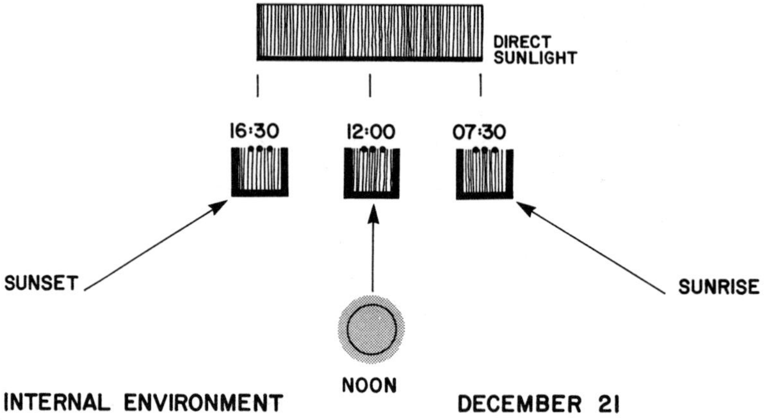

4-43. There is no sun penetration during winter solstice through a north-oriented chamber.

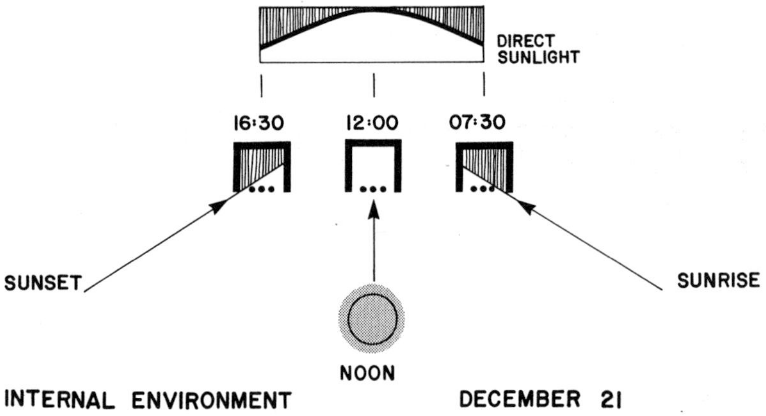

4-44. Sun penetration during winter solstice through a south-oriented chamber.

Fall Equinox	
Winter	Deepest Penetration
Spring Equinox	
Summer	Most Shallow Penetration
Fall Equinox	

Consequently, a chamber that is oriented toward true south has the most balanced solar-penetration regime (both seasonal and diurnal) which is related to a cycle of need and supply. Such a chamber receives the total available sunlight when its need is great and very little when it is least desired.

The previous figures show a maximum depth of sun penetration on five different occasions:

1. southern gate: December 21, noon (see fig. 4-44)
2. eastern gate: spring/fall equinox, sunrise (see fig. 4-36)
3. eastern gate: summer, after sunrise (see fig. 4-39); actual time is latitude specific

4. western gate: spring/fall equinox, sunset (mirror image of fig. 4-36)
5. western gate: summer, before sunset (mirror image of fig. 4-39); actual time is latitude specific

These "full blasts" of sun do not create identical sensory experiences. In regions where sunlight is a precious commodity, the designer must study these various qualities of insolation. It is good practice to draw the actual sunlight pattern along the floor and walls, penetrating the individual rooms of a building (see chapter 8). This pattern may affect the size and design of windows, the incorporation of screens, the selection of surface materials, and the layout of furniture. Although the quality of sunlight differs throughout the day and year, diagrams both of sun receivership and penetration can help the designer to develop a basic layout of a building based on a harmonious relationship between the rhythm of human activities indoors and the external rhythm of light.

Respect for Local Character

When faced with the task of developing architectural design guidelines or specifications, it is necessary to understand and incorporate a broad range of human needs, including thermal and spatial comfort, life-support systems, and the need for daylight. In northern latitudes, the sun should be the most important factor governing the design of spaces for human occupancy. Sunlight deprivation has a markedly negative impact on human health, particularly in areas with prolonged seasonal light deprivation. As a result, the maximum amount of sunshine must be allowed to penetrate living areas, especially in northern latitudes. However, the incorporation of information about the sun and climate into the design process must vary from region to region. And, in addition to the analysis and interpretation of mesoclimatic conditions, any local design strategy must consider related cultural, economic, and technological factors.

Design guidelines must vary at different latitudes and in environmental domains with different mesoclimates. Profound pressures should be anticipated from existing cultural conditioning. For example, it took a long time to develop an understanding and consensus among professionals on the subject of the passive solar house and its consequences for urban planning. After extended efforts, certain neighborhoods in North America were planned with a prevalent east/west street system. Unfortunately, houses built along these streets were not oriented torward the sun. Instead, they were oriented toward the street, resulting in only 50 percent of them, those on the south side, having their private outdoor space oriented toward the sun. The north side of an east/west street has the worst possible orientation (fig. 4-45). But cultural indoctrination prevails: the official face of the house, with its front living area behind large windows, is a sober reminder to the passerby of the owner's dignity. Image is too often placed above comfort, or even reason, regardless of cost.

In encountering such cultural design barriers, planners may do better to design a north/south system of streets in which the back and front parts of the house receive roughly equal insolation. The fluctuation of light between the eastern and western walls can be mitigated by strategically placing skylights that allow the southern sun to penetrate the center of the house at midday.

This experience points to the need to reconsider the relationships between internal layout, architectural design, and site and neighborhood planning. If the goal is to provide a stable thermal milieu at minimum cost and a high-quality, healthy environment, then there is an urgent need for a concerted, cooperative approach on the part of all the parties involved in the development process, beginning with regional urban planners and ending with interior designers.

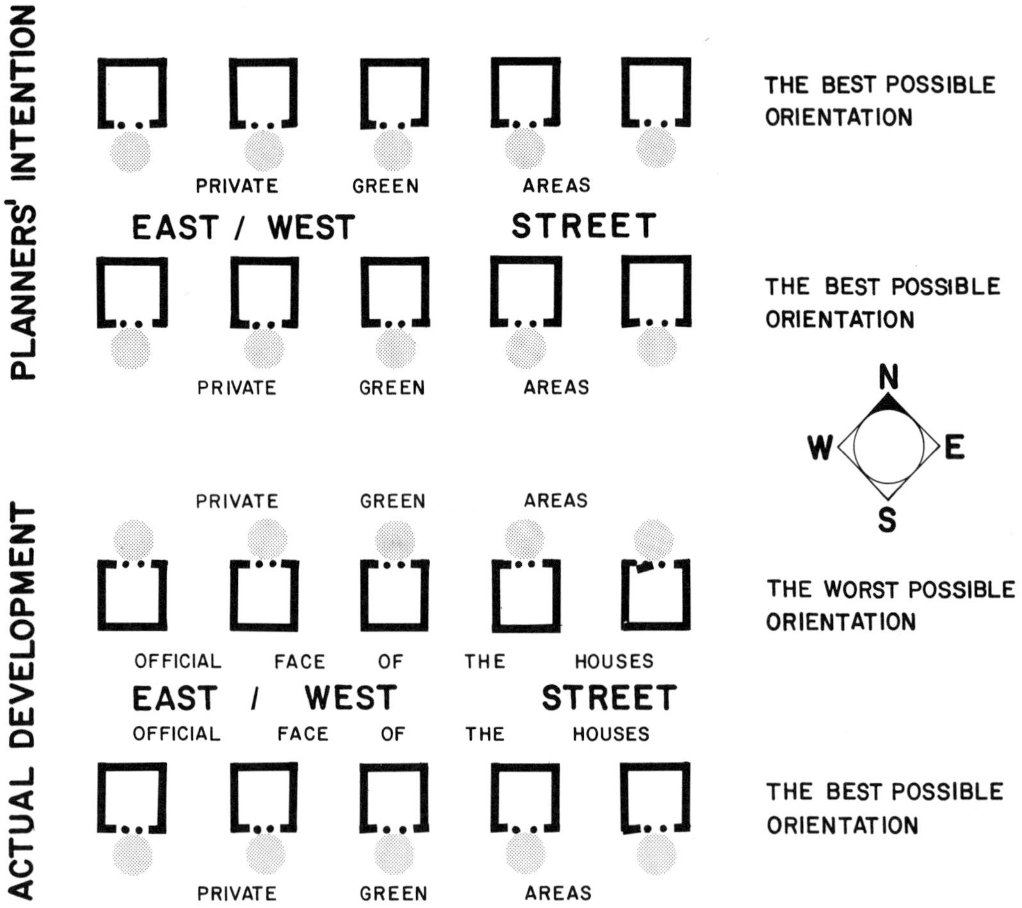

4-45. The orientation of houses on an east/west street.

THE PURPOSE AND FUNCTION OF WINDOWS

Windows have traditionally served two basic purposes: illumination and ventilation. Windows may, however, serve other purposes. The variety of shapes, sizes, locations, methods of construction, and materials used attest to the broad range of roles, both primary and secondary, that windows play.

Interior designers carve openings in walls to enhance interiors with the effects of sunlight. Architects carve openings in walls as part of their task to design a facade that reflects the owner's personality and communicates this image to passersby. A passive solar designer uses wall openings as a gate for receiving solar energy.

In northern latitudes, the importance of daylight must never be underestimated. The following review may help to illustrate the fundamental differences between northern, southern, eastern, and western windows, taking into account the path of the northern sun.

The Southern Window

The southern window, located on the warm side of the house, creates an indoor climate that is much like the warm microclimate around the house. The use of large, glazed (possibly sliding) doors extends the living area into the southern patio, an external sun pit. This indoor/outdoor link is of particular importance in northern latitudes.

The blast of the southern sun creates a dramatic visual environment—a sharp contrast between light and dark. In the summer, objects outside of the southern window are illuminated from above; their shadows look like black pedestals (fig. 4-46).

If necessary, overhangs or horizontal louvres can modify the impact of the higher path taken by the sun in summer. In winter, the low sun penetrates deeply into interior spaces, creating much needed psychological and physiological comfort. Since the winter sun path can be very short in northern latitudes, the southern wall should be oriented toward true south to take maximum advantage of this limited exposure to direct sunlight.

Each type of insolation can create a different mood which should be taken into account when allocating spaces for human activities. Southern light brings a sense of energy, triumph, joy, the mobilization of human capacities, and the drive to survive. Therefore, the room with a window facing south is the warmest, most conducive to movement, and should be allocated for family activities, particularly during the day.

The Northern Window

An opening in the northern wall introduces a different set of conditions and experiences. Generally, a source of light from the rear enables us to look into a sunlit exterior. Rooms with northern windows mainly receive reflected and diffused sunlight. Thus, the sunlight is typically less intense, but provides more uniform light throughout the day and season. The difference between a northern and southern window is comparable to the different perspectives of actors and audience members in a theater: actors are affected by light as if they were looking out of a window facing south; the audience is affected by light in a way that is comparable to a window facing north. (fig. 4-47).

In a conventional situation, the northern window carries a high energy penalty. Consequently, designers of passive solar buildings usually minimize or eliminate these windows. In certain parts of the northern latitude, however, the northern wall receives a substantial amount of direct radiation during the summer season. This allows certain systems for harnessing solar energy to be used for limited periods of time. The general performance and quality of northern interior spaces could be altered and enhanced considerably by superinsulating windows and by using a reflective vertical surface outside the window.

The soft light from a northern sky evokes peace, timelessness, dreams, contemplation, passiveness, and perhaps depression. Such internal spaces receive uniform soft light and are traditionally used as artists' studios, libraries, art exhibit rooms, and studies.

The Eastern Window

The eastern window announces the dawn of day, the cosmic trigger of daylight activities. Morning is a phase of transition in the human circadian rhythm comparable to that of spring in the annual change of seasons. The day/night cycle is part of our biological make-up and in order to remain healthy, physicians caution us not to violate this inherited cosmic pulse. In northern environments, the circadian rhythm is altered dramatically at higher latitudes, until it ultimately ceases to exist.

Our current life-style complicates the problem even more. Until the Industrial Revolution, most people were generally able to live in harmony with natural dynamics and steer their activities along the path of their circadian rhythm. But rapidly growing industry required that time be altered to fit the mechanics of engines. The twenty-four hour interval was divided and packaged to suit production efficiency and machinery maintenance schedules. This autonomous system of industrial production, with its own internal logic and timetables, began to govern most individuals' life-styles. The widespread use of technology made the basic activities, such as working, sleeping, and resting, independent of day, night, the different seasons, and the cosmic pulse.

Although the original causes of this life-style ruled by machines are becoming obsolete with the introduction of new, more flexible technology, we may continue to follow our desynchronized, uncoupled, offbeat mode of existence for some time. Consequently, the eastern window does not always announce the end of night and the beginning of day. Regular working hours rarely permit us to rise with the sun. Still, the general preference is to place bedrooms on the eastern side of the house.

During the summer, morning sunlight penetrates deeply into eastern rooms (fig. 4-48). Because it has to overcome the cold of night, there is usually no need to shade or block the sun. In the winter season, particularly where the day is very short and the sun path very low, the eastern window receives a negligible amount of energy; designers of passive solar housing in northern regions therefore usually attempt to minimize or eliminate the eastern window.

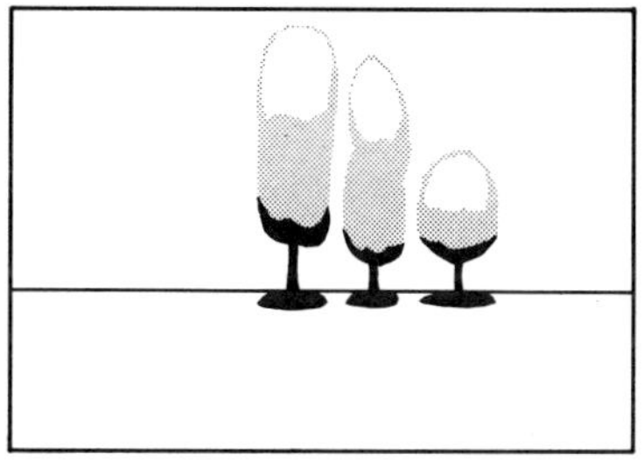

4-46. A view through the southern window.

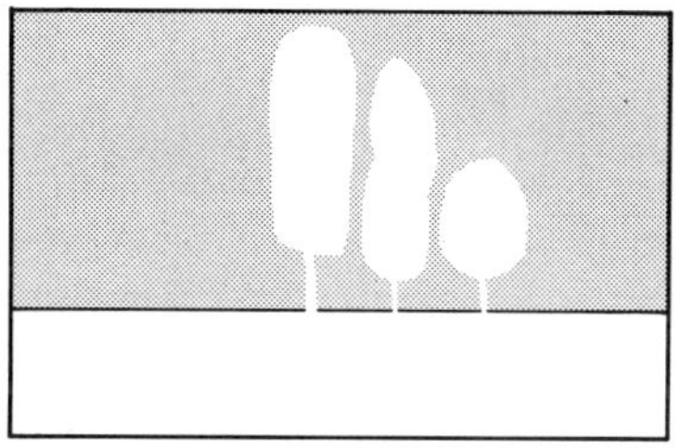

4-47. A view through the northern window.

4-48. A view through the eastern or western window.

The east and eastern sunlight are most often associated with birth, optimism, expectation, hope, and feelings of welcome. Rooms facing east are most often used as bedrooms and breakfast rooms, taking advantage of early light.

The Western Window

During the winter season, particularly in northern regions, the energy received through western windows is the same or marginally higher than that of windows facing east. In summer, however, the low sunlight can be a source of discomfort that is difficult to escape. By the time the sun appears on the afternoon side of the house, the air and ground are already well preheated; the late afternoon rays turn the western wall into an inferno. Western rooms are penetrated mercilessly by the sun's rays and looking into these rays may be as painful an experience as peering directly into the beam of a searchlight (see fig. 4-48).

Western windows are usually considered energy-inefficient propositions. Even in higher latitudes, the temptation to incorporate mechanical cooling devices to offset high summer temperatures is difficult to resist. During the winter, adjacent internal spaces must be well heated, since they are almost completely deprived of direct sunlight. Consequently, coping with these annual extremes of discomfort can be costly.

On occasion, a summer sunset can be a spectacular event. But in a cloudless sky, outdoor objects become silhouettes against a blinding source of light, with long shadows rushing toward the observer. Thus, the western window is not a very popular feature of a passive solar building, and the opportunities for design modifications or improvements are limited. The sun can be blocked, but at the cost of blocking the view. "Intelligent" windows, now emerging, promise to adjust, according to the intensity of sunlight, the light and glare absorbed.

The west and western sunlight remind one of glorious funerals with horns and trumpets, death with a promise of rebirth, falling to sleep with the promise of waking, a gateway to dreams and poetry. In the summer, rooms that receive western sunlight are warm; they enjoy very dramatic light and views of sunsets. They are excellent places in which to conduct transitional activities—ones that last into evening and eventually lead to sleep.

THE IMPACT OF NEW TECHNOLOGIES

Substantial experience has been gained by the design and construction of zero-heating-bill (ZHB) houses. The state of knowledge and technology as of 1986 permits the construction of a ZHB building up to 45°N in favorable environmental locations. Further north, the sun gradually contributes less and less to the energy budget. The goal of providing a desirable thermal environment at minimum cost and the physio/psychological need for sunlight may sometimes conflict: an increased reliance on metabolistic heat leads designers to use superinsulated walls in which openings are undesirable. This results in a dramatic reduction in window size and sometimes to the total disappearance of windows from northern, eastern, and western walls.

If technology does fulfill its promise and introduces windows with a thermal resistance equivalent to that of insulated walls, then the guidelines for the design of passive solar houses may change dramatically. Rules governing shadow casting and the geometries of impact areas would not be affected by solar access, and design practices that govern the penetration of daylight would change substantially. The window oriented toward north may even be rehabilitated.

CHAPTER 5

A CASE STUDY OF A LOW-DENSITY RESIDENTIAL NEIGHBORHOOD

In 1983, a neighborhood planning process for Ryckmans Neighbourhood was initiated by the Hamilton Planning Department. By that time, all the planning techniques and tools for a more environmentally sensitive design approach had been developed and were ready to be tested. The planning and design of Ryckmans Neighbourhood were ideal opportunities to experiment. The planners understood and accepted that a climate-sensitive approach to neighborhood planning in the north must be based on the maximum utilization of natural forces in the immediate environment. Success would be measured by the degree of dependency on an energy distribution grid (an energy supply network that includes electrical cables, gas pipes, and tranportation facilities), oil consumption, the cost of construction and maintenance, and most important, by the extent to which the quality of life has been enhanced.

The actual neighborhood planning and design process was divided into a number of steps, beginning with the construction of an Ecochart and an analysis of topography, and ending with a complete plan of the neighborhood. It was clear at the outset of the project that some previously developed techniques needed to be refined further and converted into more practical planning and design tools—tools that are easy to use. Thus, this chapter will begin with a more detailed description of the construction and interpretation of the Hamilton Ecochart. The Slope Descriptive Synthesis will be examined more closely and a pattern of Orographic Cells will be constructed and interpreted.

THE HAMILTON ECOCHART

In chapter 3 it was determined that in attempting to understand local character, planners must begin with a broader description of the environmental megaframe—the macroclimatic setting that determines the limits of extremes. The key to this information is a number placed in the upper-right corner of the Ecochart; in Hamilton it is N 43-14-00 / W 79-51-00 / 76-2, an instant address (fig. 5-1). This elongated expression simply states the three global coordinates:

Latitude	North 43° 14′ 00″
Longitude	West 79° 51′ 00″
Altitude	76.2m

As mentioned previously, the Ecochart is an ecological portrait of a given area (fig. 5-2, see insert at back of book). It is divided into first, a graphic representation and description of the surface area, which includes topographical variations, human-made features (such as roads and buildings), and natural features (such as bodies of water and vegetation). The second part comprises the overall local weather system (the dynamic part of the Ecochart).

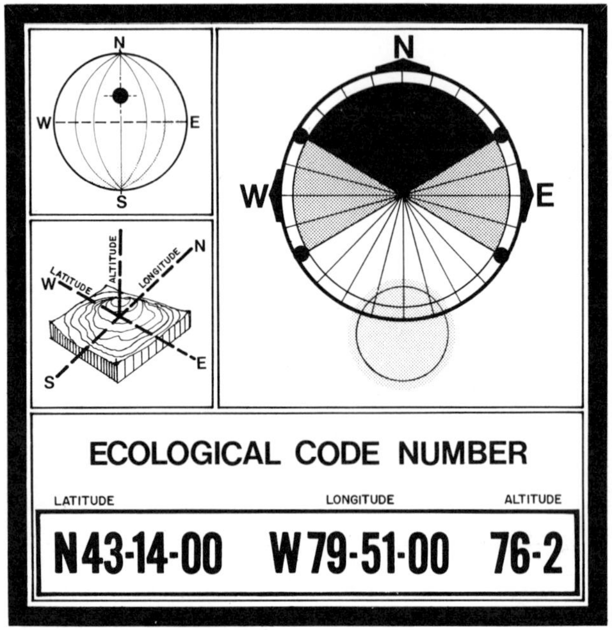

5-1. A global address.

Elements of the Local Weather System

Local climate is characterized by wide variations in *air temperature* during the year. These variations are directly responsible for differences in energy consumption. The monthly statistical maximum, mean, and minimum are recorded in Celsius degrees.

The intensity and amount of *solar radiation* depends on the atmosphere (whether there is cloud cover, the degree of obstruction on the horizon), latitude, and time of year. These measurements are absolutely essential for all systems that harness the sun's energy. Two values are recorded: the maximum possible hours and the actual hours of sun, which consider the time the sun is obscured by clouds. In addition, radiation data is recorded for specific design requirements.

Heating degree-days (HDD) is the number of degrees below a daily average temperature of 18° Celsius. For example, a day with an average temperature of 13°C has five heating degree-days ($18 - 13 = 5$). A day with an average temperature of 18°C has none. The degree-days sum for several days represents the total heating load for that period. A linear relationship exists between degree-days and fuel consumption.

The cooling degree-days (CDD) is the number of degrees above a daily average temperature of 18°C. For example, a day with an average temperature of 23°C has five cooling degree-days ($23 - 18 = 5$).

Precipitation is the water that falls on the earth's surface in the form of rain, snow, or sleet. The amount of precipitation affects the design and construction of buildings (water-proofing, insulation, and so on) and the municipal system for handling storm water. In addition landscape design is, to a great extent, determined by precipitation. Snow can have both a positive and a negative effect on energy consumption; on the positive side, it reflects sunlight and heat and creates a blanket of thermal insulation; on the negative side, it increases the roof load and incurs additional costs for snow removal and melting. Rain is expressed in millimeters (mm), snow in centimeters (cm), and total precipitation —the water equivalent of snow and rain—in millimeters.

Relative humidity (R/H) is the percentage of moisture in the air at a particular temperature. Since the amount of moisture in the air is directly related to human thermal comfort, R/H data is important in designing and setting performance specifications for mechanical systems that handle and condition air. R/H is expressed as a percentage and is recorded twice daily at 08:00 and 19:00 hours.

Soil types and thermal characteristics vary. These properties are of interest to landscape architects, builders, and engineers when specifying insulation for subsurface construction. This Ecochart measures soil temperature 10 cm. below the surface in Celsius degrees at 10:00 hours.

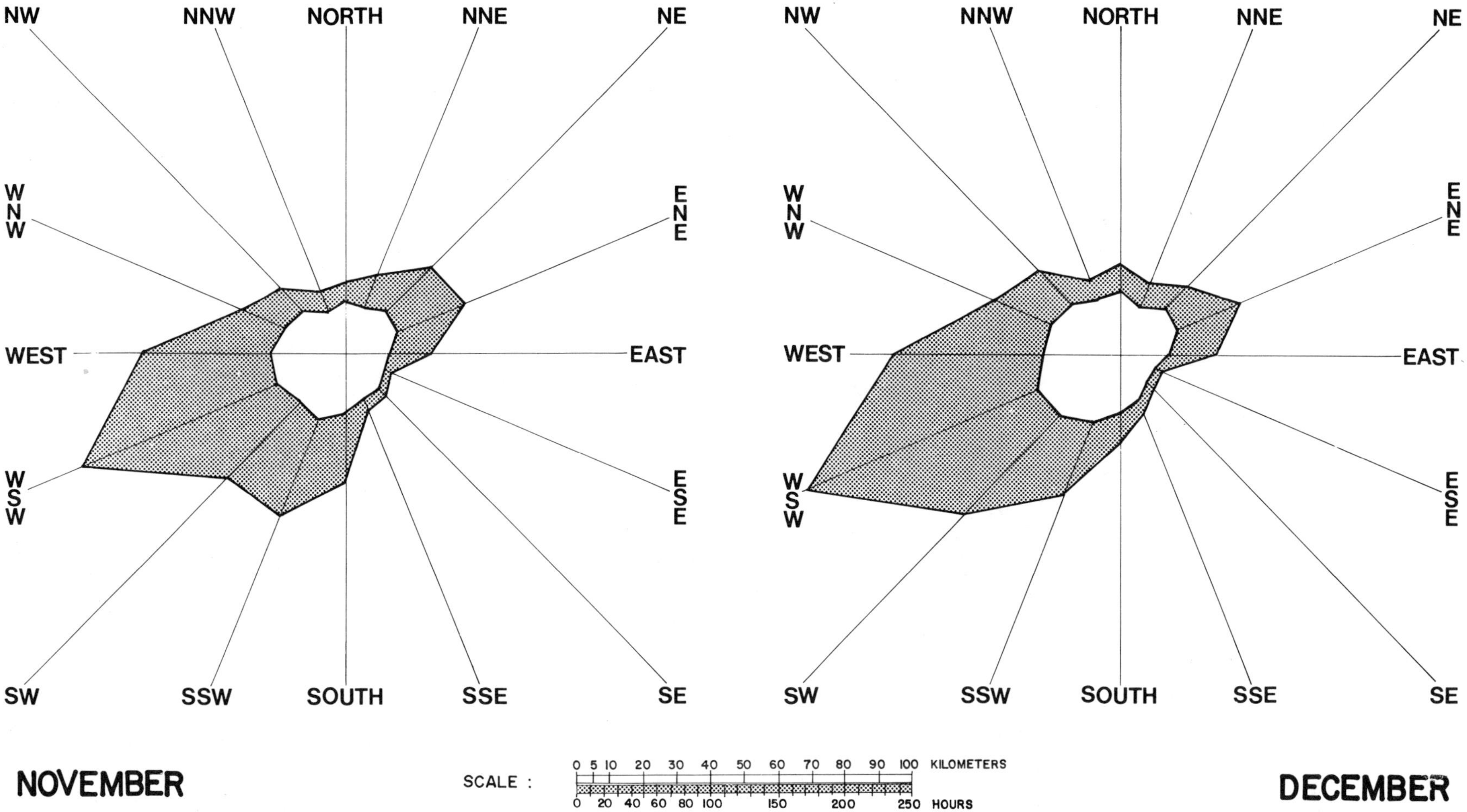

5-3. In air movement data, frequency of wind is indicated with shading, and the white area at center indicates speed.

Air movement data, or the speed, direction, and frequency of seasonal winds affect the energy requirements for indoor spaces as well as the comfort of outdoor spaces (fig. 5-3). Three variables are recorded —speed (km/hr), frequency, and the prevailing direction (16 nominal directions).

Angles of and *shadows* cast by the sun constitute another aspect of the overall local weather system (fig. 5-4). The sun's position in the sky is defined by two angles. One is determined by the sun's height above the horizon and is called *solar altitude*. The position of the sun from east to west is called *solar azimuth* and is measured clockwise from true north to the projection of the sun on the horizon. It equals 0° at north, 90° at east, 180° at south, 270° at west, and 360° at north to complete the circle.

Sun angles determine shadow length and direction. (See chapter 8 for techniques to construct shadows cast by the sun.)

The Hamilton Ecochart tabulates and illustrates the

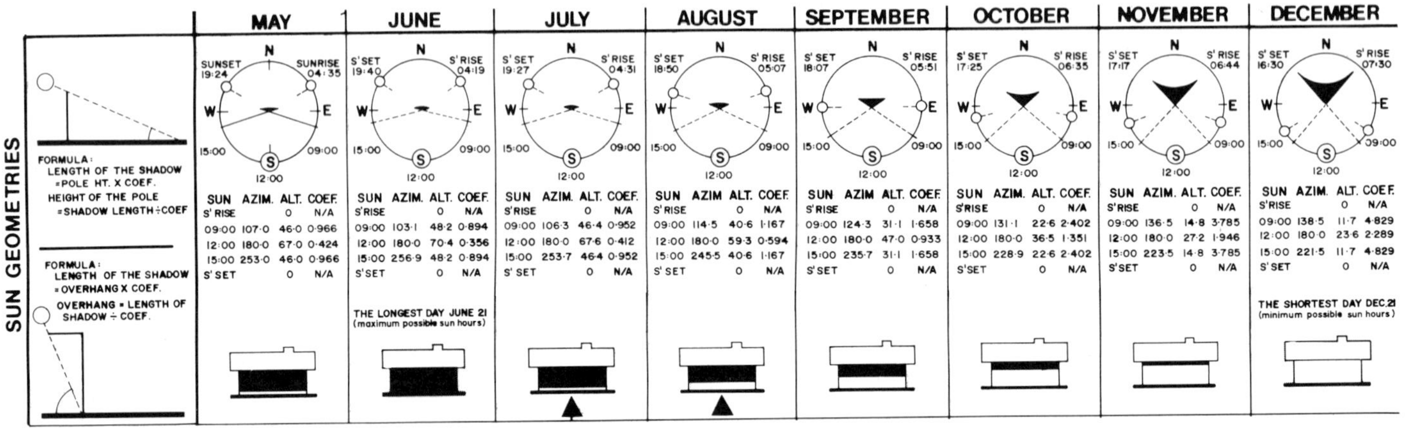

5-4. Effects of sun angles and shadows on horizontal and vertical surfaces.

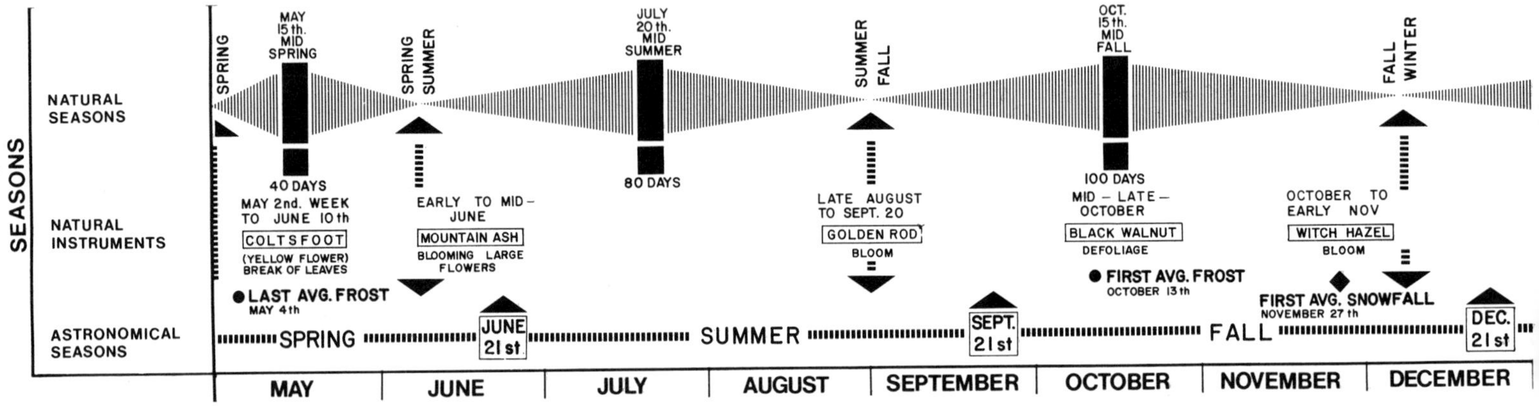

5-5. Natural and astronomical seasons.

following: 1. the composite shadow pattern cast by a pole on horizontal surfaces between 09:00 and 15:00 hours on the twenty-first day of each month; 2. the shadow cast on a vertical surface, for example, by an overhang on the southern wall of a house, at 12:00 hours on the twenty-first day of each month.

Astronomical seasons are a product of the earth's movement around the sun (fig. 5-5). These four seasons are of uniform length, and their interval is dictated by four important dates—summer and winter solstice and spring and fall equinox.

Natural seasons and *natural measuring instruments* must also be examined (see fig. 5-5). The actual or natural seasons are the product of complex and ever-changing environmental forces. They are not of uniform length and thus do not coincide with the astronomical seasons. Natural seasons differ from region to region; even the number of seasons in a year may vary.

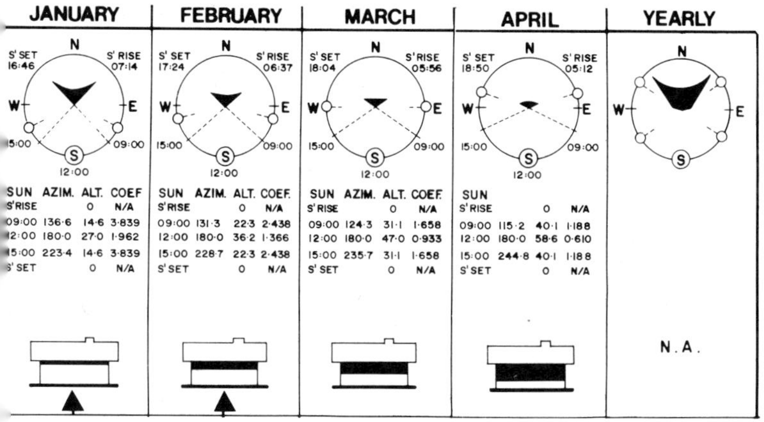

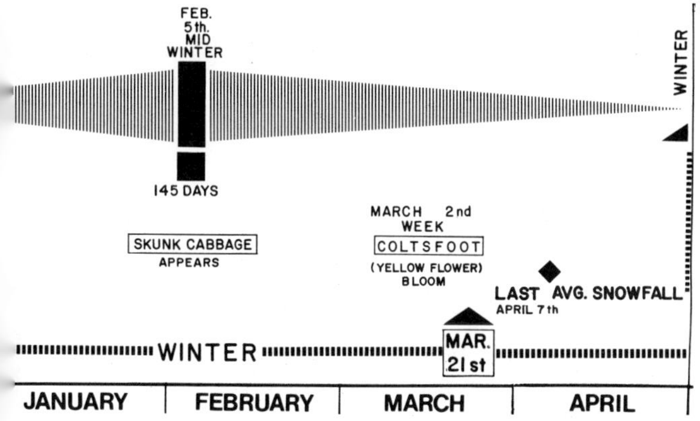

Living systems are the only devices able to indicate the beginning and duration of a natural season. Plants, for example, are highly sensitive meteorological instruments with a special capacity to integrate a number of weather factors, such as sunshine, air temperature, precipitation, humidity, and air movement. Plants exhibit the measurements of the composite environmental force with a variety of growth responses. Animals also reflect changes in the environment with changes in their behavior (nesting, mating, migrating, hibernating, and so on), and in their appearance (changes in fur density and color, fat deposits, and so on). The peak of a season is easier to identify. The beginning and the end are blurred and may vary significantly from year to year.

Analysis of the Ecochart

The Ecochart indicates that Hamilton is a northern city. The general character of the climate is continental with certain modifications due to Lake Ontario, which is adjacent. The prevailing southwest winds supply the city with fresh air and disperse industrial emissions over the lake. Although not the sunniest region in Canada, Hamilton does enjoy a considerable amount of sunshine. December has the least hours of sunshine (69 out of the 280 available hours), but is fortunately not the coldest month. The maximum heating requirement comes in January (762 HDD), followed by February (661 HDD), but by then hours of actual sunshine jump to 95 hours and 107 hours respectively.

The heating and cooling degree-days indicate that heating and cooling are not of equal importance in this region. And, although maximum available sunshine *should* be sufficient to supply heat throughout the cold season, clouds make this availability irregular. Consequently, the clearest message this Ecochart imparts is that heating is more energy devouring than cooling, and planning for the cold period of the year should be the major consideration. More emphasis should therefore be placed on exposure to the sun than on providing shade. Guaranteeing access to the sun is a prime task for an urban planner, who is directly involved in designing neighborhood plans, zoning, and other planning regulations, whereas providing shade is a prime task for an architect and landscape architect, who determine site design, building design, and landscaping.

The Ecochart also indicates that there are significant opportunities for reducing space-heating costs. This,

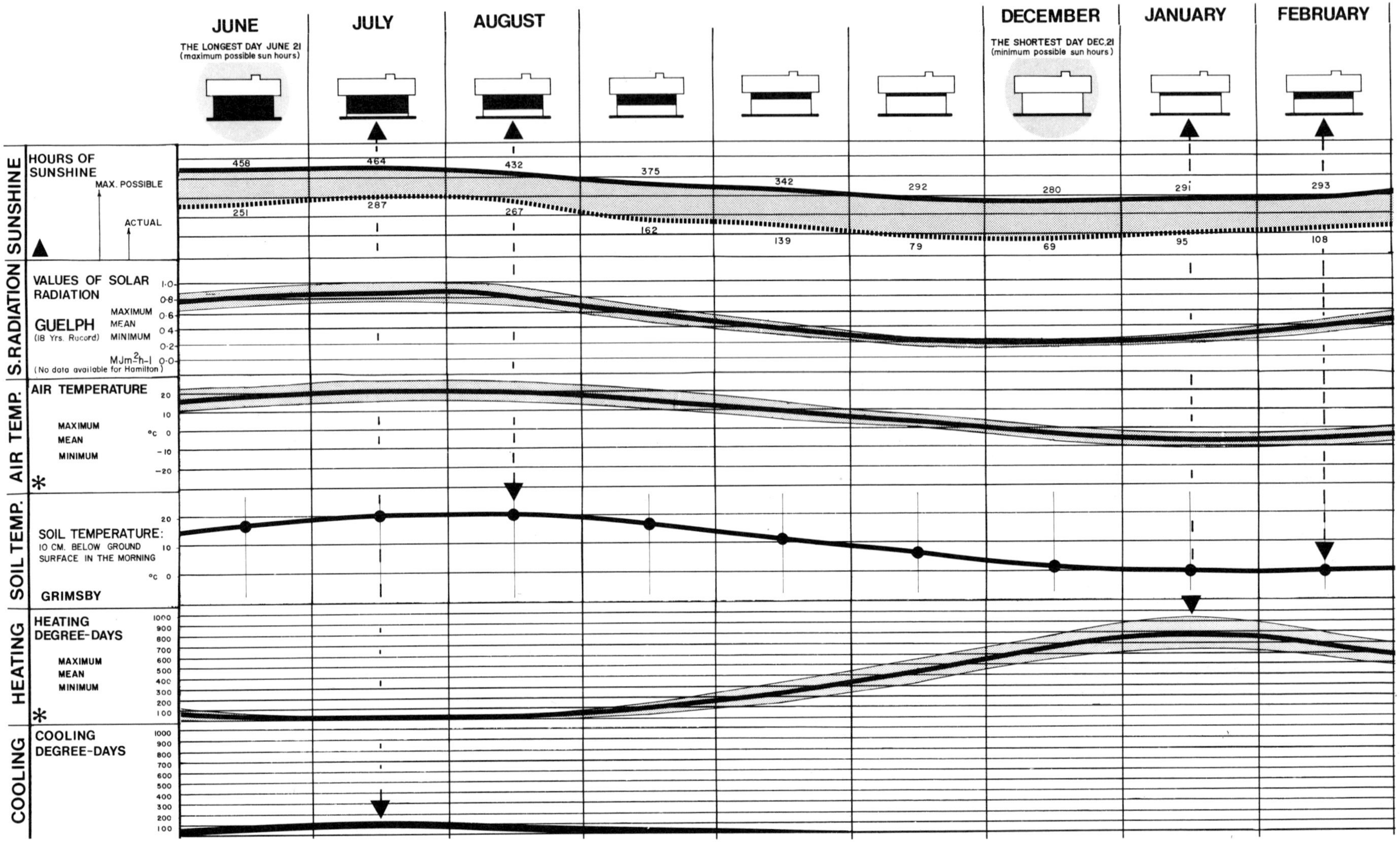

5-6. Stimulus-response intervals.

however, requires extensive knowledge of the local environment.

Time Lag

An interesting phenomenon, which occurs in any terrestrial setting but whose extent and magnitude differs from locality to locality, is the discrepancy between the astronomical and actual seasons, or the delay before planetary mechanics have a full effect on a given area (fig. 5-6). This is clearly illustrated in the Ecochart.

The shortest day of the year is the day of maximum sun deprivation; the longest day offers the maximum possible sunshine at a given latitude. Astronomically, these two annual shocks take place on December 21 and June 21. However, it requires time for the earth-atmosphere system to respond to changes in solar-energy input. The air temperature cycle lags behind the solar cycle by approximately thirty days in the Ecochart area. This reaction time, the stimulus-response interval, or simply the time it takes to recover from the previous hit, delays the consequences of the solar cycle by approximately one month. This, of course, complicates a soft path strategy. For example, in the case of the performance characteristics of fixed, self-shading, architectural devices such as overhangs, the best possible geometries of an overhang would totally expose the windows on a southern wall to the sun on December 21 and completely screen them on June 21. Such an arrangement would perform flawlessly in a vacuum. Unfortunately, because of the lazy response of air mass temperature, the actual need for total exposure and total screening takes place four to six weeks later. This difference must be compensated for by installing movable screens or shading devices.

Another striking example of the time lag is in the underground temperature profile (fig. 5-7). In the area covered by the Ecochart, the maximum and minimum temperatures of soil ten centimeters (four inches) below the surface exhibit a delay of eight weeks. The time lag continues to increase up to a depth of six meters (19.7 ft), where the lag is six months. This means that summer on the surface coincides with winter underground. In other words, a six-month delay phase juxtaposes surface summer and underground winter and vice-versa at the given depth. Engineers can use this information to design heat pumps and thermal storage underground.

In principle, this time lag is a result of the combined effects of the thermal properties of matter (soil, air, water) and the annual solar radiation profile. It is a result of the complex interaction between a large number of factors and is always locale specific. The Ecochart neither attempts to nor is able to explain this phenomenon; it only records and displays its existence and pattern of behavior.

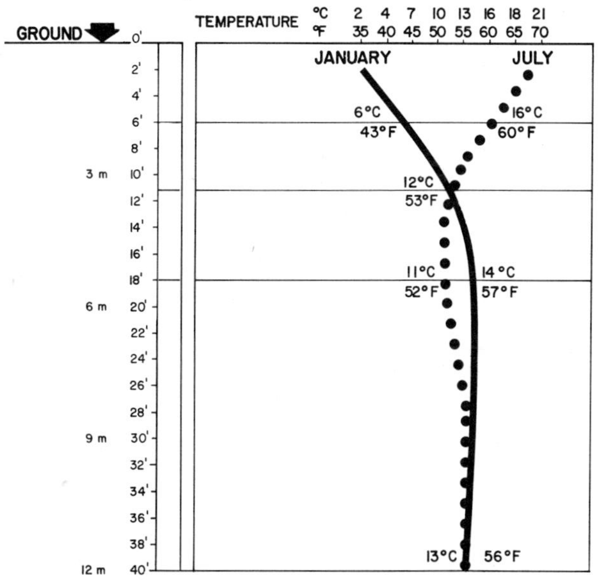

5-7. The measurement of variations in soil temperature at McMaster University Campus, Hamilton, Ontario, from October 1964 to September 1966.

Climate and Local Topography

The local topography is, of course, an outstanding modifier of the regional weather machine. For example, the upper part of the Ecochart indicates that the city of Hamilton is divided by the Niagara Escarpment. A mere drop of approximately 100 meters (328 ft.) creates distinct weather patterns on the upper and lower levels. The difference, measurable by conventional instruments, shows a noticeable deviation from the mesoclimatic data as stipulated by the Ecochart.

For the architect and urban planner primarily interested in human comfort, measurements of individual weather elements by conventional instruments are of limited value. The human comfort zone is the elusive product of the synergy of several environmental ingredients, and no single instrument can measure it. However, the ability to understand instrument readings in conjunction with the natural responses of plants and/or animals leads the professional to a better assessment of outdoor human comfort in a given location at a given time.

This approach is common in other scientific fields. The Ontario Ministry of Agriculture has been publishing maps indicating corn heat units (CHU). These are values that express the relationship between the environment and corn development. The rate of progress to maturity is related to the optimum thermal conditions for corn. The higher the number of CHU, the faster the plant grows. Hamiltonians may be interested in knowing that the lower bench of the city is rated at 3,300 CHU, whereas the upper plateau is only 2,700 CHU.

A topographical variation, such as a sudden drop of one hundred meters, modifies the general weather pattern. But even less profound, gradual, topographical variations, sometimes barely noticeable, can and do create unique climatic conditions.

Planning Considerations

Taking into consideration the length of the cold season, the unpleasant weather, and the amount of precipitation as shown by the Ecochart, neighborhood shopping and other services and community amenities should be located to minimize travel during bad weather. An aging population would expect a northern neighborhood to be interwoven with covered walkways to shopping and other communal facilities. Sidewalks and roadways should be wide enough to accommodate vehicles that remove snow, and there must be provision for snow storage. Street alignment should be derived primarily from the geometries of solar access. Under certain circumstances, however, streets might be placed perpendicular to the prevailing winter winds, as indicated by the Ecochart.

Every attempt should be made to reward people who leave their cars at home. This can be achieved by designing exciting walkways that take advantage of or enhance the natural topography, by creating convenient public transit, by minimizing the effects of extreme weather conditions with shelters or canopies, by introducing pedestrian amenities, such as benches, garbage cans, drinking fountains, street lights, sculptures, and fountains. The warm season can be dramatically extended with wind protection, berms, fences, evergreens, and by controlling the roof slope, the overall shape and layout of the house, and by landscaping.

It might be necessary to introduce more specific guidelines and new ordinances in order to extend the warm season. In areas suitable for passive solar housing (flats, south slopes), solar access must be protected. The most convenient tool is the Solar Envelope, whose height is determined and regulated by the maximum permissible size of the building's impact area (shadow). To screen private green spaces and to shade eastern and western exposures, flexible fence ordinances are needed (ordinances are often based on southern aesthetics and are not necessarily suited to an area where increased microclimatic control is desired). Furthermore, it would be advantageous to have flexible siting provisions to achieve maximum exposure to the sun, regardless of street orientation and the orientation, shape, and size of lots, front yards, side yards, and backyards. Earth forms that are naturally shaped to reflect the sun in the winter should be used wherever possible. This can be emphasized with reflective paving.

A northern climate requires a compact neighborhood design and urban form. This could result in more clustered houses, houses with a common wall, and attached, compact buildings, since a low surface/volume ratio is an efficient way to save energy. Planning for the cold requires that individual buildings be airtight and compact in form. Sheltered sun traps, (comfortable even in very cold weather) should be created wherever possible. Sun traps can be included in the design of individual houses (atriums in U-shaped, L-shaped homes) as well as in neighborhood design (groups of buildings forming sunny, wind-sheltered public spaces).

The Hamilton Ecochart indicates that some areas in the region, such as southern slopes, may benefit from a favorable microclimate and have enough sun to provide year-round thermal comfort. A passive solar strategy in combination with an intermediate thermal-storage system for sunless winter days can therefore be combined with the use of metabolistic heat. Should bathing individual buildings in sun be preferred, then they would have to be spread sufficiently apart to respect the boundaries of impact areas.

The climate of this region does not dictate a single planning and design solution but permits a variety of strategies. On southern slopes, flats, and shallow eastern and western slopes, year-round thermal comfort can be achieved at minimum cost by building low-density urban forms. On northern slopes and on more dramatic eastern and western slopes, the surface/volume ratio would have to be lowered. This would lead to larger, more compact buildings.

Housing Design

The Hamilton Ecochart provides eloquent instructions for house design and construction. Two rather pronounced annual blows require that any structure in this area be built to absorb the effects of these blows. Six meters of soil cover would be ideal (see fig. 5-7), but even a meter of soil prevents water from freezing. Because earth sheltering reduces heat loss as well as heat gain, this strategy would be worth considering in this area.

The winter sun must be treated as a precious natural asset. Northern countries in particular must take full advantage of the sun to ensure physical and psychological well-being. A sheltered, internal court can stretch the warm season throughout most of the year, particularly if it is combined with a mobile glass roof and mobile shades. To enhance the warmth and light provided by a weak winter sun, receiving surfaces could be painted in dark colors. Also, the south side of structures could be equipped with greenhouses or sunspaces that have movable insulation to prevent overnight heat loss.

The general layout of a house is also of great importance. A cold climate requires that living areas are positioned on the south side of the house. Indoor living activities should be planned adjacent to an outdoor private green space. Heat-producing spaces with low occupancy, or spaces that require little light should be located on the north side of the building. Bedrooms should be located on the east to receive the morning sun.

The Ecochart records the frequency of wind, rain, and snow. Wherever possible, the entrance should be located on the sheltered side of the house. In any case, this regional prototype should have an entry that is well protected and has an air lock.

When designing energy-efficient houses, careful attention must be paid to window location, size, construction, and shading. There should be maximum fenestration on the south—the sun gate—a minimum on the east and west, and no windows on the side facing north. If northern windows cannot be avoided, then the glassed area should be reduced and assembled from triple and quadruple panes. Northern windows should be fixed rather than openable and equipped with movable, high-performance insulation, for example, indoor or outdoor shutters. High-quality weather stripping and caulking should be applied to all windows, especially since the Ecochart shows that wind is a permanent feature. If the window is to be used for ventilation as well, the upper part should be openable, allowing for very efficient winter ventilation.

During the summer, all windows should be shaded by a combination of fixed and movable overhangs and exterior movable insulation. Shading reduces summer heat gain by as much as 80 percent. Exterior shades are approximately 35 percent more effective than interior shades. Eastern and western exposures usually require special shading devices to block the low sun.

Every participant in the planning, design, and development process can extract much useful information from this Ecochart. It offers a variety of solutions and a means for making intelligent trade-offs and compromises.

Climate and Construction

The construction of buildings is a weather-dependent activity and can have serious economic consequences. Careful planning is therefore required, particularly if the construction site is unfamiliar. The Ecochart is an excellent source of information for local as well as out-of-town construction managers and contractors. Its data is relevant for most phases of building construction.

Minimum *excavation* depths for burying pipes, municipal services, and building foundations are determined by the degree of frost penetration in a given area and are usually specified by law. A cold season could be of some advantage here because trenches cave in less easily and it is often easier to handle material. To break through the solidly frozen surface, however, requires the use of heavier equipment, and in this particular area, the surface could be as deep as thirty centimeters (11.8 in). Another problem is that frozen material must never be used to backfill.

During the cold season, *concrete* must be treated with chemical additives and/or covered with thermal blankets, or warmed by an external source of heat, usually hot-air blowers. Concrete poured outdoors requires waterproof blankets, since the impact of rain can be devastating. A layer of snow, on the other hand, could act as insulation during an extremely cold period.

Pouring concrete below grade (in trenches) is possible in the area covered by this Ecochart throughout the year. However, a contingency for special treatment must be provided for the period between December and March. Above-ground pouring is very susceptible to environmental variations and is more difficult to protect. A minimum temperature of 4°C (39.2°F) is needed for curing without additives. It would then take twenty-four hours for concrete to be 50 percent cured and seven days to be 95 percent cured. It is therefore safer to schedule this type of pouring before December or after March. Pouring in enclosed buildings is possible throughout the year, since heaters can maintain the desirable temperature.

Sudden swings in temperature can adversely affect the curing process. For example, large concrete walls cure too quickly in temperatures above 25°C (77°F), and their structural properties may be affected. Rain, by washing away not yet hardened mortar, can be even more damaging than frost. In general, masonry contruction should be scheduled for above-freezing temperatures; special precautions that must be taken when the temperature is below freezing, such as heating the sand, water, and other materials, as well as the building of protective shelters can escalate costs. The Ecochart indicates that masonry construction in this area should not be scheduled from December through March.

Any process involving water, for instance, external stucco or painting, must proceed when the temperature is above freezing. Both rain and sudden swings in temperature are risky.

When creating a flat *roof,* felt and gravel construction require a 1- to 2-percent slope. The minimum temperature required for application is 4°C (39.2°F), lasting at least forty-eight hours. The same parameters apply to membrane construction.

Given the climatic profile of the region as described in the Ecochart, roofing projects should not be scheduled for January and February. At the appropriate time of the year, it is best to wait for favorable weather and then to proceed as quickly as possible. Similar considerations apply to paving and certain landscaping projects.

Wood-frame construction and prefabricated steel construction could be scheduled for any time of year according to the Ecochart for this area. Both types of construction are practically unaffected by weather. Wind, however, could be destructive. Any large, tall, multistory construction, or warehouse block walls, require special bracing to withstand the wind. Moreover, cranes cannot operate when it is very windy.

Any construction planning and trades scheduling within the Ecochart area should be guided by the following figures, which should be considered optimistic:

January and February: three days of construction time in a five-day week. The speed of construction is reduced by nearly 50 percent.
December and March: four days of construction time in a five-day week.
The rest of the year: five days of construction time in a five-day week.

Rain, averaging two days of every thirty throughout the year, should also be considered.

TOPOGRAPHY—THE CONSTRUCTION OF OROGRAPHIC CELLS

Two methods of illustrating topographic variations based on a square grid were described in chapter 3. As previously discussed, the Slope Gradient Description contains a number of instructions for the planner. Figure 5-8 summarizes these instructions for optimum allocation of land with respect to slope gradient. The Slope Orientation Description relates topography to the cycles of the sun. It is constructed by assigning a nominal direction to each square. Figure 5-9 indicates preferable land use with respect to the different degrees of energy received from the sun.

Both ways of describing land (by slope gradient and orientation) treat land as a natural solar collector with an uneven pattern of absorbing sun energy. In order to extract the instructions necessary for plan design and to make the Slope Descriptive Synthesis a useful design

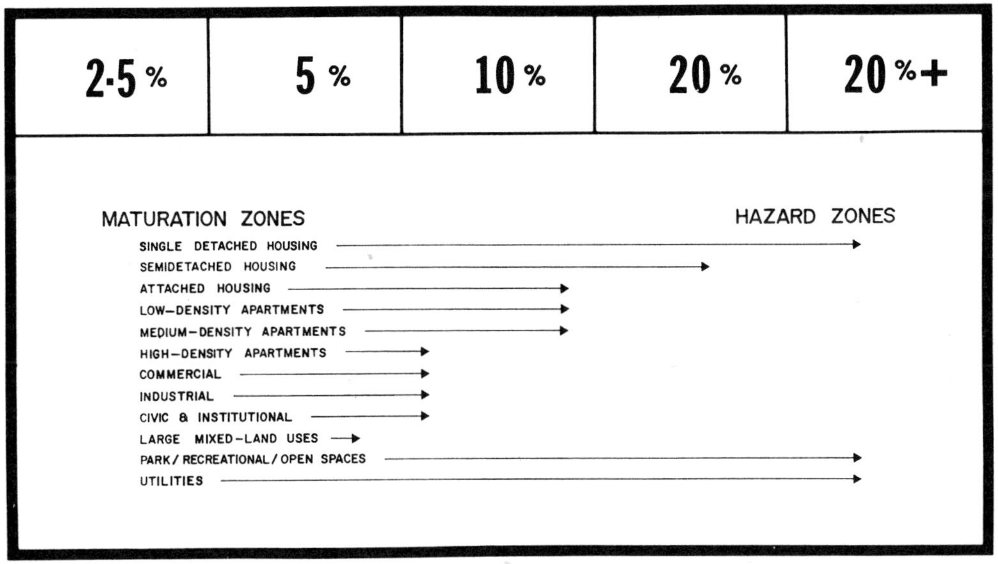

5-8. Slope gradient and land use allocation.

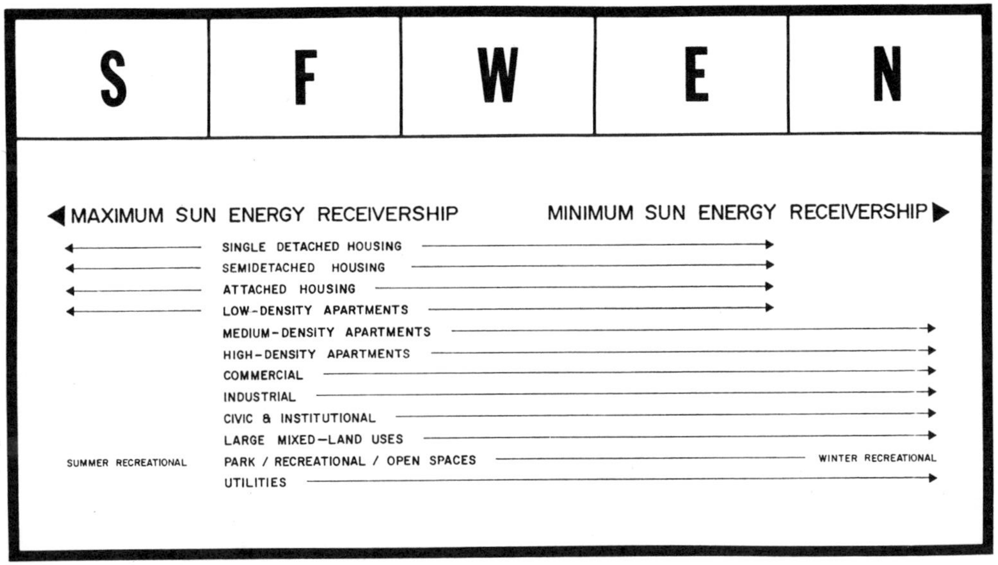

5-9. Slope orientation and land use allocation.

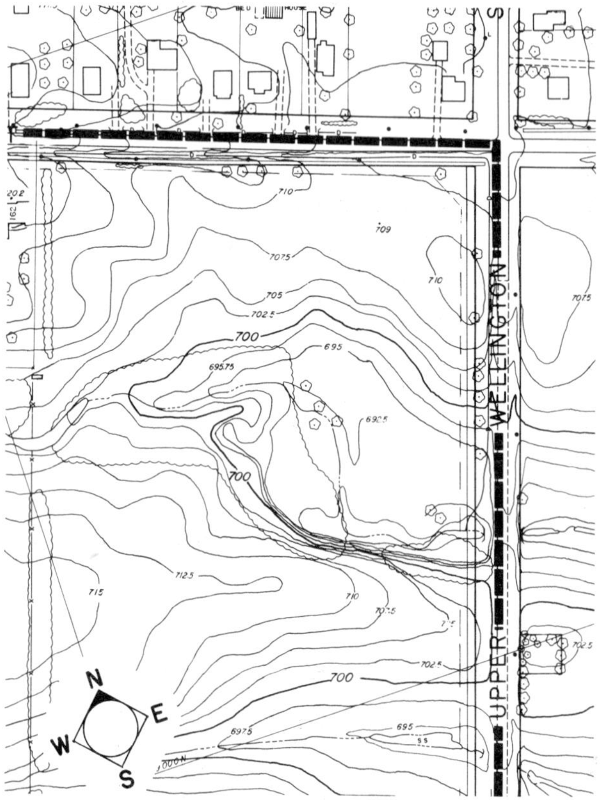

5-10. Conventional topographic contours.

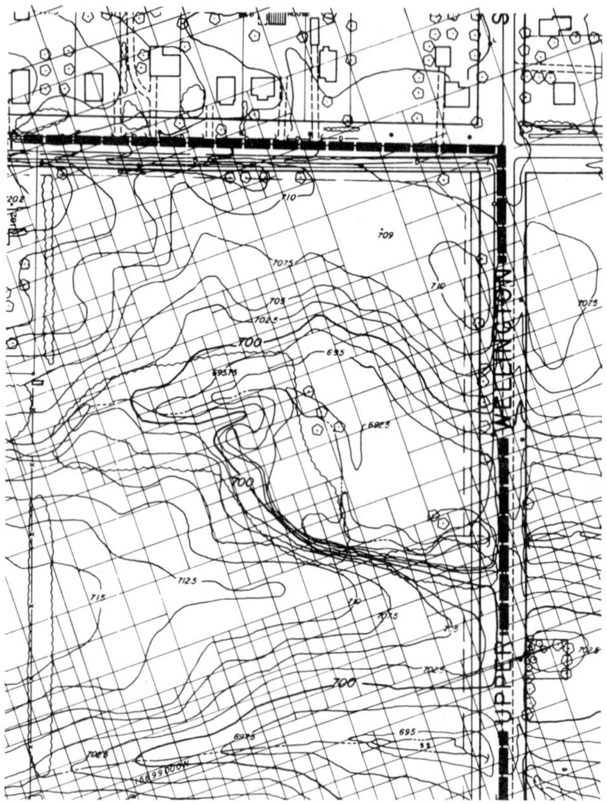

5-11. The basic grid superimposed on contours and subsequently divided to ensure that each square is intersected by a single contour.

tool, a new practice must be adopted, that of constructing orographic cells.

Orographic cells are derived from the Slope Descriptive Synthesis (S.D.S.). Both methods of S.D.S. construction allow conversion to a pattern of orographic cells. The following is a review of the first method of construction, where the basic grid is subsequently divided into smaller squares to describe land form. This fast and simple method was the one chosen by planners for the development of Ryckmans Neighbourhood Plan.

Figure 5-10 shows the topographic contours. In figure 5-11 the basic grid, drawn within true north/south and east/west coordinates, is divided into smaller squares that represent the slope gradient. Figure 5-12 shows the grid with contours removed. One of four nominal directions (south, west, east, north) are assigned to each square in figure 5-13. This figure is an S.D.S. for the given area. It shows two variables: slope gradient and slope orientation.

In the next step, grid squares having the same slope

5-12. A description of land form with contour lines removed. The size of each square indicates the degree of slope.

5-13. Arrows are used to indicate four nominal categories of slope orientation.

gradient and orientation are grouped into cells and labeled accordingly. For example, W5 represents an area or cell on a western slope with a gradient of 2.5 to 5 percent (figure 5-14). The actual process of converting the Slope Descriptive Synthesis into a pattern of orographic cells requires that two design criteria be balanced.

1. The minimum practical size of cells are determined by the type of development or land use. For example, a $10m^2$ (107.64 sq ft) cell would be meaningless to a land-use planner.
2. The maximum permissible level of impurities must also be established. In order to ascertain the minimum practical size of a cell, it may be necessary to absorb and integrate squares with different slope gradient and orientation. For example, an E5 cell may include several squares with a 10 percent and 20 percent slope and with a northern and western orientation. These would be the impurities within a

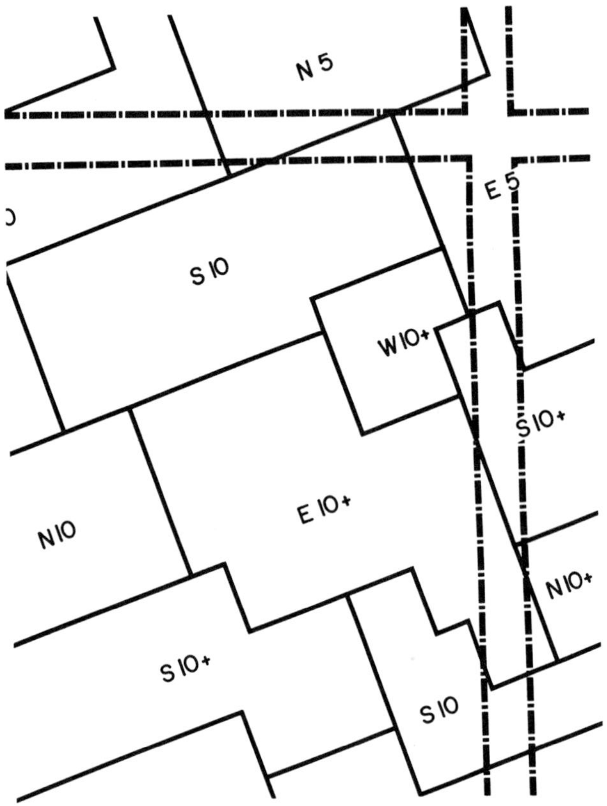

5-14. Squares that have a similar slope orientation and gradient are grouped together and labeled accordingly.

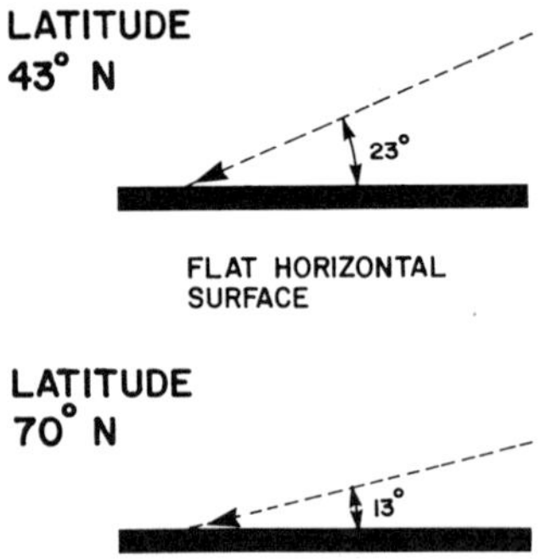

5-15. Comparison of sun altitude at 42°N latitude and at 70°N latitude at the same time of day.

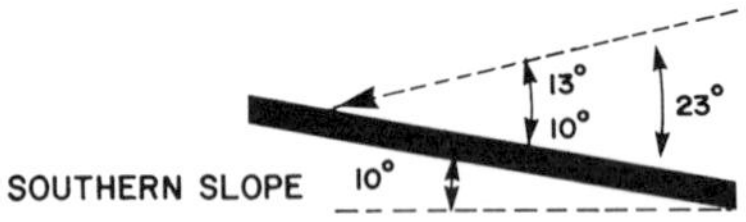

5-16. Southern slopes of 10° at 70°N latitude and flat surfaces at 43°N latitude have a 23° angle of solar beam incidence at the same time.

given cell. However, practice shows that the level of impurities can be held below 15 percent of a given cell area.

The tapestry of cells represents a variety of microclimatic conditions, areas that have a unique climatic profile and length of seasons. In fact, a special Ecochart for each cell can be constructed, and the recorded weather elements in some areas may resemble more closely the climatic profiles of far-away places than those found in the mesoclimatic Ecochart of the given region. One outstanding reason for this is the variation in solar radiation density. (See chapter 4 for a detailed discussion of the relationship between orientation and solar geometries.) Barring atmospheric phenomena, it is the angle of solar incidence that determines the amount of energy received. If, at a given hour, solar altitude at 43°N latitude is equal to 23°, then, because of the curvature of the earth's surface, the solar altitude at 70°N latitude would be equal to 13° at the same time (fig. 5-15). The only way to achieve the same radiation density at 70°N latitude would be to tilt the horizontal surface 10° toward the sun. A flat horizontal surface at 43°N latitude and a 10° southern slope at 70°N latitude would receive the same amount of solar radiation (fig. 5-16). This would not only make the average soil temperature equivalent, but also, to a large extent, the microclimate. These distance relationships for 43°N can be calculated and tabulated as follows:

North or South Slope (%)		Flat (0%) Site
(43°N LATITUDE)	EQUIVALENT TO	(KM NORTH OR SOUTH)
5 percent		319 (198.2 miles)
10 percent		634 (394 miles)
20 percent		1250 (776.7 miles)

As a result, a cell described as S10 (southern slope, 10 percent gradient) would receive the same radiation and have roughly the same microclimate as a flat site 600 km (372.83 miles) farther south. The opposite would apply to a N10 cell (northern slope, 10 percent gradient). To illustrate this point, a conventional pattern of cells is described in terms of their related geographic locations (fig. 5-17).

In summary, by virtue of its unique properties, every orographic cell contains special constraints and offers a variety of opportunities for the development and use of land.

Information Derived from Orographic Cells

Information derived from orographic cells can be applied to land use, angles of street alignment, space-heating modes, and density. The Slope Gradient Description (see fig. 5-8) and the Slope Orientation Description (see fig. 5-9) are merged to create preferred *land-use* types (fig. 5-18).

Since the division of land into individual lots, site design, and the placement of structures is, to a great extent, determined by the street pattern, it is essential to develop a design methodology based on the knowledge gained in chapter 4, where angles of *street alignment* ensure maximum solar exposure for structures and buildings. In fig. 5-19, a four-sided hill with a flattened top represents five nominal topographical variations: northern slope, southern slope, eastern slope, western slope, and a flat horizontal surface. Five identical structures (cubes) constructed in these five areas cast five different annual composite shadow patterns due to five topographical configurations (see also fig. 4-12).

In figure 5-20, a nominal hill is described with uniformly distributed, identical cubes. If on the flat top the distance between the rows of structures is such that it ensures solar access to southern walls (if each row of houses is built just outside the shadow impact area of

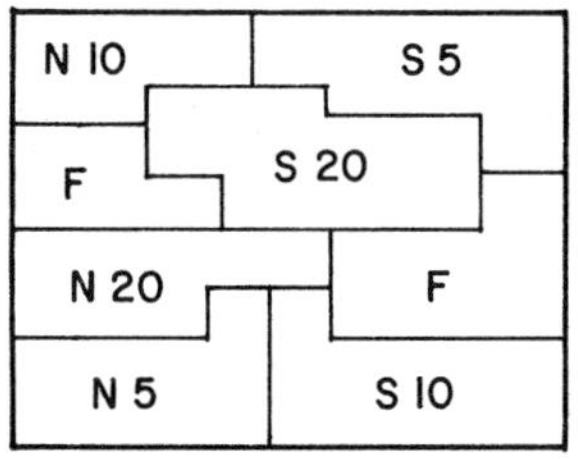

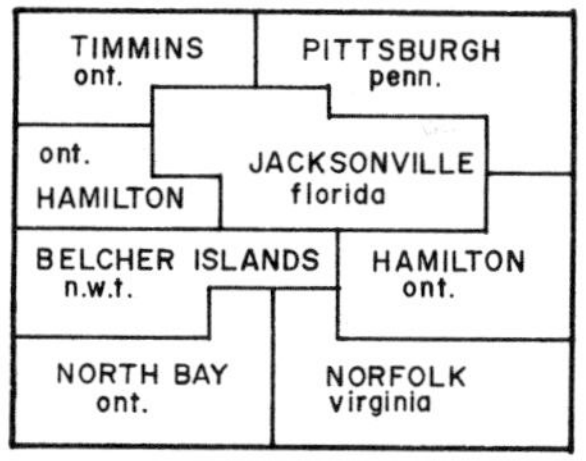

5-17. Orographic cells and geographic areas with sunbeams of the same angle of incidence.

<table>
<tr><th>SLOPE % →
ORIENTATION ↓</th><th>2·5</th><th>5</th><th>10</th><th>20</th><th>20 +</th></tr>
<tr><td>S</td><td colspan="2">SINGLE DETACHED
SEMIDETACHED
ATTACHED HOUSING
LOW DENSITY APTS.
PARK/REC./OPEN SP.</td><td>SINGLE DETACHED
SEMIDETACHED
ATTACHED HOUSING
PARK/REC./OPEN SP.</td><td rowspan="3">SINGLE DETACHED
SEMIDETACHED
PARK/REC./OPEN SP.</td><td rowspan="3">SINGLE DETACHED
PARK/REC./OPEN SP.</td></tr>
<tr><td>W</td><td colspan="2" rowspan="2">SINGLE DETACHED
SEMIDETACHED
ATTACHED HOUSING
LOW–DENSITY APARTMENTS
MEDIUM–DENSITY APARTMENTS
HIGH–DENSITY APARTMENTS
COMMERCIAL
INDUSTRIAL
CIVIC & INSTITUTIONAL
LARGE MIXED–LAND USES
PARK/RECREATIONAL/OPEN SPACE
UTILITIES</td><td rowspan="2">SINGLE DETACHED
SEMIDETACHED
ATTACHED HOUSING
LOW–DENSITY APTS.
MED–DENSITY APTS.
PARK/REC./OPEN SP.</td></tr>
<tr><td>E</td></tr>
<tr><td>N</td><td colspan="2">MEDIUM–DENSITY APARTMENTS
HIGH–DENSITY APARTMENTS
COMMERCIAL
INDUSTRIAL
CIVIC & INSTITUTIONAL
LARGE MIXED–LAND USES
PARK/RECREATIONAL/OPEN SPACE
UTILITIES</td><td>HIGH–DENSITY APTS.
PARK/REC./OPEN SP.
UTILITIES</td><td colspan="2">PARK/REC./OPEN SPACE
UTILITIES</td></tr>
</table>

5-18. Preferred land use for different orographic cells.

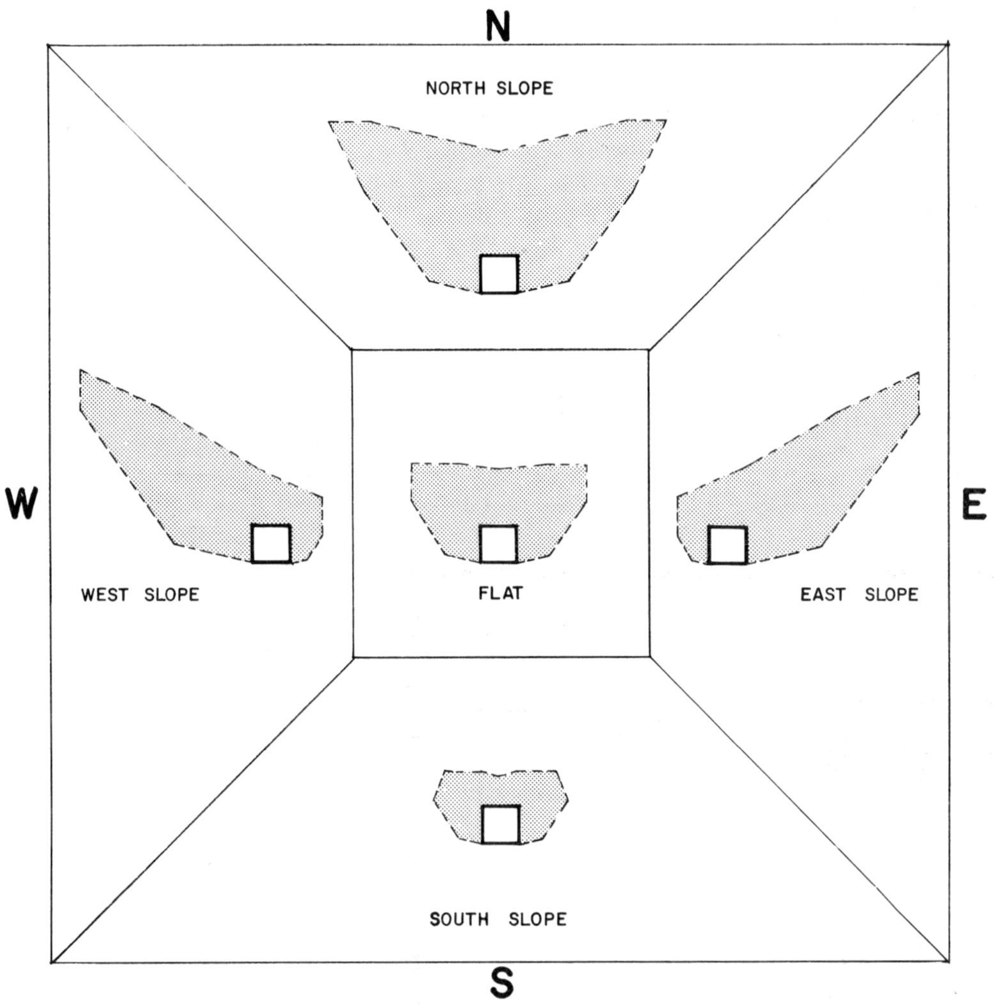

5-19. The distortion of impact areas in a nominal hill with four slopes and a flat top.

the previous row) maximum development would be possible on the flat top. On the southern slope, however, the same distance between the rows of structures would be too generous; whereas structures on the northern, western, and eastern slopes would be overshadowed and any solar harnessing opportunities significantly limited or impossible.

In figure 5-21, structures are rearranged to ensure the insolation of southern walls—all structures are built just beyond the impact areas. The density on the flat top therefore remains the same. On the southern slope, density would increase; and density would be profoundly lower on the eastern, northern, and western slopes.

Since the arrangements on eastern and western slopes are strikingly cumbersome, a second look has to be taken at the nature of shadow distortion. It is evident from figure 5-22 that the degree of shadow distortion is directly related to the degree of slope. In order to minimize the distance between rows of structures, it is necessary to align these structures according to shadow distortion characteristics. As a result, the row of structures cannot be aligned from true east to west. The degree of deviation from the cardinal east/west direction would be directly related to the degree of east or west slope (fig. 5-23). This relationship is latitude specific; for 43°N latitude, it was calculated that the degree of deviation from the cardinal east/west direction equals the percentage of slope (east or west) multiplied by two. Thus, 1 percent (east or west) slope requires a deviation of 2°, 2 percent slope, a deviation of 4°, 3 percent, a deviation of 6°, and so on (fig. 5-24). At 43°N latitude this system works only up to 10 percent of an east or west slope. Inclined planes over 10 percent are not insolated by the low winter sun within the given sun window (09:00 to 15:00 hours). The structures are submerged in the shadow of the hill itself, and any change in alignment is therefore meaningless.

Lines can now be drawn into the individual cells to indicate recommended alignment (fig. 5-25). This allows incorporation of the maximum possible recommended alignment angles into the actual neighborhood design. Through the inclusion of slope orientation and gradient considerations, this method improves on the first generation of subdivision designs to conserve energy. These promoted a standard east/west alignment for solar access, regardless of topography.

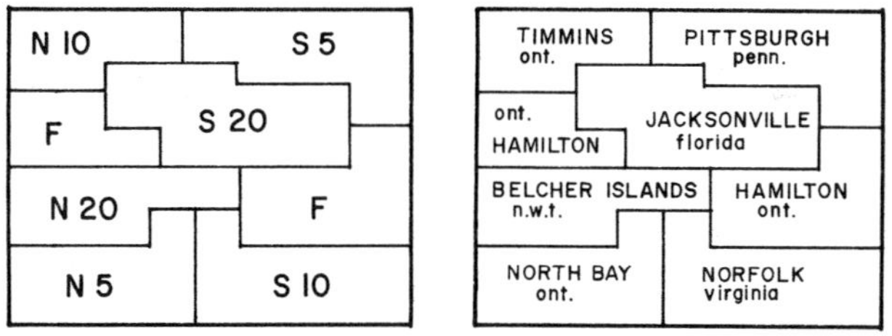

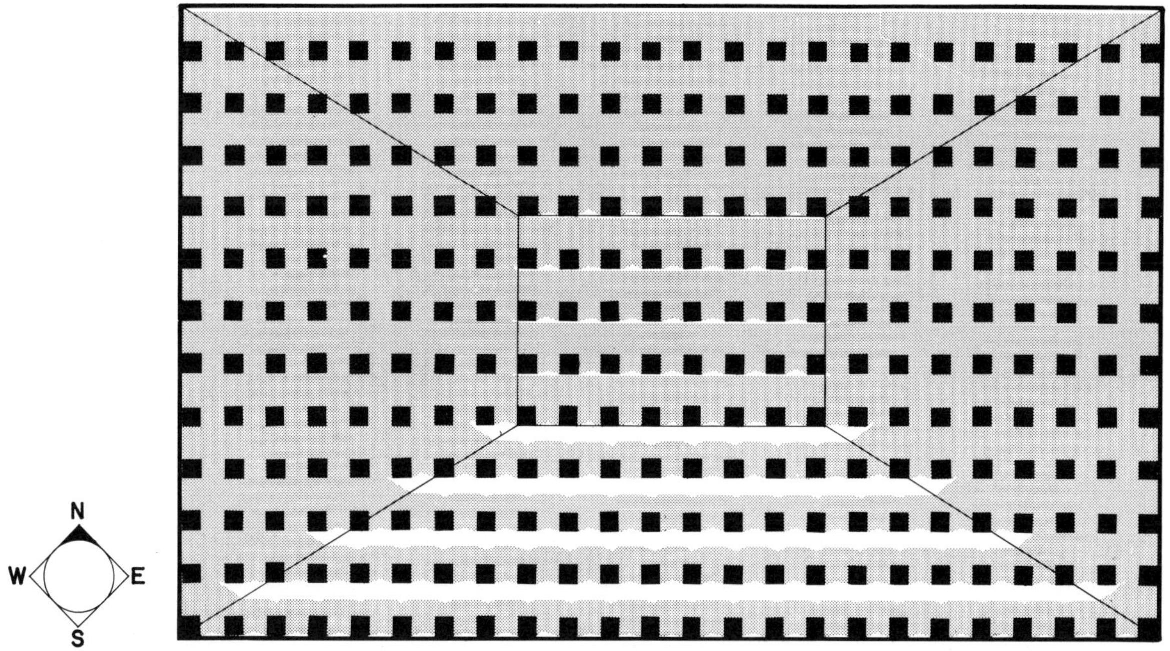

5-20. A nominal hill uniformly developed with identical structures; the extent of the impact area equals the distance between the rows of structures only on the flat top.

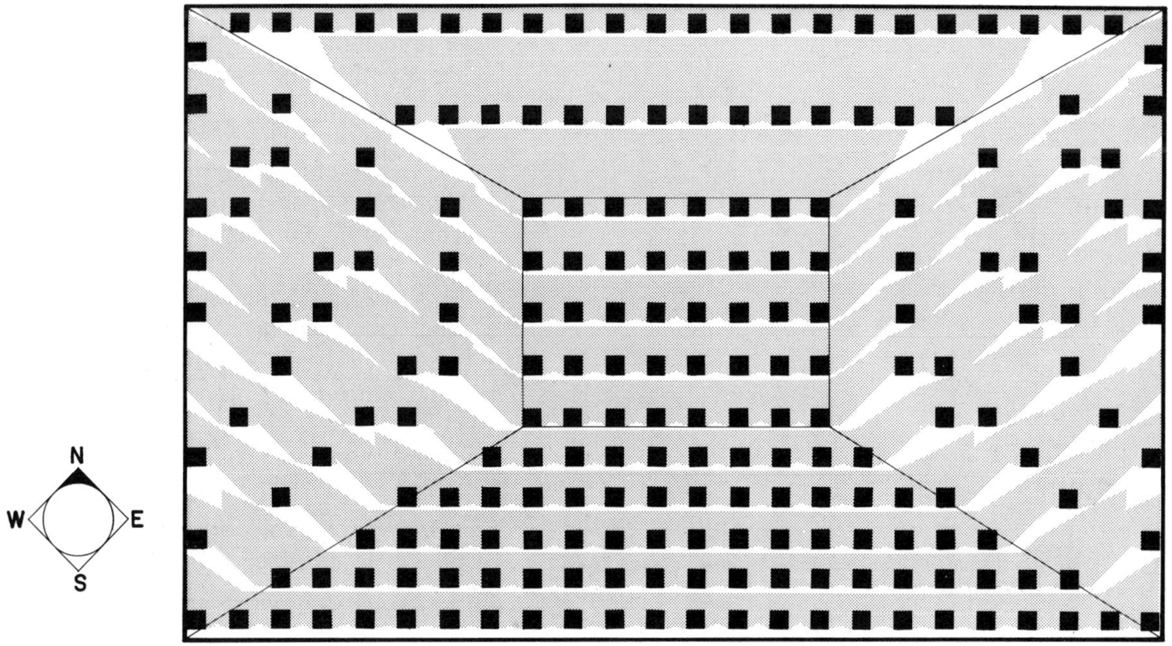

5-21. The densities of structures are modified to accommodate solar access on different slopes.

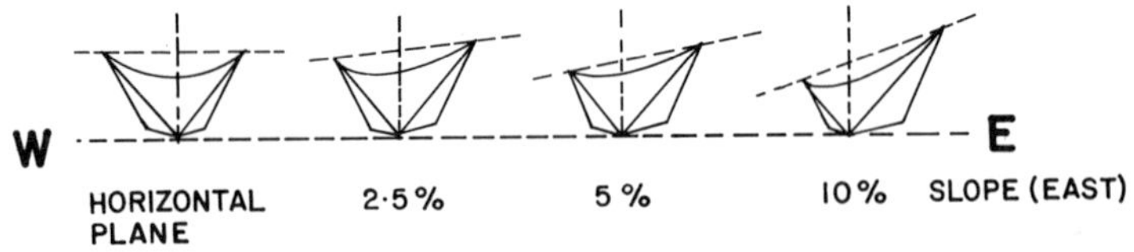

5-22. The asymmetrical distortion of the Solar Pyramid on eastern slopes relates to the gradient of that slope.

The preferred *space-heating mode* varies in response to an individual structure's exposure to sun. Figure 5-26 shows the opportunities for a variety of space-heating modes in relation to different degrees by which external surfaces are insolated. Increases in the amount of solar energy used for space heating will result in corresponding decreases in the amount of conventional

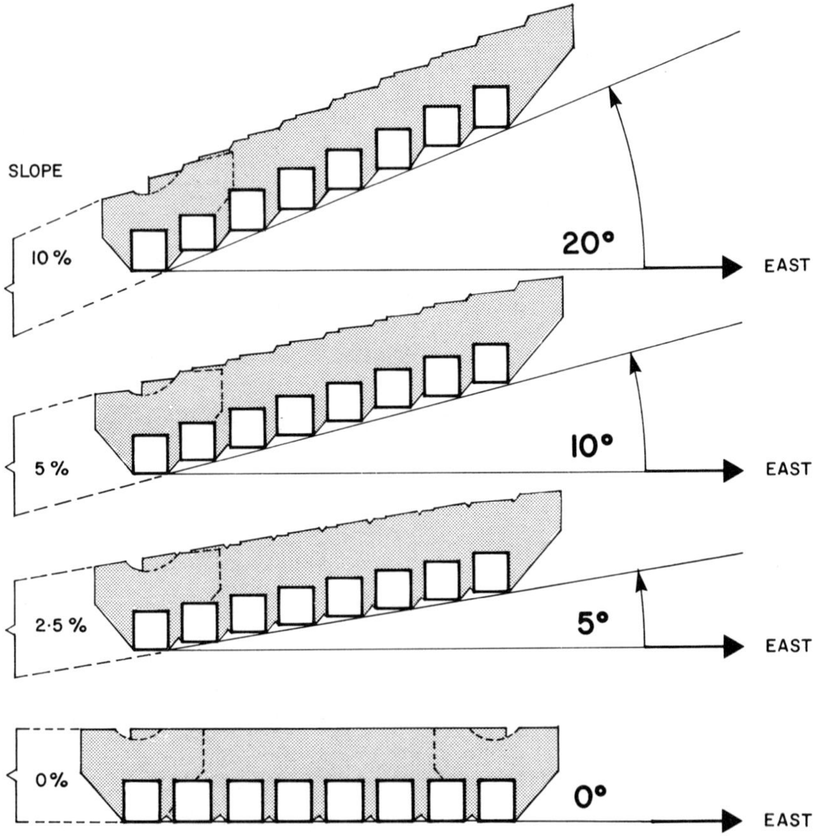

5-24. Street alignment deviates from true east/west on eastern slopes to accommodate impact area distortions.

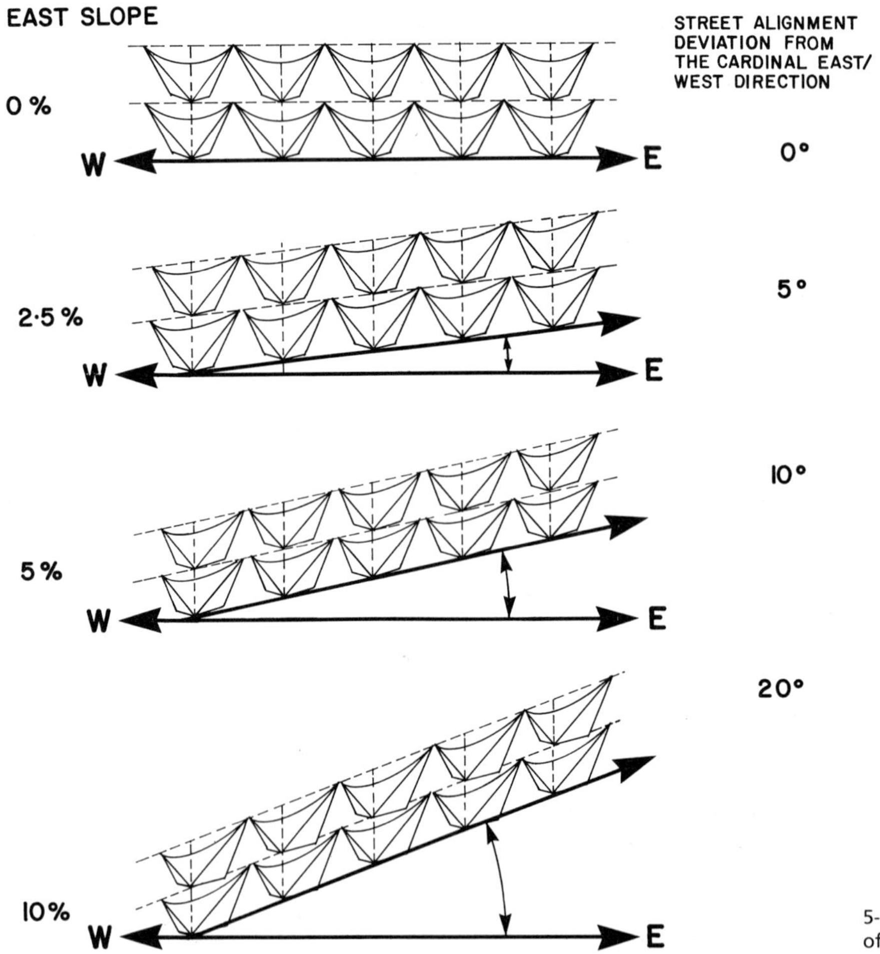

5-23. The relationship between a variety of eastern slope gradients and distortion of the Solar Pyramid.

energy used. To determine the appropriate space-heating mode for a given cell, a table is constructed to relate the slope degree and slope orientation description—inherent levels of insolation with categories of space-heating modes (fig. 5-27). It may be of interest to note that areas suitable only for conventional heating modes (category 5) are also the best areas for superinsulated houses. Both conventional and superinsulated houses are environmentally independent; however, the former is totally dependent on the input of delivered energy, whereas the latter is autonomous.

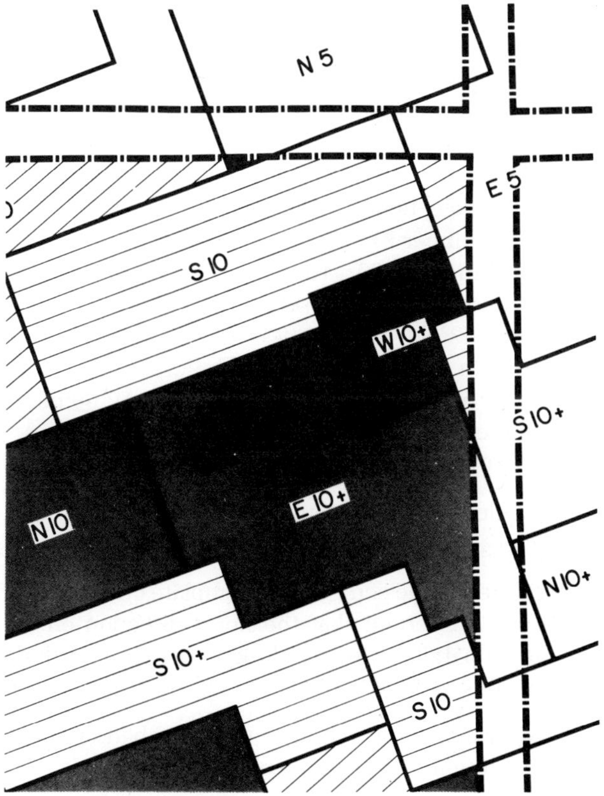

5-25. Recommended street alignments drawn into the orographic cells.

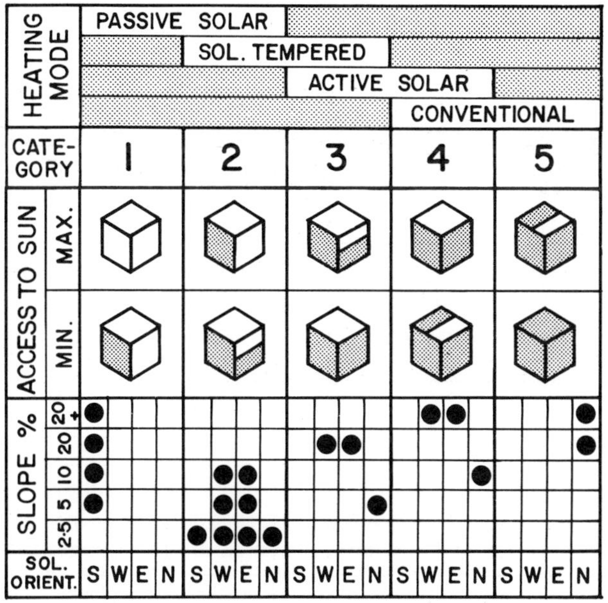

5-26. Space-heating modes and a variety of topographic locations.

SLOPE % ➡ / ORIENTATION ⬇	2·5	5	10	20	20 +
S	2	1	1	1	1
W	2	2	2	3	4
E	2	2	2	3	4
N	2	3	4	5	5

5-27. Recommended space-heating modes for different orographic cells.

The selection of preferred land use, street alignments, and space-heating modes is based on the maximum possible intensification of permitted land uses. Orographic cells indicate maximum permissible densities for specific land uses. For example, E5 and W5 cells have a wide range of optimal land uses, ranging from low- to high-density residential use as well as mixed use; cells N20, N20+ are best suited for open spaces, winter and summer recreation, parks and utilities (see fig. 5-18).

RYCKMANS NEIGHBOURHOOD

Once the development and refinement of the above planning and design tools were completed, Ryckmans Neighbourhood became the first experimental application of this new planning methodology in 1983. The neighborhood, a rectangular area of approximately 79 hectares (195 acres), is located near the southern limits of the city of Hamilton (fig. 5-28). It was intended that the neighborhood plan would provide for approximately one thousand single-family and multifamily dwelling units; other land uses to be included were commercial, institutional, and recreational spaces. The plan had to adhere to the established pattern by requiring four major gates or entrances and a few auxiliary access and egress points to and from the four major arterial roads.

The actual planning process was organized and carried out according to established planning procedures with the following new steps. Data was collected that included information on current land use, physical characteristics of the topography and existing human-made structures, population, environmental concerns, municipal services, planning control, land ownership, and historical sites and buildings. Weather data was tabulated, and the Ecochart was constructed and interpreted (see fig. 5-2). The topography was analyzed by dividing Ryckmans Neighbourhood into approximately eighty cells, varying in size from approximately 0.37 hectares (0.9 acres) to 1.9 hectares (4.6 acres). Alignment angles were drawn into each cell. Cells containing northern, western, and eastern slopes over 10 percent were darkened, since they have no bearing on alignment (fig. 5-29).

Design Alternatives

Four alternative neighborhood layouts were developed to accommodate a variety of opportunities and constraints and to produce compromises among a multitude of conflicting requirements, interests, and pressures.

Option A was designed without any concern for solar access, winter insolation, space-heating costs, or energy efficiency (fig. 5-30). In this conventional planning strategy, the street layout did not minimize overshadowing between buildings, nor did it maximize solar access. The street pattern reflected the curvilinear fashion of other neighborhood blueprints developed in the 1970s. A primary collector loop was branched into a series of cul-de-sacs and crescents. The total street length was 7.173 km (4.46 miles) with forty-two intersections, the majority being three-way. This option also had the lowest overall residential density, approximately 16.2 dwelling units per gross hectare of residential land.

Option B represents the first systematic planning effort to integrate a conventional street layout with instructions derived from the orographic cells (fig. 5-31). The interior street pattern conformed to the existing institutional property boundaries; modified street alignments were included wherever possible. Since solar access was incorporated to a certain degree, the number of opportunities to use systems for harnessing solar energy and for passive and active solar space heating were increased. Cul-de-sacs and crescents were attached to a primary collector loop. The total street was 6.695 km (4.2 miles) long. There were forty-one three-way inter-

5-28. *(Opposite)* Ryckmans Neighbourhood area.

UPPER WELLINGTON STREET
MELISS
UPPER JAMES STREET
RYMAL ROAD EAST
STONE CHURCH ROAD EAST
CEMETERY

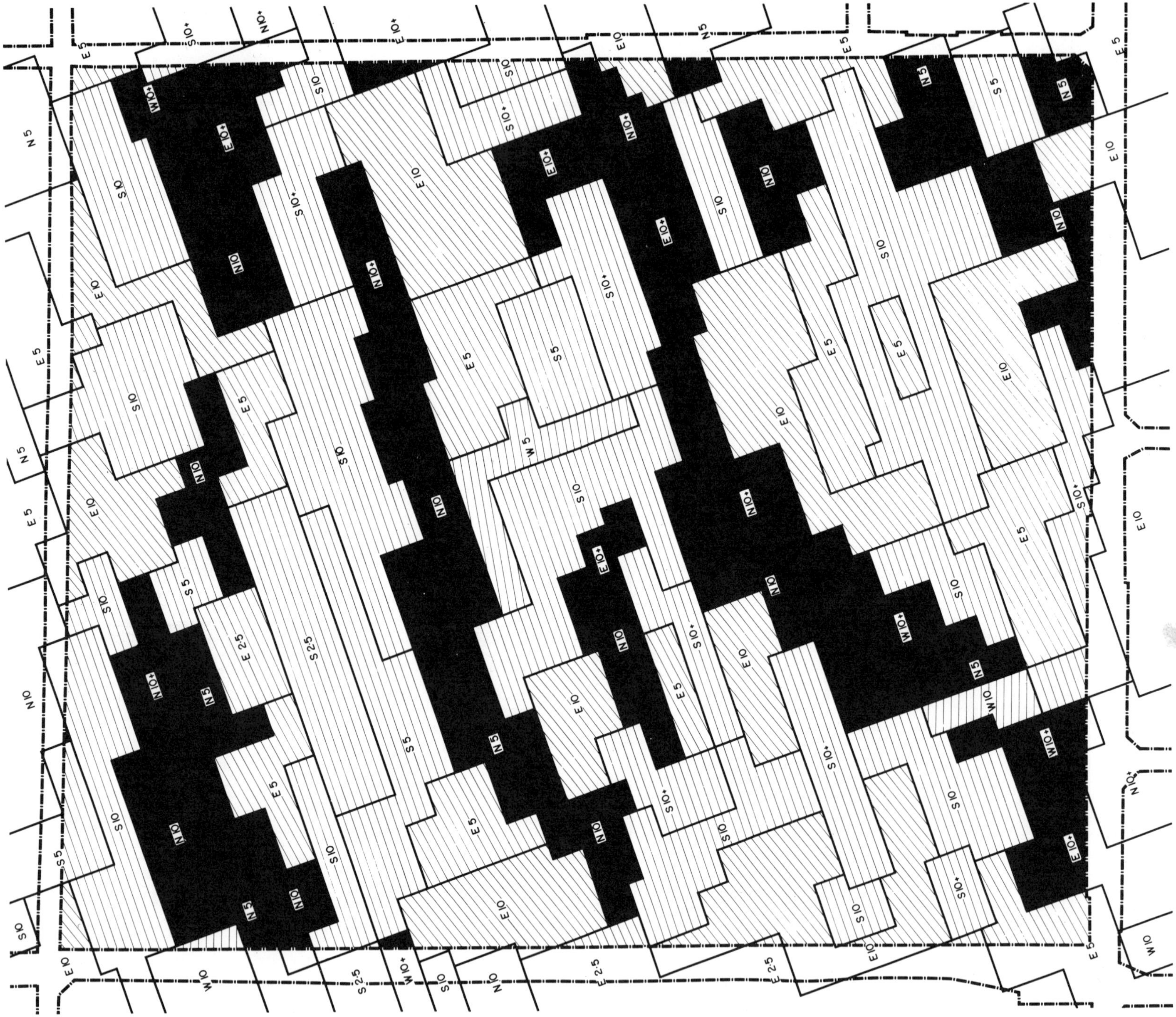

5-29. Patterns of orographic cells for Ryckmans Neighbourhood.

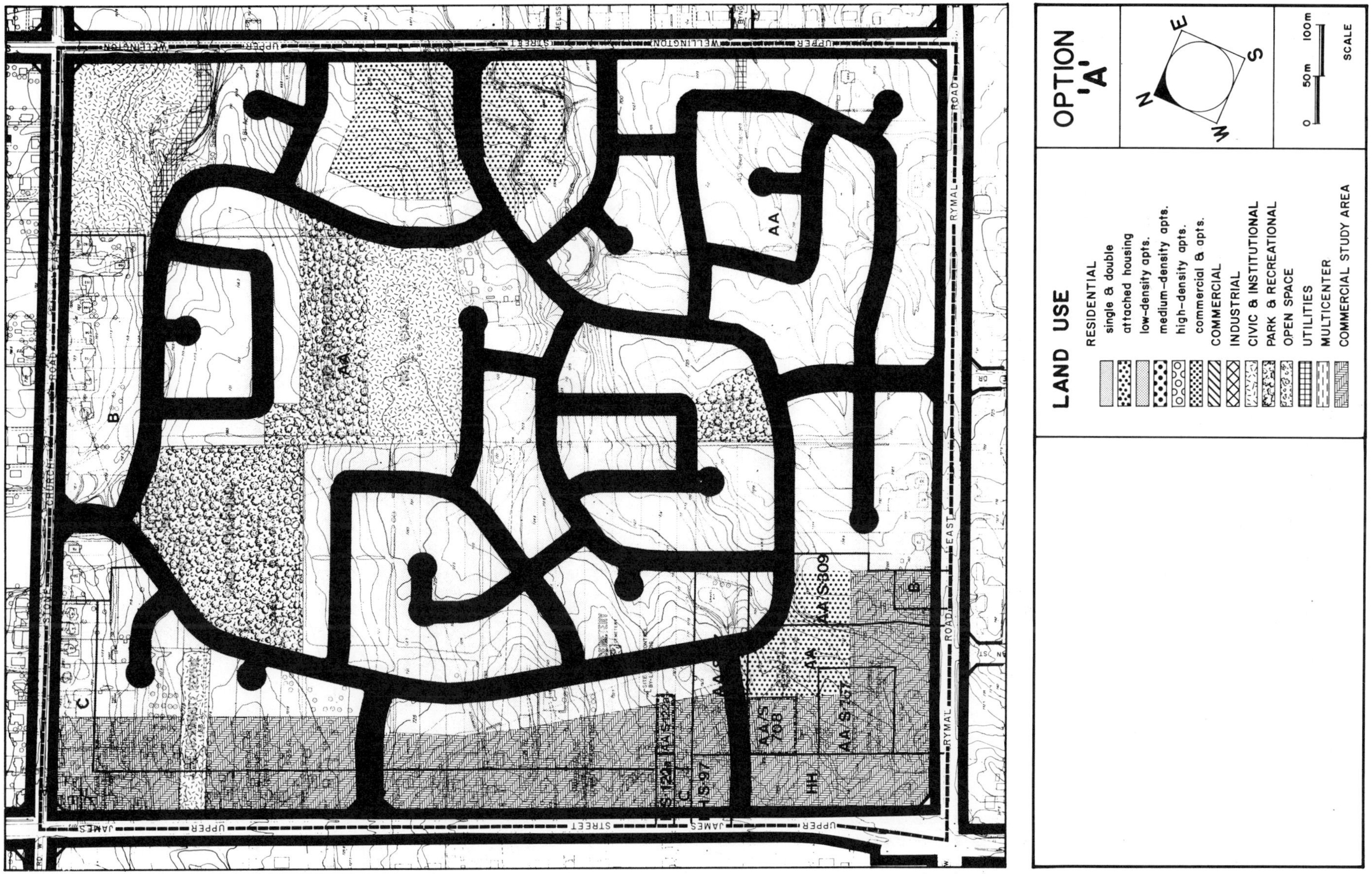

5-30. Design option A.

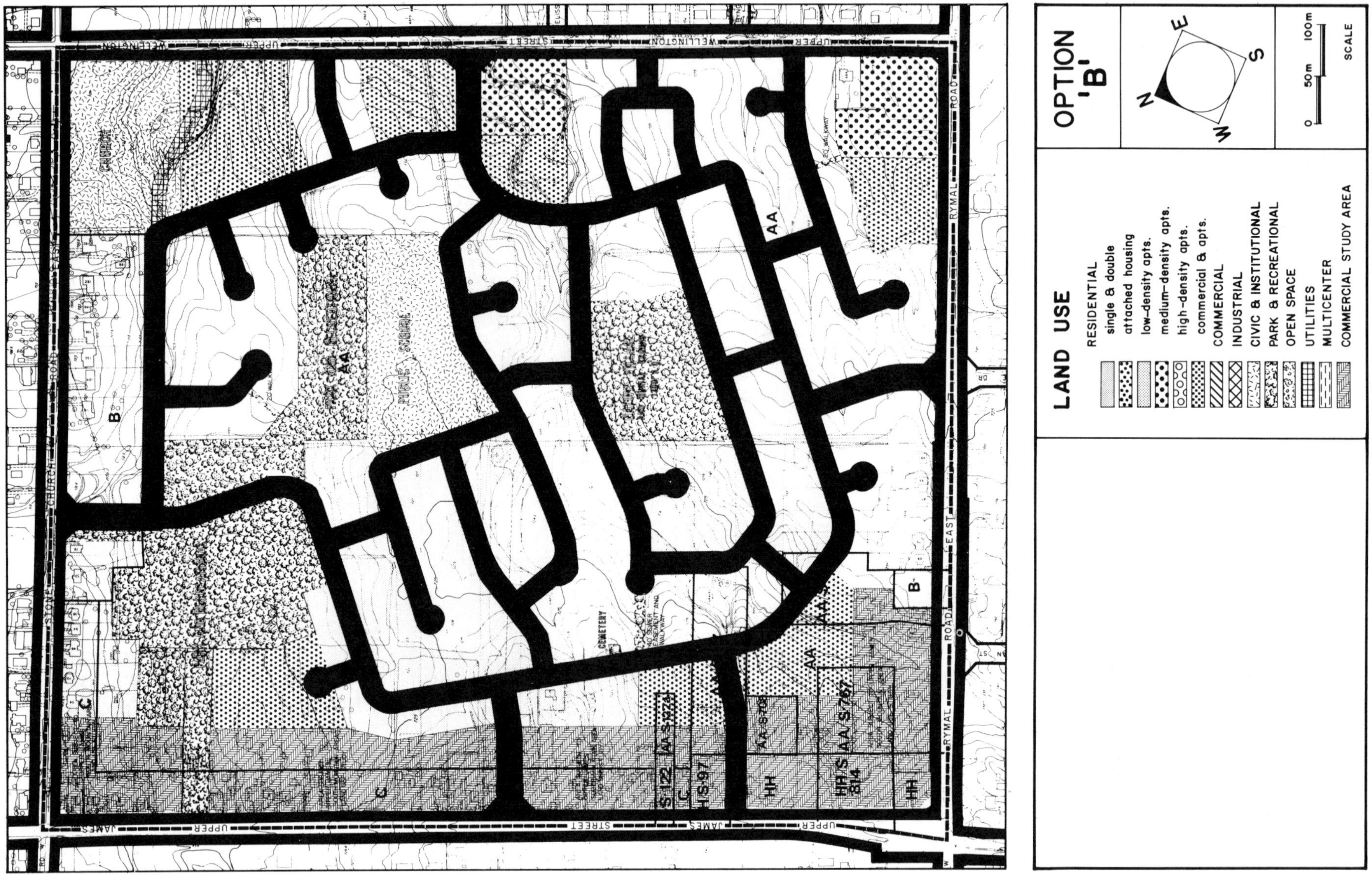

5-31. Design option B.

sections. The residential density was 21.0 dwelling units per gross hectare of residential land.

Option C offers improved solar access by further modifying the internal street system (fig. 5-32). Street alignment followed more closely the instructions provided by the orographic cells. There was a well-defined hierarchy of streets—a primary collector loop, looped streets, and cul-de-sacs. The total number of intersections was thirty-one; two were four-way and the rest, three-way. The total length of the streets was 6.490 km (4 miles). In addition, a partial pedestrian and bicycle pathway was established through the center of the neighborhood along the north/south axis. The residential density was 22.6 dwelling units per gross hectare of residential land.

Option D is based on the best possible application of the new planning method in terms of street alignment, land use, density, and mode of space heating (fig. 5-33). It is a well-balanced solution that resulted from sensitive compromises to conflicts. In this optimum neighborhood layout, opportunities for solar access, energy-efficient land use, open spaces and recreational areas, and other urban design features were maximized. Near the northern and southern gates, two rondels were included on the primary collector loop to enhance aesthetically the space design as well as to help orient visitors to the neighborhood. A linear recreational open space constituted a north/south spine; it included a continuous pedestrian and bicycle pathway and treed boulevards on the outlet roads. Thus, all the open spaces were integrated into one linear, uninterrupted recreational area in the center of the neighborhood. Constraints, such as land ownership, both institutional and private, existing buried municipal services, housing market pressures, and so on, were accommodated with the maximum possible flexibility. The collector loop was well defined and easy to read; the intersections totaled thirty-seven, and the total length of streets was 6.914 km (4.3 miles). Residential density was 21.6 dwelling units per hectare of residential land.

Computer-assisted Cost Evaluation

All four design alternatives of the Ryckmans Neighbourhood Plan were evaluated by two cost-analysis programs developed by Curtis and Schlosser at the University of Regina. The first analysis was the Residential Area Cost Analysis Model (RACAM), and the second was the Residential Area Traffic Analysis Model (RATAM).

The RACAM was used to estimate the construction, operation, and maintenance costs of water and sewer systems, roadways, sidewalks, street lights, gas, electrical power, telephone systems, parks, and buffer systems. In addition, the model calculated the present and annual worth of all four alternative plans based on a given interest rate and the life expectancy of the neighborhood.

In summary, the RACAM results were as follows:

1. subdivision construction costs in $Canadian
   option C $12,146,000.00 (BEST)
   D $12,699,000.00
   B $13,004,000.00
   A $13,048,000.00 (WORST)
2. subdivision operation and maintenance costs in $Can./year
   option D $83,990.00 (BEST)
   C $84,970.00
   B $86,405.00
   A $91,360.00 (WORST)

The RATAM estimated the automobile fuel consumption for work- and nonwork-related trips based on the street layout of each alternative. In summary, the RATAM results were as follows: Compared to option A, option B is 6.6 percent more fuel efficient; option C is 4.6 percent more fuel efficient; option D is 6.1 percent less fuel efficient.

In order to provide more useful feedback to the neighborhood planner, these results require a more detailed explanation. Option B is most fuel efficient be-

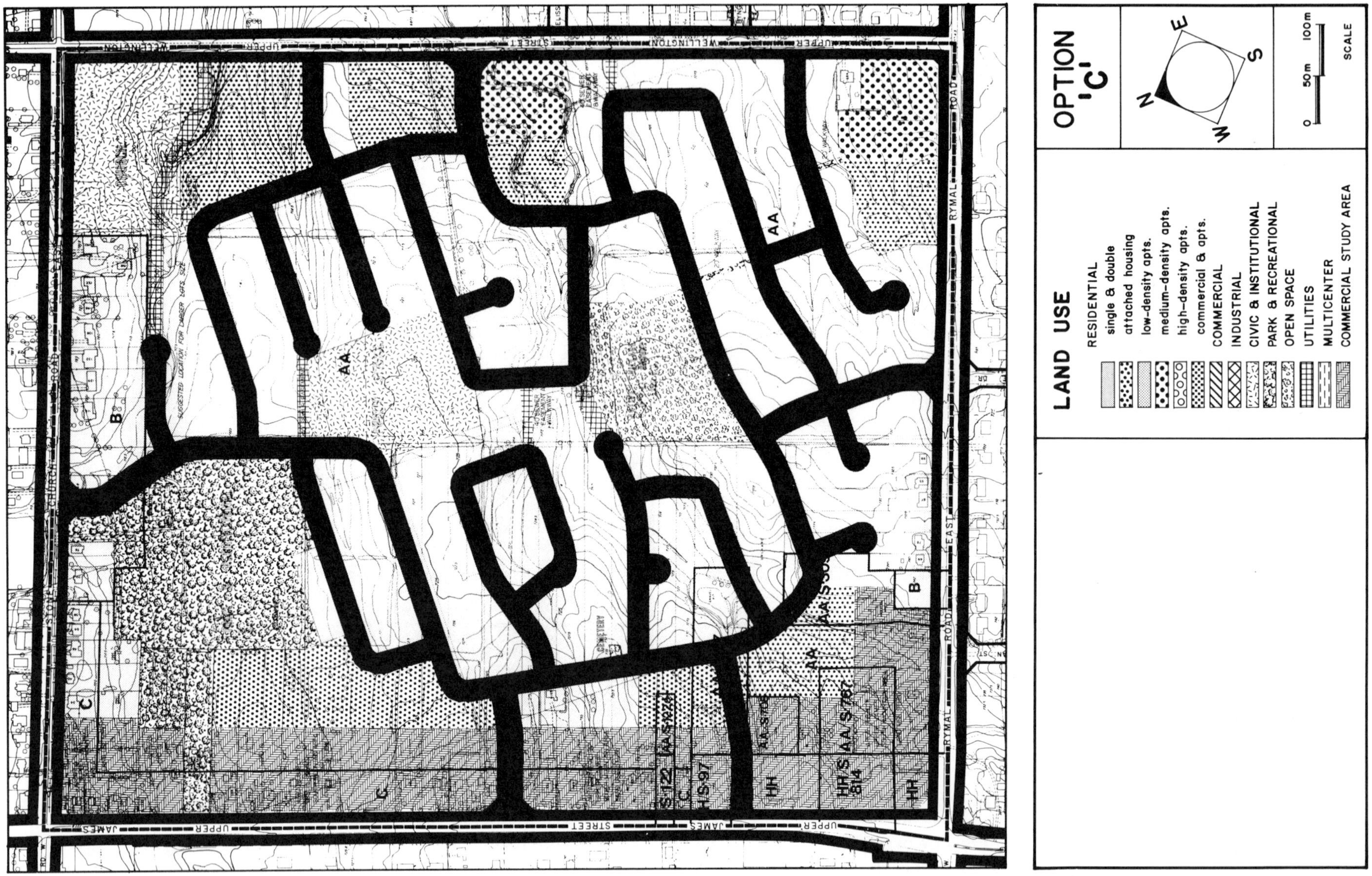

5-32. Design option C.

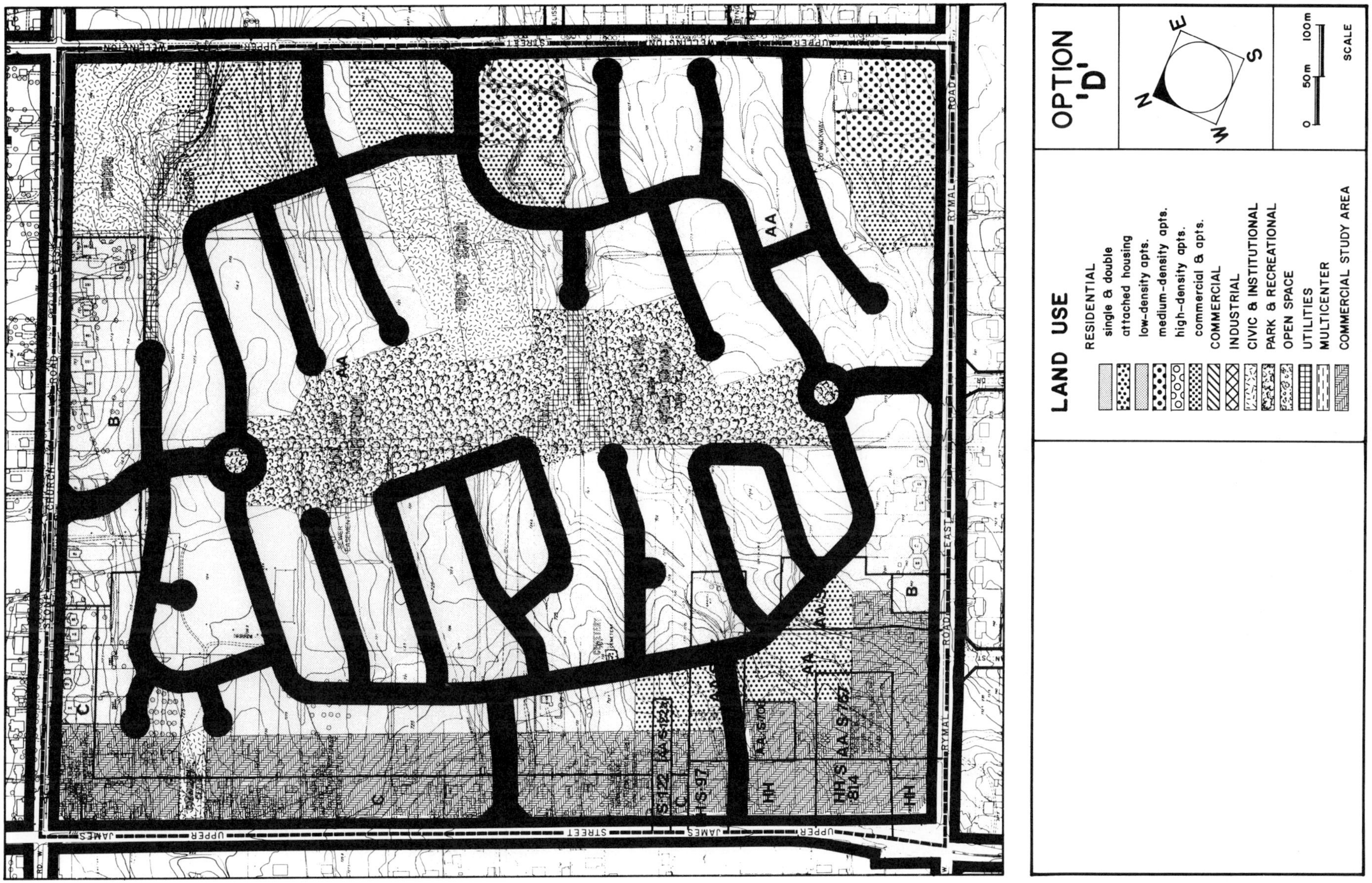

5-33. Design option D.

cause of (1) the decreased total street length; (2) the ability to traverse the collector loop; (3) fewer curvilinear east/west streets. This results in an increase in the average speed of vehicles, since it minimizes the need for acceleration and deceleration.

Option C is the second most fuel-efficient alternative, because the street layout allows a 20-percent increase in the average speed of vehicles and because there is a 26-percent decrease in the number of intersections in comparison with option A. Option C is not as fuel efficient as option B, because the collector loop cannot be traversed, and the consequent backtracking increases the average length of a trip.

Option D is the least fuel efficient for several reasons. The street layout requires the most backtracking, thus increasing the length of the average trip. The collector loop has the largest diameter and cannot be traversed. The model anticipated congestion at the loop, which would result in a reduced average speed and increased travel time.

Combining the results of both the RACAM and RATAM analyses and taking into consideration solar potentials and potential revenue based on saleable frontage ratios, the overall rating of the four alternative plans would be as follows:

1 option C (BEST)
2 option D
3 option B
4 option A (WORST)

The computer-assisted process of cost evaluation confirmed that this planning and design methodology, one goal of which is to promote maximum harmony between the natural and the human-made environment, also leads to a very cost-efficient layout in terms of construction, operation, and maintenance (options C and D). The differences between options C and D illustrate the conflict between a car-oriented and a people-oriented neighborhood layout. If the planner takes into consideration generally accepted population, economic, and technological trends and forecasts (including the advancing age of baby boomers with their proven political clout), then the planner might assume that the role of the passenger car may be diminished because

1. the elderly drive less often
2. fewer people would travel to work because of increased computer networking
3. a rebirth of the corner store and/or teleshopping is anticipated
4. recreational opportunities within walking distance, as well as home gym and recreational facilities, would be more readily available
5. low-density residential neighborhoods will gradually be restructured into higher-density, complex, mixed land-use organisms, and thus, walking and using public transit may be the preferred mode of locomotion

In addition, neighborhood planning practices often support rather than oppose efforts to slow down and impede vehicular traffic in order to increase safety.

On the other hand, the same inquiry into statistical forecasting may lead to a different conclusion, one where the role of the automobile will become more significant because of modifications permitted by technological advances.

1. The automobile would not damage the environment, and its use would therefore be encouraged.
2. The car might become a semiintelligent extension of a human that is directed by simple finger or eye movements.

Favoring option D over option C could create a type of future that would be preferable to many planners. Market preferences, the community's involvement in the planning process, and the present economic climate, however, will continue to determine today's plans for tomorrow's neighborhood.

PART 3

THE CITY

CHAPTER 6

THE GENESIS OF URBAN CLIMATE

THE NATURAL STATE AND HUMAN INTERVENTION

We have thus far concentrated on tools to aid in the planning and development of individual shelters and low-density residential neighborhoods. But on a larger scale, in densely developed urban environments, unique climatic phenomena emerge. These require closer examination in order to address intelligently the problems of planning and designing high-density areas, especially in northern latitudes.

Because of the great number of isolated structures, the generally poor insulation of buildings, and the presence of human activities, cities are like radiators that heat the sky. A layer of warm air envelopes every city and the weather pattern within this warm dome, or *heat island,* can be markedly different from that in the countryside. Recent thermographic analyses have confirmed that a substantial amount of heat in urban areas escapes and thus affects the urban weather pattern.

A reciprocal relationship exists between climate and human activities: activities are modified by climate and climate is modified by human activities. The presence of hunting and harvest seasons demonstrates how the climate modifies human activities. Early, relatively primitive, actions, such as clearing fields, slash-and-burn practices, the introduction of monocultures, and the draining of fields and wetlands, have a marked influence on local weather patterns. Building a shelter, village, or town, paving surfaces, constructing transportation systems, creating industry, and developing a metropolis, however, have an even more pronounced impact on the general mesoclimatic weather system.

Within the heat island, the urban climate and its unique characteristics are the product of a large number of factors. Even if we were to stop heating the buildings or if we were to superinsulate all buildings so that not a single Btu was allowed to escape, there would still be a heat dome hovering above the city. To better understand these complex mechanics that affect planning and design results significantly, the genesis of the urban heat island should be reviewed. The process begins with an examination of a flat, horizontal surface and its microclimate. New elements will be added, and the resultant microclimates examined. These various microclimates, together with human activities, generate the urban heat island and a unique urban climate.

A FLAT, HORIZONTAL SURFACE

A horizontal plane that does not support any three-dimensional object can create distinctive microclimatic conditions. Since shadows are not cast on such a plane, the impact area includes only the air space immediately

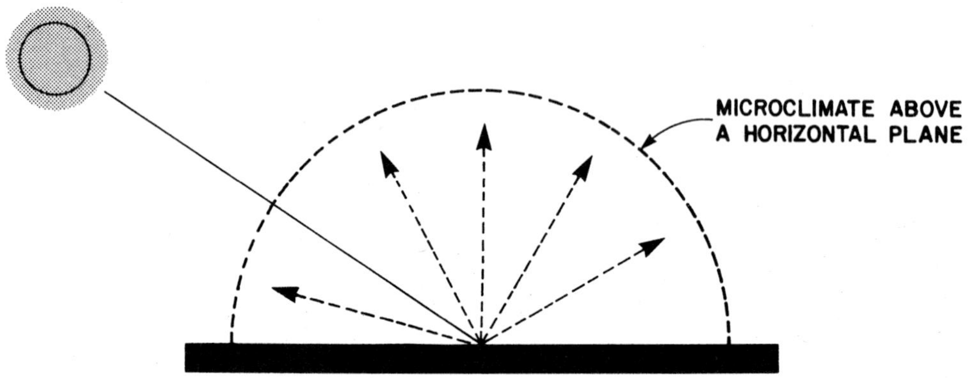

6-1. The microclimate generated by a horizontal surface.

above the surface (fig. 6-1). Interaction between the rays of the sun and the surface properties generates the sheath of a unique microclimate. An example of a natural steady state is a horizontal surface covered with grass—the temperature and moisture are more or less equally distributed. This relationship might be modified by human intervention (for instance, ploughing), a change in the color and texture of the original surface, the introduction of materials with different absorbtion/reflection qualities (dark and rough surfaces), paving and roofing materials, light- and warmth-absorbing surfaces, as well as water, reflecting pools, horizontal mirrors, white, light surfaces, and smooth, shiny surfaces that reflect light and warmth. In addition, because of direct exposure to precipitation, a variety of materials can disturb and alter moisture distribution. In an urban context, horizontal surfaces play an important role in shaping urban microclimatic conditions, reflecting light and warmth to the vertical surfaces of surrounding structures and thus contributing to the generation of an urban heat island.

Sun Radiation Received

Sun radiation is an important element in the formation of microclimatic situations. Horizontal as well as vertical surfaces receive different levels of direct sun radiation. The radiation density varies with the angle of the beams' impact. One of the best ways to illustrate the amount of direct sun energy received by a surface is to draw the surface from the sun's point of view. The size of the area affected by the sun is directly related to the sun energy it receives.

A square horizontal plane is shown in figure 6-2 as it is seen from the angle of the sun. The broken line represents the actual surface area affected by the sun's rays. The dark area takes into consideration the effect of atmosphere: the morning and evening sun has to penetrate a thicker layer of air, water vapor, and pollution than does the noon sun. Values of surface area or sun energy received (indicated by the broken line) are multiplied by the sine of solar altitude in order to accommodate the effect of atmosphere.

The Skywindow—a Further Elaboration

In low-density residential design, the Skywindow is used primarily to determine simple solar access/impact configurations. Since sun radiation is an important factor in the formation of the urban microclimate, the Skywindow must be explained in greater detail within the context of a complex, densely built urban environment. By definition, the Skywindow is an area of sky representing in size the sum of the daily and annual variations of the sun's position in the sky. There are three different types or categories of Skywindow.

1. The NATURAL SKYWINDOW is the largest (fig. 6-3, see also fig. 4-1). It spans the distance from natural sunrise to natural sunset and from June 21 (the highest sun path) to December 21 (the lowest sun path). In reality, such a window can be seen only from midocean in calm waters.
2. The DESIGN SKYWINDOW represents the smallest possible area of the unobstructed sky necessary to provide sun radiation for a given purpose (fig. 6-4).

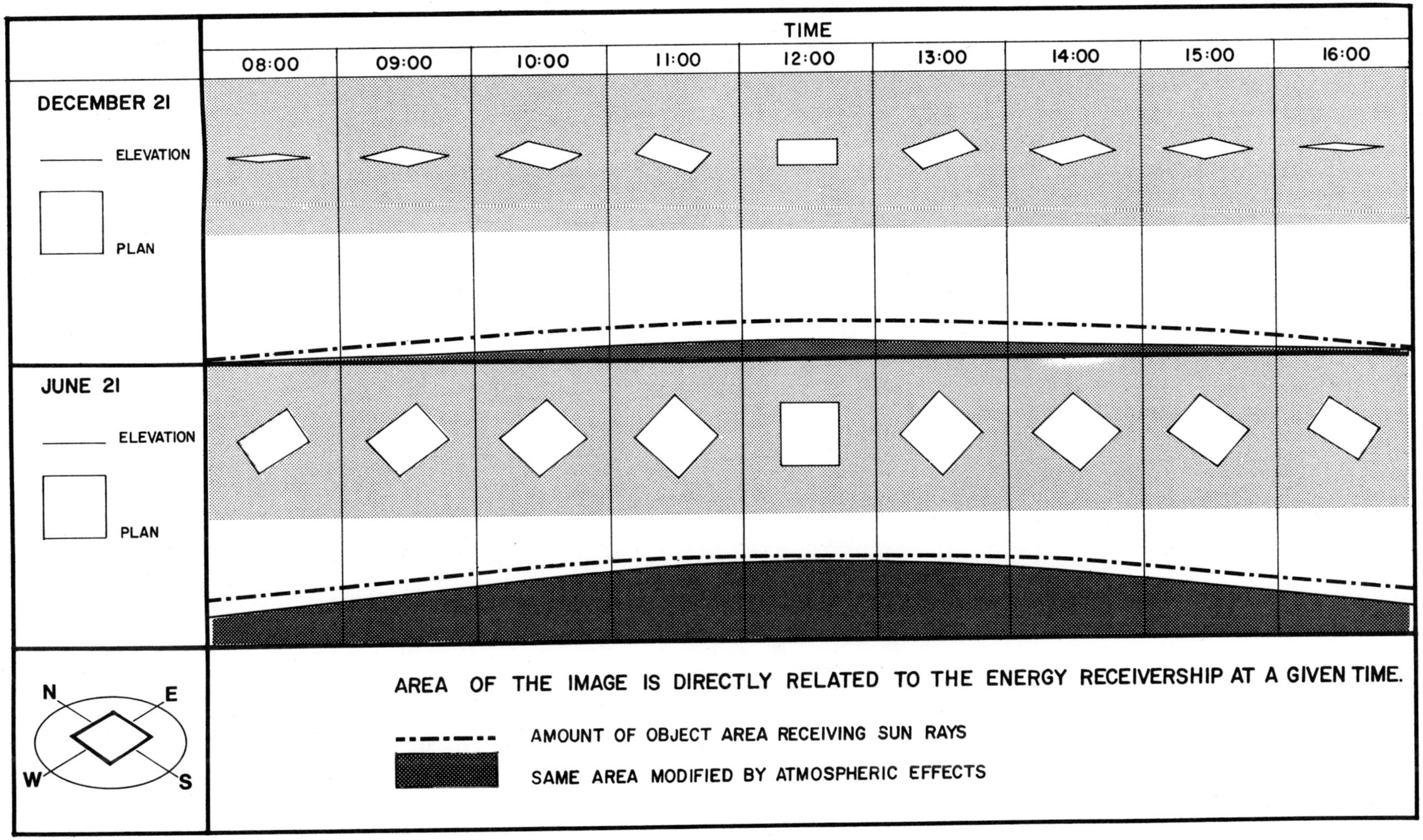

6-2. The image of a horizontal surface generated by the sun's rays.

Its size may differ for the requirements of a photovoltaic collector from that of a passive solar house.

3. The REAL SKYWINDOW is like the natural Skywindow except that it incorporates natural and/or human obstructions (fig. 6-5). The size of the real Skywindow varies. Whether on a prairie, in a mountain valley, or at the bottom of a bear pit, an observer would see Skywindows of different sizes under different conditions.

For purposes of this analysis, it is assumed that the horizontal plane does not lie within the impact area of any surrounding object; the natural Skywindow is therefore equivalent to the real Skywindow.

Exposure to Air Movement

The microclimatic sheath above a horizontal surface would be subject to the direction, speed, and frequency of air movement as described by a regional Ecochart

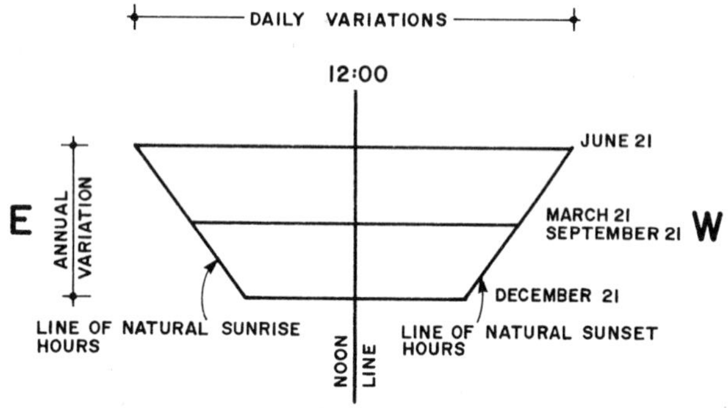

6-3. The natural Skywindow.

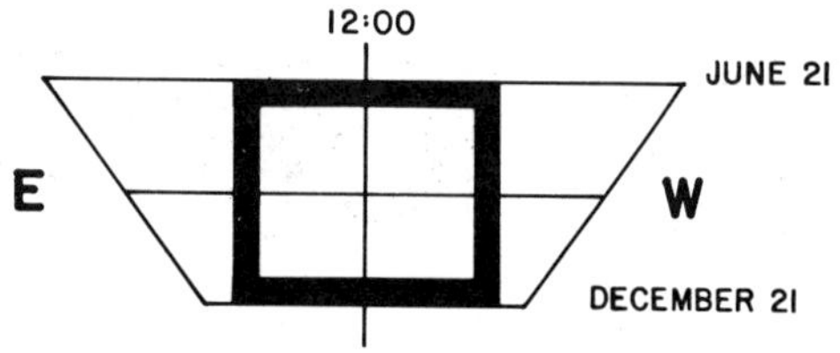

6-4. The design Skywindow.

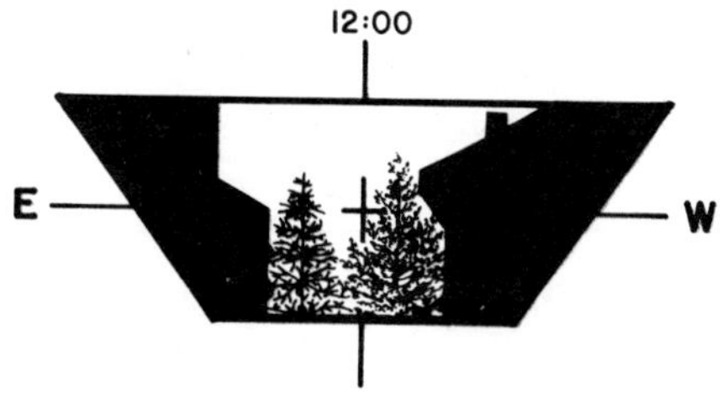

6-5. The real Skywindow.

(see chapter 3). This information will enable the designer to assess the contribution of wind to the creation of a particular microclimate.

A WALL—EAST/WEST ORIENTATION

By introducing a new element, a wall oriented toward east and west, to a flat, horizontal surface, a new set of environmental circumstances will result. The east/west-oriented wall (fig. 6-6) has a northern and southern face. Each face possesses a unique microclimatic sheath.

The Impact Area

Two types of impact areas are created with the introduction of a wall. Within the context of a cold northern environment, a shadowed impact area inhibits life-supporting conditions and is therefore considered negative. An impact area lit by reflected sunlight, on the other hand, enhances life-support conditions and is therefore considered positive. Figure 6-7 illustrates both positive and negative impact areas within the 09:00 to 15:00 hours time frame on June 21, March/September 21, and December 21.

Sun Radiation Received

Both the northern and southern faces of the wall receive direct rays in certain locations at certain times. The High Arctic receives direct sunlight during the season when sun altitudes change but never sink below the horizon. Direct sunlight is received on the equator, where overall annual insolation is equal on both faces of the wall (insolation values from March through September are equivalent and symmetrical to those from September through March). However, with the exception of these two extremes, the northern face generally remains cold and dark, receiving very little or no direct sun during the winter; the southern face receives direct solar radiation throughout the year. The latter is thus warm and sunny. Figure 6-8 shows the sun radiation

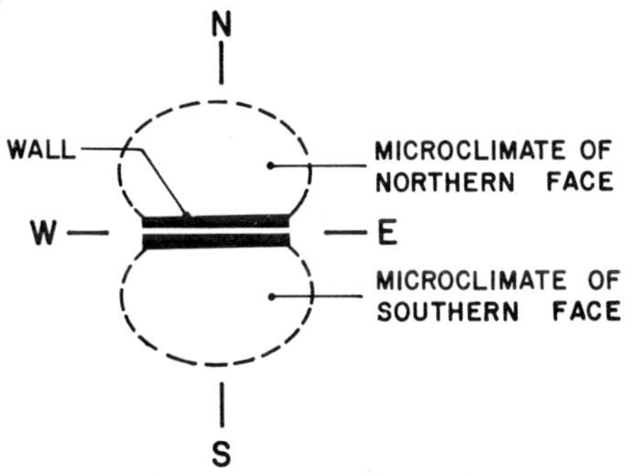

6-6. Microclimates generated by an east/west wall.

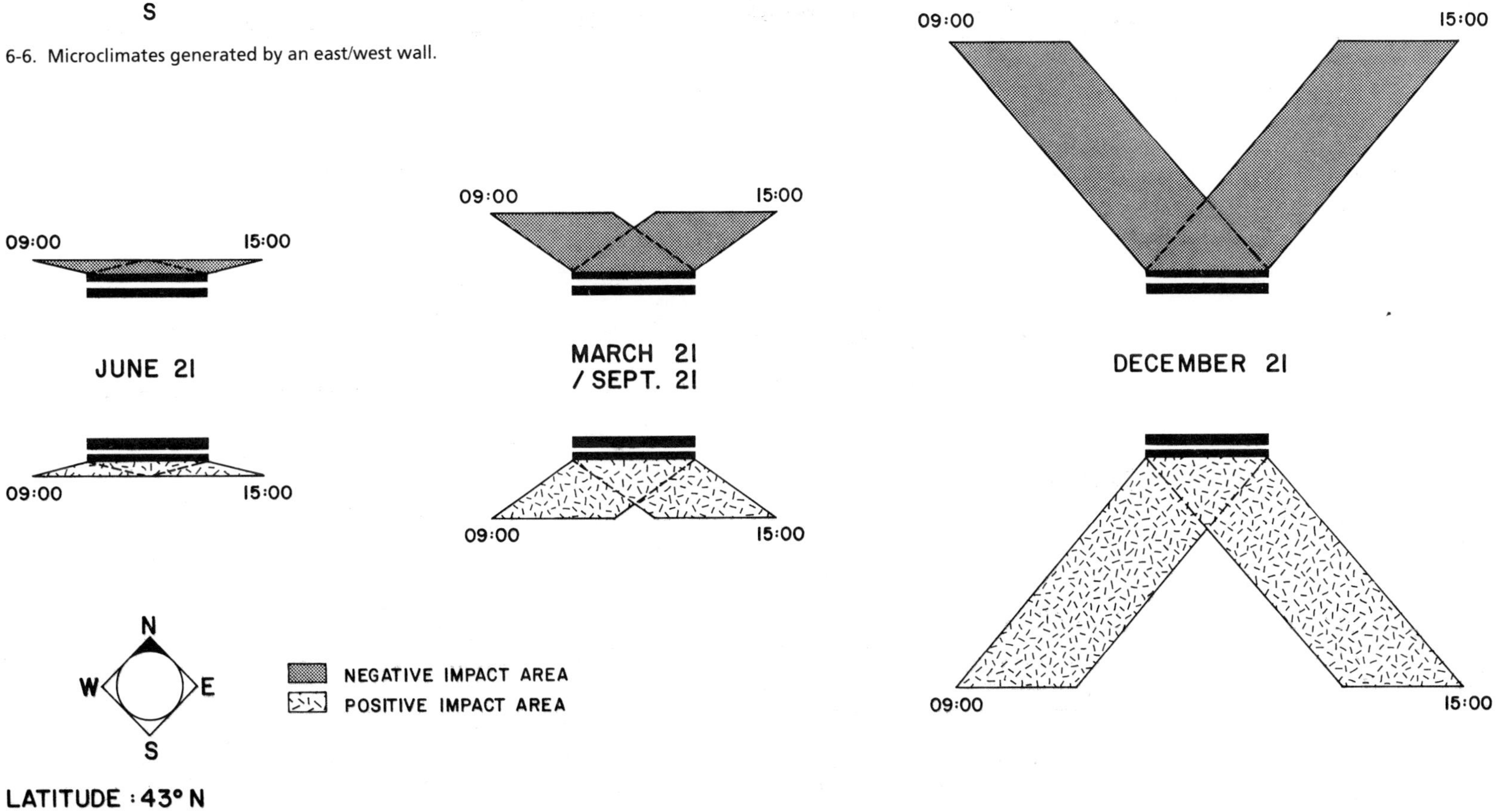

6-7. The negative and positive impact areas of an east/west wall during summer solstice, equinox, and winter solstice.

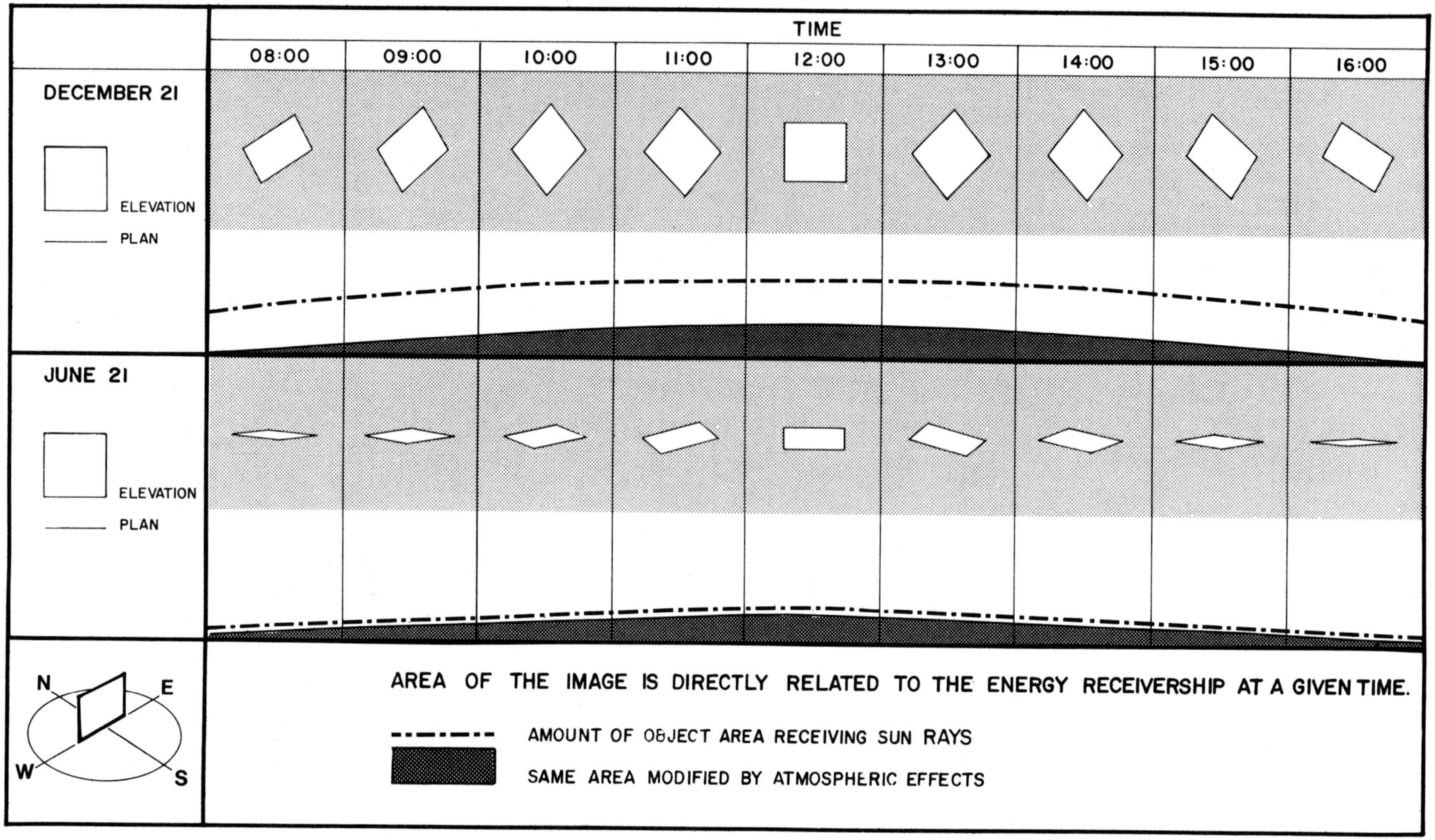

6-8. The image of the southern face of an east/west wall generated by the sun's rays.

received by the southern face within the 08:00 to 16:00 hour time frame.

Skywindow

The real Skywindow for the northern face of an east/west wall is shown in figure 6-9. Between the spring and autumn equinox, this wall has two sunrises and two sunsets a day. During the period between September 21 and March 21, the sun is never visible.

At 42°N latitude, these are the selected sunrise/sunset data:

	A.M.		P.M.	
DATE	SUNRISE	SUNSET	SUNRISE	SUNSET
April 21	05:12	07:00	17:00	18:50
June 21	04:19	08:00	16:00	19:40
August 21	05:07	07:00	17:00	18:50

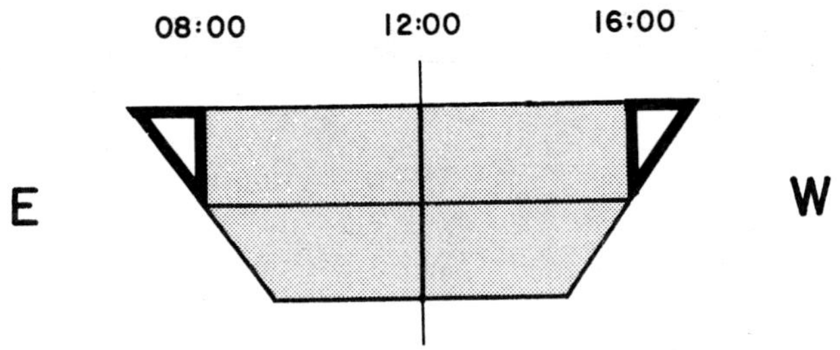

6-9. Skywindow for the northern wall.

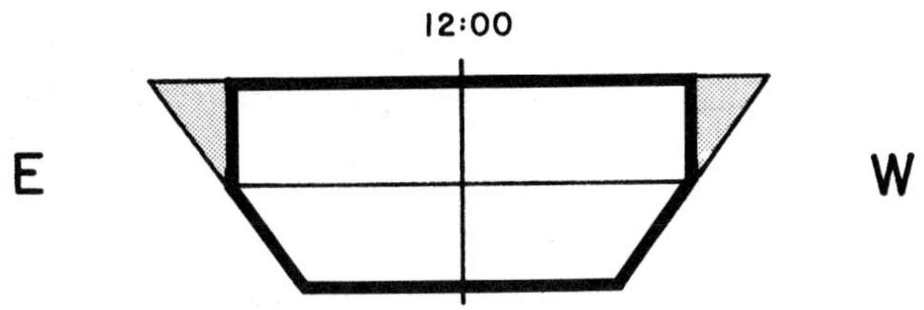

6-10. Skywindow for the southern wall.

The real Skywindow for the southern face is shown in figure 6-10. During the period between spring equinox (March 21) and autumn equinox (September 21), the "real" sunrise is at 06:00 hours and the "real" sunset (one that an individual can actually see) is at 18:00 hours. Natural (astronomically calculated) sunrises and sunsets take place during the rest of the year, for example, at 07:30 and 16:30 hours on December 21.

Exposure to Air Movement

Monthly averages of selected wind exposure are shown in figure 6-11. These charts are derived from a regional Ecochart and thus represent local mesoclimatic data. Neither gradient effect nor the effects of possibly interfering structures or objects in the vicinity were taken into consideration.

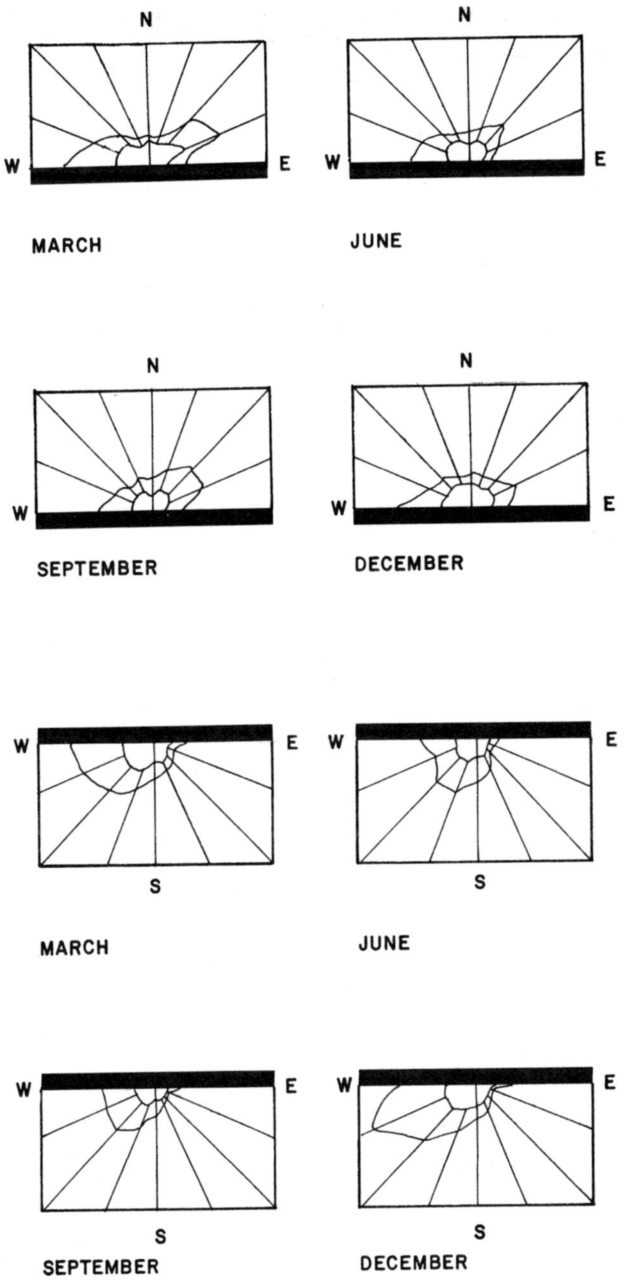

6-11. Northern and southern winds.

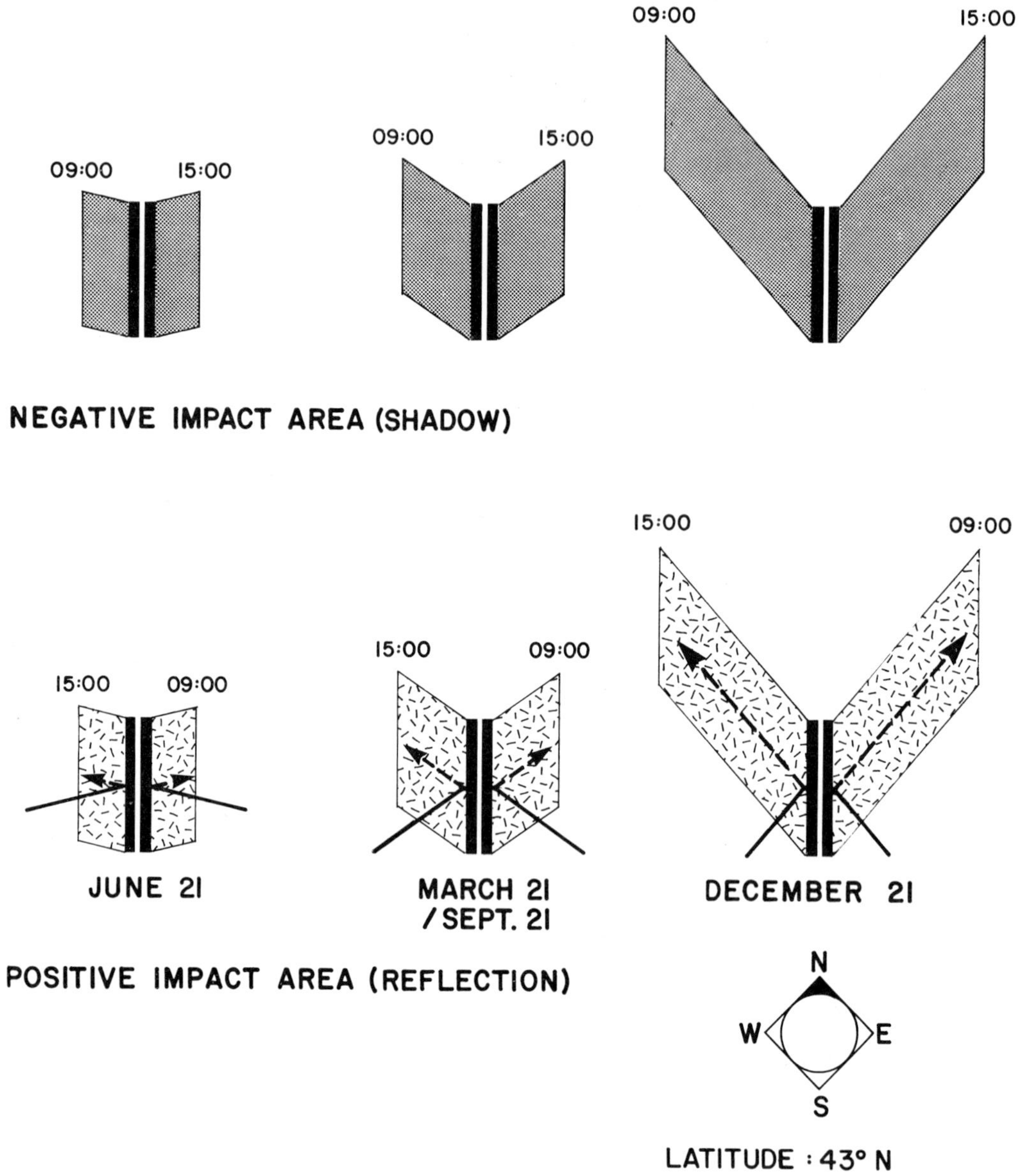

6-13. Negative and positive impact areas of a north/south wall during summer solstice, equinox, and winter solstice.

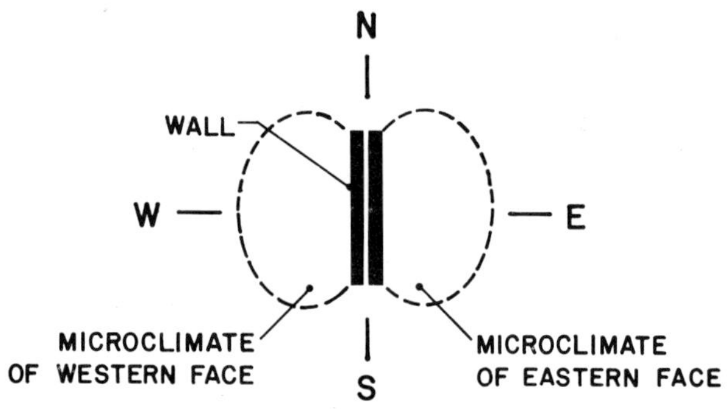

6-12. Microclimates generated by a north/south wall.

A WALL—NORTH/SOUTH ORIENTATION

A wall that is oriented north and south has eastern and western faces with adjacent, unique microclimatic conditions (fig. 6-12).

The Impact Area

Figure 6-13 shows both the positive and negative impact areas of such a wall. Note the significant difference between north/south and east/west walls: the positive and negative impact areas are transposed in a north/south wall orientation; an east/west wall has negative and positive impact areas permanently attached to each face of the wall (see fig. 6-7). Within a given time frame, the two impact areas of the east/west wall do not overlap.

Sun Radiation Received

Both the eastern face (morning side) and the western face (afternoon side) receive the same amount of direct solar radiation daily and annually (fig. 6-14). Geographic location affects the overall amount of radiation

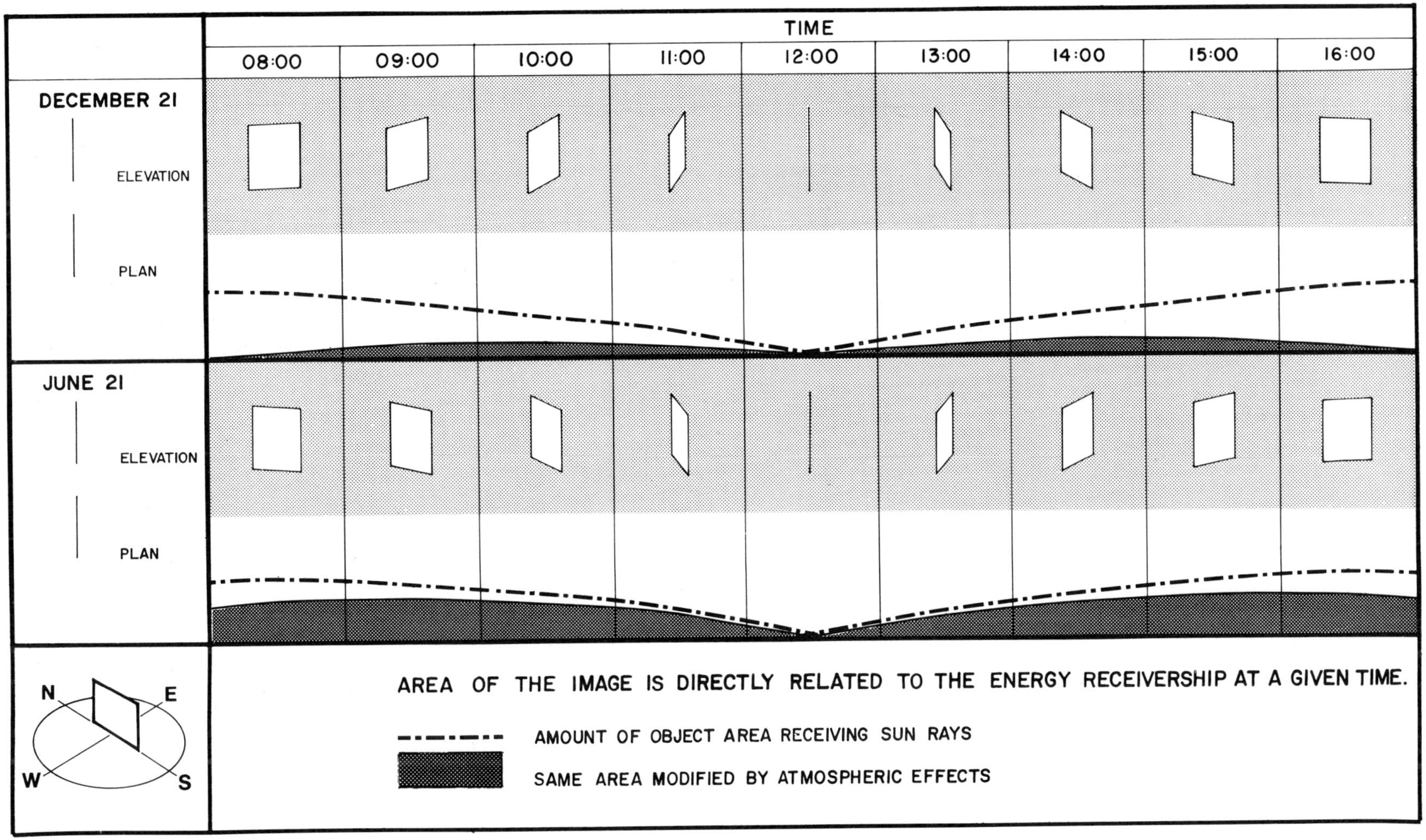

6-14. Image of a north/south wall generated by the sun's rays.

received but not the equal distribution between the eastern and western side. Noon is the axis of symmetry.

It should be noted that equal insolation does not produce equal microclimatic sheaths. Due to the delayed response of air mass to the day/night temperature cycle, the afternoon side is substantially warmer than the eastern side.

Skywindow

The real Skywindow for the eastern face is shown in figure 6-15. Sunrise is natural throughout the year, for example, on June 21, it is at 04:19 hours, and on December 21, it is at 07:30 hours. Sunset is at 12:00 hours throughout the year.

Figure 6-16 illustrates the Skywindow for the west-

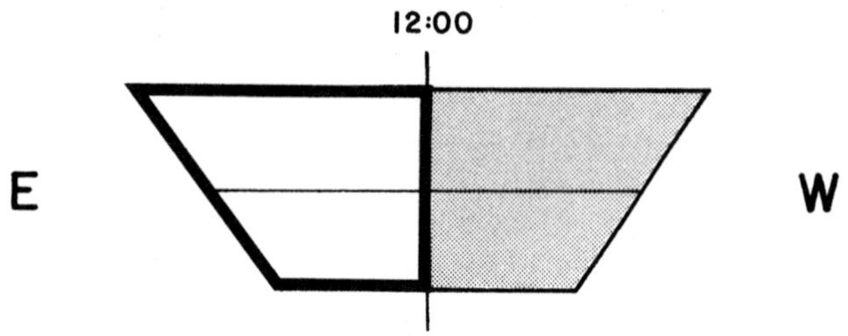

6-15. Skywindow for the eastern wall.

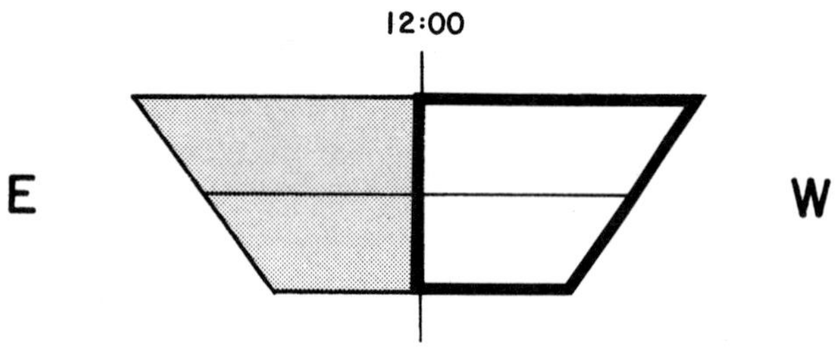

6-16. Skywindow for the western wall.

ern face. The real sunrise takes place at 12:00 hours throughout the year, and the sunset is natural throughout the year, for example, on December 21, it is at 16:30 hours, and on June 21, it is at 19:50 hours.

Exposure to Air Movement

Figure 6-17 shows the monthly averages that have been transferred from the regional Ecochart for the direction, speed, and frequency of wind. This information can assist the designer in assessing the impact of wind on the microclimate.

Application

Both east/west and north/south walls can be imaginatively combined in landscape and urban design to enhance certain microclimatic conditions, such as sun, shade, wind, or calm. Figure 6-18 shows a variety of wall configurations that can result in an array of microclimatic conditions by reflecting light, casting shadows, shielding structures against and exposing them to the wind.

A BUILDING AND ITS SURROUNDINGS

Once the relationships between individual walls and the resultant microclimates are understood, the presence of a different element, a cube, can be analyzed. A cube can be viewed as a representation of a building. Picture a flat, green meadow with no major human-made or natural obstacles nor any profound topographical variations in the immediate vicinity. The weather pattern is as indicated by the Ecochart. The ground temperature, moisture content, and living strata are more or less uniformly distributed. This pattern is dramatically altered when a building is constructed, since a building upsets the existing equilibrium. Although nature seeks a new state of balance, the new situation would differ considerably from the original one. The following is a partial list of anticipated changes.

1. The building would develop its own climatic sheath.
2. The hours of sunrise and sunset for different areas around the building would change.
3. The building would become both absorber and radiator, and thus a change in the radiation regime would take place.
4. The wind pattern would be disturbed.
5. Moisture would be redistributed; depending on the roof drainage system, the moisture regime would be
   - very wet in the vicinity of downspouts (10–100 times the normal amount of precipitation, usually at the corners of the building)
   - normal to very dry at other locations.
6. Snow would be redistributed so
   - windward side would be relatively free of snow

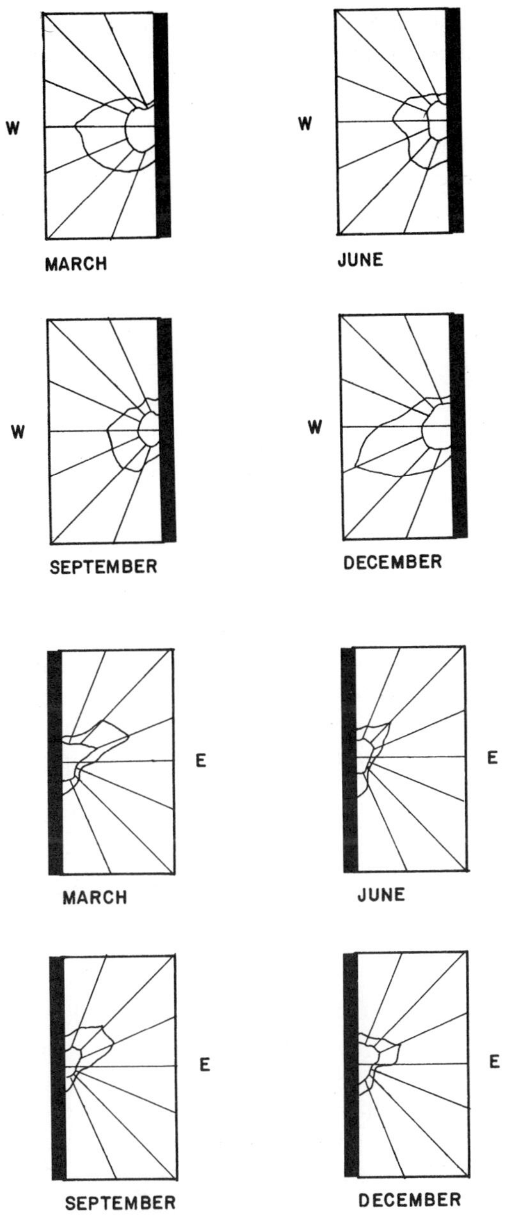

6-17. Western and eastern winds.

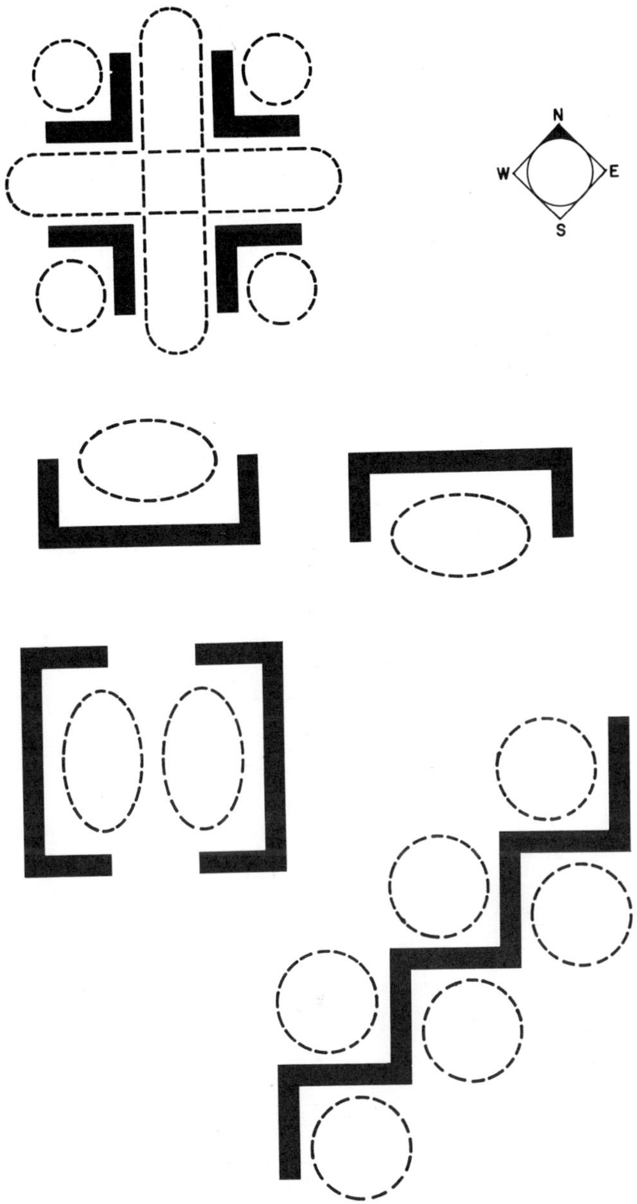

6-18. A variety of wall configurations that create a desirable microclimate.

- leeward side would have a high accumulation of snow.

Clarification of the nature and mechanics of the complex relationship between a building and its surroundings necessitates a closer look at what is actually happening inside the climatic sheath, since each wall of the building is responsible for the unique microclimate in its vicinity (fig. 6-19).

Southern Wall (facing true solar south)

The southern wall has hotter and drier ground; the winter season is shorter and snow melts sooner. In the winter, the warmth of the direct and reflected sun can be enjoyed here. In the spring, insects and plants will appear earlier. This is where birds find their first feeding ground. The hotter and longer summer may be too hot and dry for certain life forms. Its microclimate is similar to that of steeper, southern slopes in the same area. Summer extremes, and direct radiation in particular, however, can be mitigated with shade. The southern wall is the easiest to shade. Both fixed (self-shading) and/or mobile shading devices can be used.

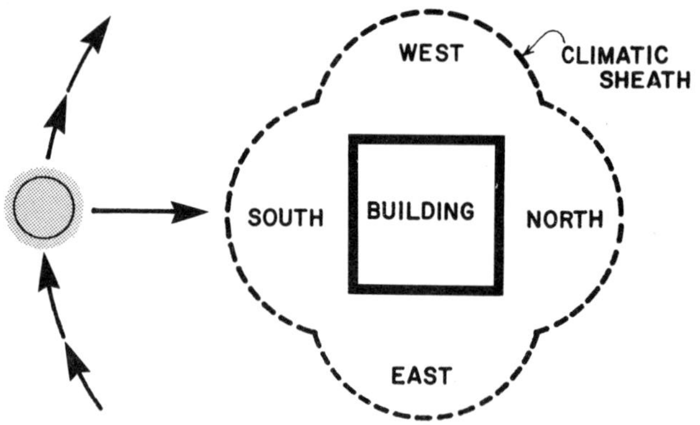

6-19. Four microclimates.

Western Wall

In the summer, the western wall bears the full impact of the afternoon sun; intensive radiation is absorbed and radiates heat both to the ground and the air for prolonged periods of time. The ground is hot and dry. Self-shading is possible, but is more complex than shading the southern wall, because devices like overhangs do not block the low, western sun. Certain living species may therefore find the summer microclimate beyond their tolerance limits. During the winter (at 43°N latitude), only a small amount of direct radiation is received from the low winter sun between noon and sunset, not nearly as much radiation as is received at the southern wall during the cold period.

Eastern Wall

Although the eastern wall is exposed to intensive morning sunlight, it does not have a microclimate identical to that of the western wall. Most of the direct and reflected energy is exhausted to overcome the night cold. The temperature dip usually occurs between midnight and one hour before sunrise. The microclimate of the eastern wall is sheltered against the western winds. Still, it is the coolest place next to the northern wall area.

Northern Wall

The area around the northern wall is characterized by a darker, cooler, and wetter environment. The winter season is longer and the accumulated snow is still melting as blossoms are appearing on the southern side. In certain areas, the microclimate of the northern wall is similar to that of the northern slopes in the region. Direct sun is received only between spring equinox (March 21) and autumn equinox (September 21) during a limited time of the day. Thus, during the extremely hot summer period, this mostly shaded environment provides a cool, relaxing place for insects, birds, beasts, and humans. This single quality is obviously of limited value in northern latitudes.

Figure 6-20 shows both the negative and positive impact areas of a building for each hour during the winter solstice, equinox, and summer solstice at 43°N latitude. Figure 6-21 illustrates the diurnal composite pattern and figure 6-22, the annual pattern of these impact areas.

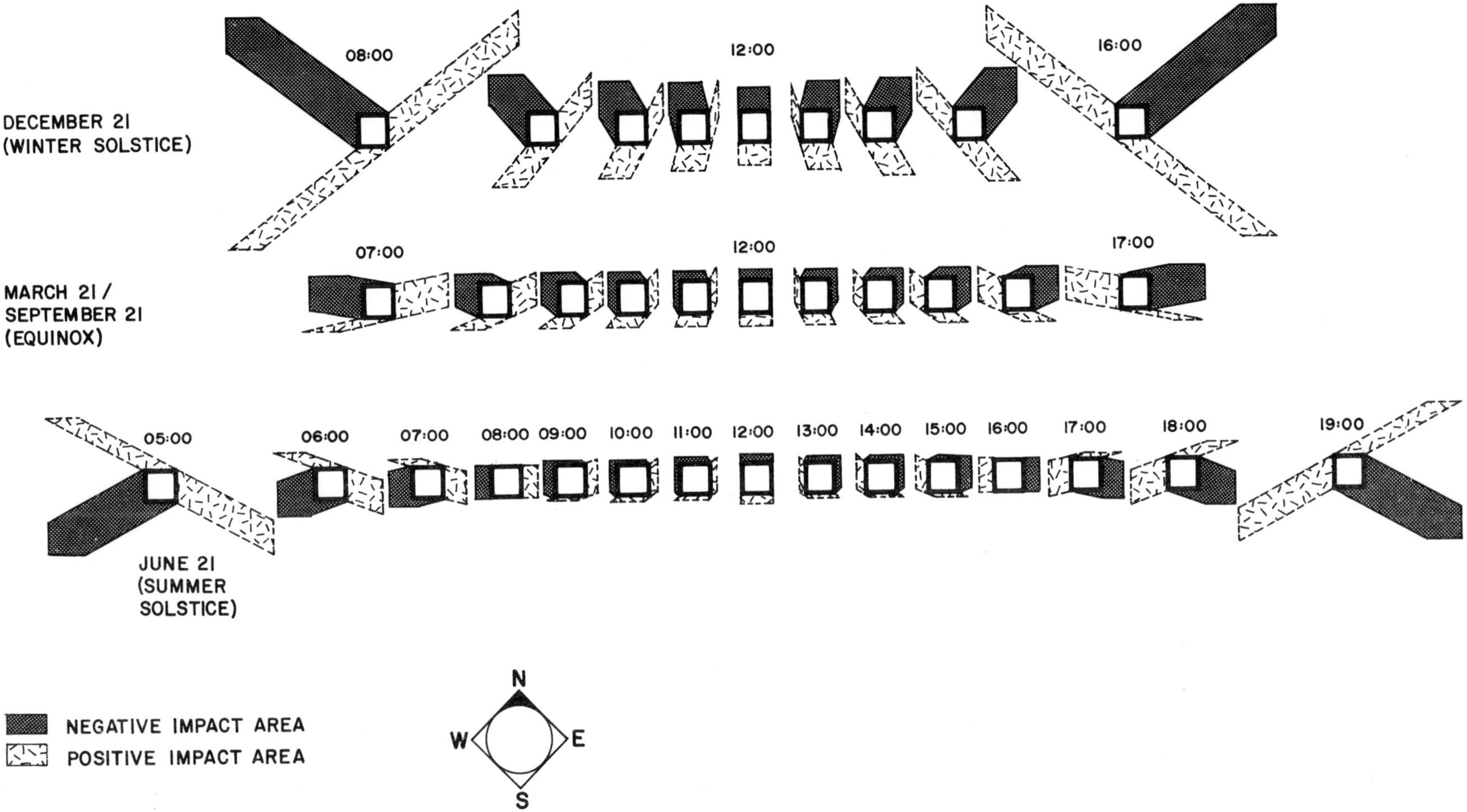

6-20. The positive and negative impact areas of a building for each hour during winter solstice, equinox, and summer solstice.

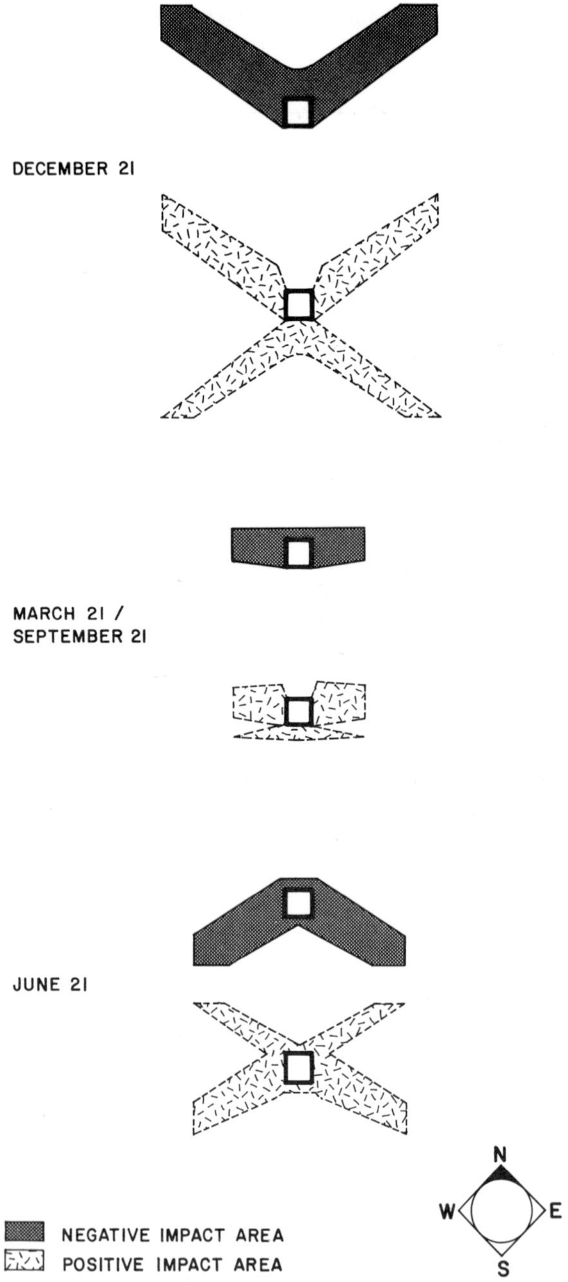

6-21. The diurnal composite pattern of the negative and positive impact areas during winter solstice, equinox, and summer solstice.

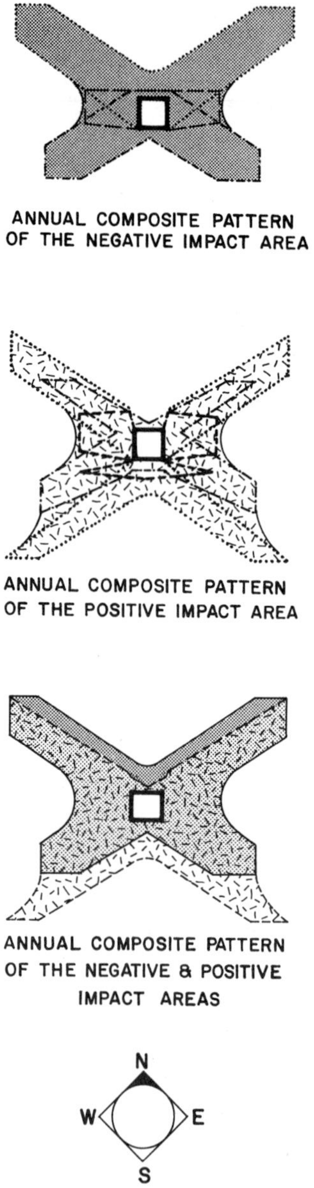

6-22. The annual composite pattern of the negative and positive impact areas.

THE SPACE BETWEEN TWO BUILDINGS

The next step is to place two buildings side by side. The walls of each building affect the microclimate in the same way as the walls of a single building, but a new element would be present—the space between the buildings (fig. 6-23). The major factors determining the character of the microclimate would be the proportions, that is, the width of the alley and the height of the buildings.

A narrow north/south-oriented alley would lack sun and thus be rather cool (fig. 6-24); the ground would have a high moisture content, especially if the downspouts were located here. The growing season would be short, and snow would melt very slowly. This space might also accelerate winds by creating a funnel (fig. 6-25).

Direct exposure to the sun would be limited to a short period at noon (see fig. 6-24). By widening the gap between the two buildings, however, the microclimate can be gradually improved. The period of access to direct sunlight would be extended, and the surrounding walls would reflect the morning and afternoon sun.

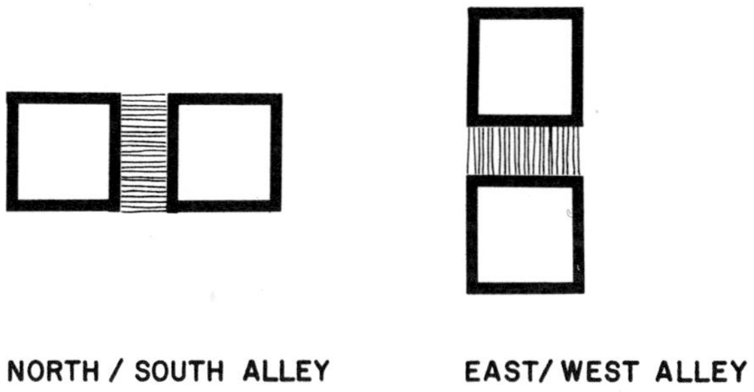

6-23. A space between buildings.

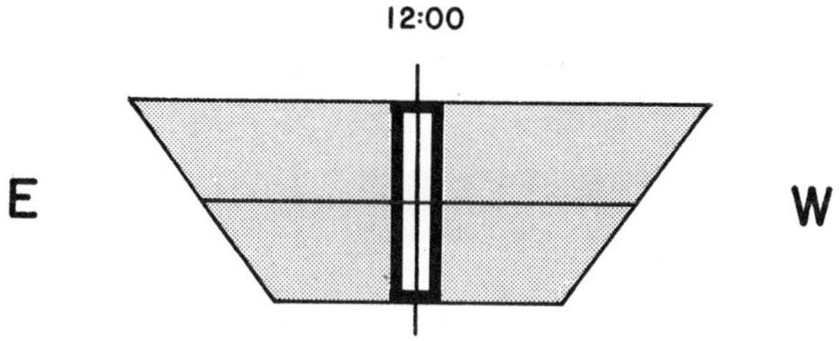

6-24. A Skywindow in a north/south alley.

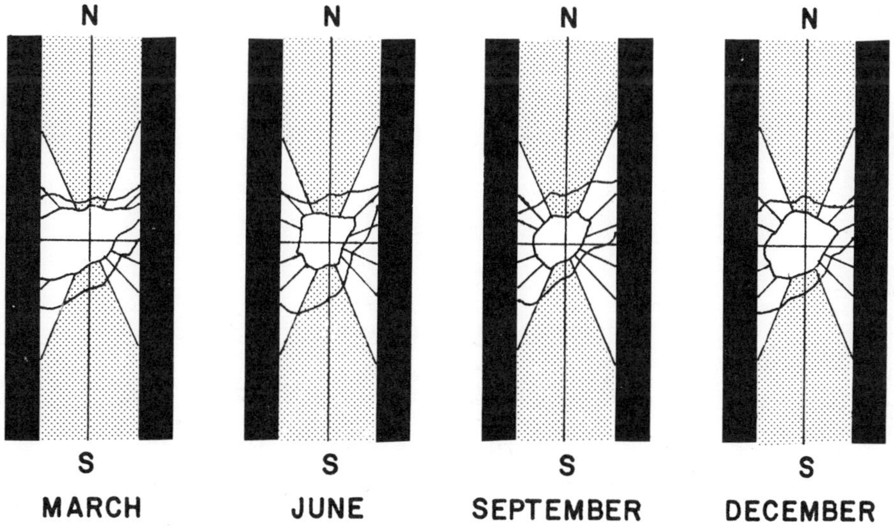

6-25. North/south winds.

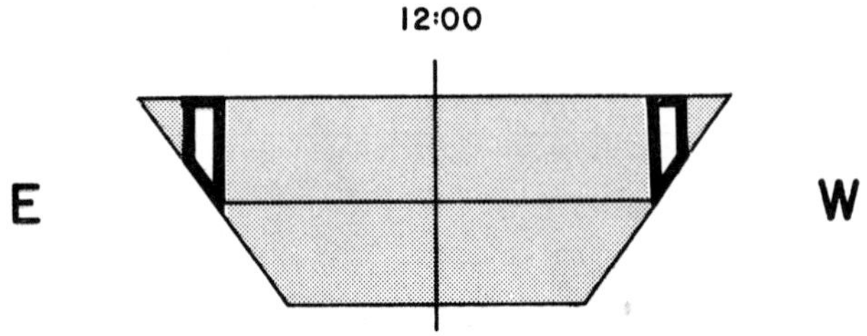

6-26. A Skywindow in an east/west alley.

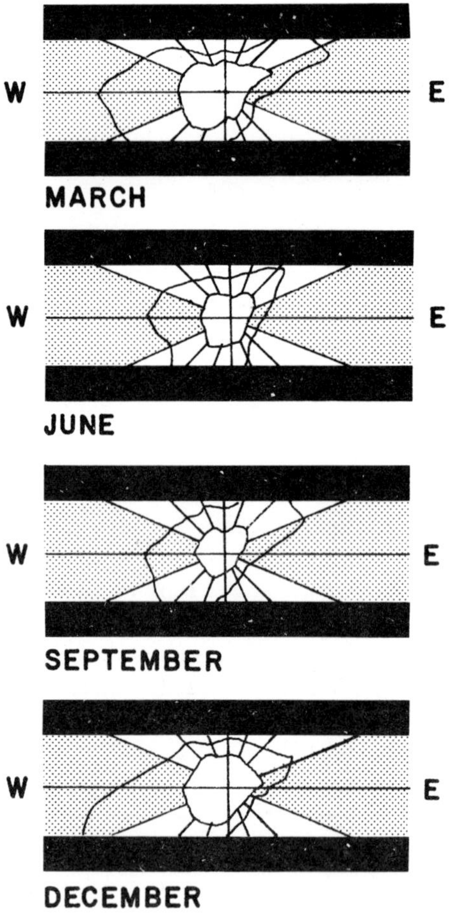

6-27. East/west winds.

Further widening would develop different microclimatic characteristics near each wall, which means that the microclimate of the alley would not be uniform. When the two buildings are sufficiently far apart, each wall would create its own microclimate, one distinctly eastern and one western. The alley and its unique microclimate would cease to exist.

A narrow alley with an east/west orientation would receive no direct sun radiation during the winter season. A limited amount of direct exposure does take place between spring equinox (March 21) and autumn equinox (September 21), once in the morning and a second time in the later afternoon (fig. 6-26). Frequent westerly winds would most likely whistle through at an accelerated speed (fig. 6-27); consequently, only a few life forms would exist in this location, unless they were seeking refuge during a summer heat spell. Living conditions in this alley could be less bearable than those of the northern wall.

Once the east/west alley is widened, the microclimatic conditions dramatically improve; at a certain height/width ratio, they become reversed. The fortuitous combination of direct and reflected sun radiation makes the creation of a microclimate superior to that of the southern wall possible on the northern side.

Any further widening of the alley, however, would result in the gradual development of two distinctive microclimatic environments: one that is associated with the northern wall, and a second that is associated with the southern wall. The alley and its unique environment would cease to exist.

INTERNAL COURTYARDS

An atrium, or interior courtyard, can be either carved out of a single building or created with several buildings. The microclimate within this space is unique, depending (as in the case of the alley) on specific proportions (fig. 6-28).

Except in equatorial regions, where the sun's rays can hit the ground for a short period during the noon hours, a very small courtyard surrounded by tall walls can efficiently block direct sun access. This practically sunless space is an excellent cooling device that can promote drafts and cross ventilation through adjacent rooms. In northern latitudes, such a slim void in a building mass would be like an icicle pierced through the heart of a building. A wider courtyard, on the other hand, allows for a direct penetration of the sun's rays. Thus, a gradual metamorphosis can be seen: a cold pit becomes a warm, solar pit.

Larger, shallower atria built in hot climatic zones actually escalate the temperature. This is why a fountain is placed in the center, airy and shady arcades are constructed on the perimeter, and why, in order to enhance air movement, decorative grids and louvres are used instead of glass. An atrium can be a powerful amplifier of temperature: shielded on all sides from the wind, the four walls and the ground receive, absorb, and reflect direct sun radiation continuously.

The ideal proportions of a courtyard designed to achieve a desirable temperature vary according to latitude. In southern Ontario, a properly designed atrium, one derived from the geometries of the sun at that particular latitude and from a survey of weather conditions as shown by the local Ecochart, can double the length of the summer season and easily provide an environment for sunbathing from April to November.

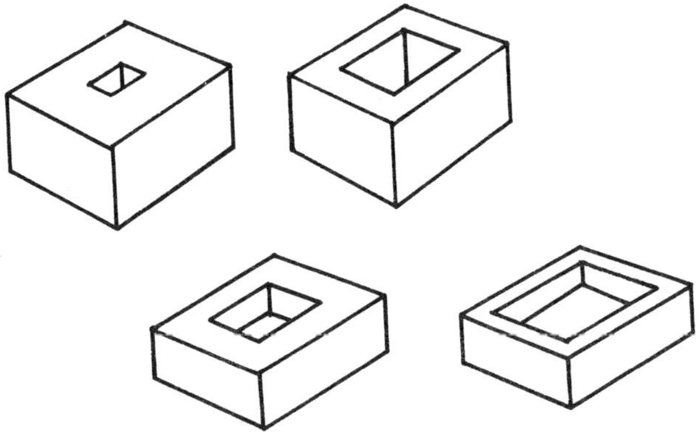

6-28. Internal courtyards.

By further modifying the proportions and extending the distance between the walls of the courtyard, the unique microclimate created by each wall would become more pronounced. The ultimate result would be four walls with four distinct microclimates. This phenomenon is similar to the climate created within the space between two buildings, which was explored earlier.

An understanding of how a microclimate is generated along the four walls of a building, the unique microclimatic properties of north/south and east/west alleys, and the principles governing courtyard performance should lead to a more efficient development of impact analyses, to general improvements and possible remedial actions in the urban environment. These various microclimates help generate the phenomenon called the *urban heat island.*

FROM VILLAGE TO METROPOLIS

After this brief excursion to the complex world of external microclimates, it should be no surprise that a small village on a windy plain, or one built on a frosty mountain, can maintain a comfortable environment on the sunny and leeward side of individual buildings. Since the only reliable comfort indicators are living organisms, one would find that these places are usually favored by children, the elderly, animals, and plants.

Gradually, as the village becomes a town, the town a city, and the city a metropolis, certain microclimatic conditions in positive impact areas may improve considerably. For example, buildings can create calm, windless areas. The most deeply felt improvements would be experienced in the winter, since the urban heat island creates higher temperatures. On the other hand, comfort gains are mitigated by, say, a higher density of buildings, which can create wind turbulence powerful enough to impede human movement, with up drafts and down drafts caused by the unequally heated walls of the high-rises. The worst penalty of all may be pollution, a direct and indirect product of human activities and a major contributing factor to the generation of the urban heat island.

The climates of individual urban heat islands inevitably differ, since each city has unique physical properties. Broadly speaking, the magnitude of a heat island is roughly related to the size of the city. The heat island is created from:

1. the compact grouping of buildings
2. the concentration of human activities

Five factors determine the specific character of a heat island.

1. SURFACE MATERIALS: Stone, concrete, and brick conduct heat three times more effectively than wet soil; as a result, the city accepts and stores a large amount of energy in a very short period of time.
2. SHAPE AND ORIENTATION OF SURFACES: Incoming energy is reflected by surfaces (it, in effect, bounces between horizontal and vertical surfaces). All surfaces, even those not directly exposed to the sun, therefore receive and store heat. (In the countryside, only the upper layer of vegetation receives energy).

   The regional airflow pattern (see the Ecochart) is slowed by buildings; on the other hand, buildings may be arranged to create bottlenecks (reduced volumes of passing air), the Venturi effect (a wind funnel that results in increased wind speed), turbulence, up and down drafts, and pockets of absolute calm.
3. HEAT SOURCES: A city generates excessive heat as a result of poor insulation, industry, and transportation and as the by-product of space-heating buildings.
4. MOISTURE SOURCES: City roofs, pavements, and drainage systems redistribute rain water and snow. (In the countryside, moisture is partially retained and partially evaporated, cooling the environment.)
6. AIR QUALITY: The city emits a large amount of pollution. This dome-shaped blanket of pollution is visible from a long distance or from an aircraft. The dome contains dust particles, water vapor, and carbon dioxide, which reduce the amount of direct sunshine, but reflect the upward radiation down, heating the air and ground, similar to the greenhouse effect.

DYNAMICS OF THE URBAN HEAT ISLAND

Buildings are constructed to provide an optimum thermal environment, to shelter a range of human activities and needs. However, by satisfying human requirements, buildings also generate a new set of external conditions. In investigating how the immediate environ of a single building is affected, it has been found that within the "climatic sheath," a number of anomalies related to temperature, humidity, and air movement can be detected. Obviously, in cities, a large number of buildings generate a large number of modifications.

The sun is, of course, a prime force behind any climatic situation. The urban form is exposed to two basic types of radiation: direct radiation from the sun and diffused radiation from the sky. Human-made features, such as horizontal and vertical surfaces, absorb and reflect both types of radiation; urban surface materials, their reflectivity, and their other physical properties are therefore important.

The performance of the urban heat island is not uniform throughout the year; it changes according to the season. In the summer, the walls of tall buildings, pavement, concrete, stone, and brick absorb and store a large amount of solar radiation, more than vegetation and soil. The thermal admittance is, of course, directly related to the geometries of the urban form. Very little energy is consumed for evaporation, since municipal engineering systems are efficient in getting rid of precipitation. A large amount of stored heat is therefore given up at night, making urban night air warmer than air in the countryside. During the winter, on the other hand, solar radiation is less intensive. Vertical urban surfaces reflect the low sun, providing light and warmth to open urban spaces. During the cold season, heat produced by sources such as combustion for home heating, the power-generation industry, transportation, and human as well as animal metabolism with their associated by-products contributes significantly to the thermal budget. Some of these energies warm the urban air directly, whereas others do so by filtering through poorly insulated walls and roofs. The removal of snow allows direct and reflected radiation to be absorbed by horizontal surfaces.

The urban heat island maintains its form and position in windless or mildly windy conditions (fig. 6-29). Increased wind may deform its characteristic dome shape, but each heat island is unique, and a definite critical air speed is required in order to eliminate a particular heat island. For example, in the period between 1955 and 1966, this wind speed was 22–30 km/hr. in Hamilton, Canada.

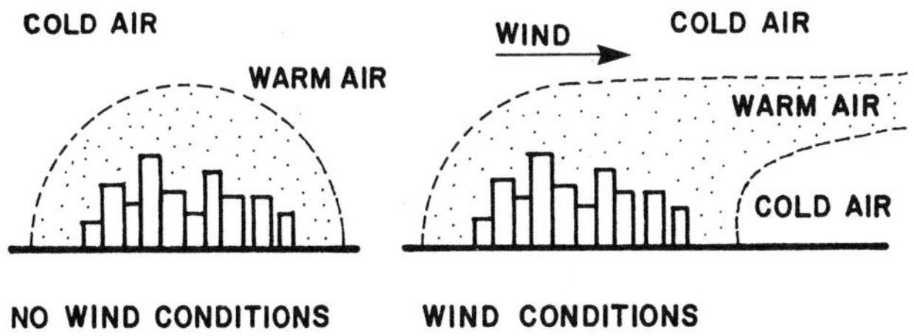

6-29. An urban heat island.

Table 6-1 is a description of some of the mesoclimatic changes that may be expected as a result of the urban heat island.

In the final analysis, the urban heat island must be considered by architects and urban designers. The tangible benefits associated with living and working within the urban dome—warmer city environment and fewer heating degree-days—lead to lower heating bills for citizens and the municipality. The reduced amount of snow and the softened impact of environmental extremes result in lower municipal costs. In addition to

TABLE 6-1
Mesoclimatic Changes Caused by the Urban Heat Island

ELEMENT		CHANGE
Radiation	on a horizontal surface	less 15%
	ultraviolet, winter	less 30%
	ultraviolet, summer	less 5%
Temperature	annual mean	more 0.7°C
	winter maximum	more 1.5°C
	freeze-free season	more 2–3 weeks
Wind speed*	annual mean	less 20–30%
	extreme gusts	less 10–20%
	calms	more 5–20%
Relative humidity	annual mean	less 6%
	seasonal mean, winter	less 2%
	seasonal mean, summer	less 8%
Cloudiness	cloud frequency and amount	more 5–10%
	fog, winter	more 100%
	fog, summer	more 30%
Precipitation	amount	more 5–10%
	days (with less than 5mm)	more 10%
	snow days	less 14%
Dust particles		more 10 times

* The wind speed data indicates changes in the values of wind speed on the Ecochart. It must be understood that the urban form, and high-rises in particular, introduce new elements; some of these, if large enough in magnitude, can be hazardous to human life (sun induced currents, vortices, eddies, rotors, and so on).

the direct economic benefits, plants have an extended growing season and many animals require the warmth of city living to guarantee survival through the winter.

For cities and communities located in northern latitudes, the hot season is of little concern. Extremely high temperatures are rare and last for short periods of time; hot spells are usually perceived as pleasant exceptions to the rule. The high summer sun is easy to shade and induced drafts help to keep the temperature within the human comfort zone. The challenge therefore lies in designing for the three remaining seasons. Since the urban heat island is generated not only by the sun's interaction with the urban form, but also by heating and other by-products of human activities, the contribution of this metabolistic heat in winter should not be overlooked, since it can be equal to or even greater than the contribution of the sun.

The years of the energy crisis taught us that it is costly to allow a building to heat the sky. In order to provide internal thermal comfort at minimum cost, precious heat had to be stored. Buildings thus began to be superinsulated. Similarly, if any industrial process or mechanical device is emitting Btus, fumes, gases, or solid particles, it is most likely the result of an archaic and inefficient technology. In our slowly evolving post-industrial society, the performance, economy, and consumption of critical resources are of utmost concern. The present wasteful processes will be less and less affordable.

Despite all the technical and technological improvements, however, the city is still a living organism that must release some metabolistic heat into its surroundings. An urban heat dome generated partially by the sun and partially by the city itself will probably continue to exist, but heat from the sun will exceed urban metabolistic heat.

A design strategy aimed at maximizing the advantages of the heat island should be twofold:

1. Excess metabolistic heat should be captured and channeled into specific locations to improve the local microclimate (building entrances, sidewalks, and so on).
2. The urban form should be regulated to allow maximum use of the sun's radiation for improving pedestrian comfort in public spaces.

This brief review of urban climatology urges rational urban design and planning. The cumulative effect of microclimatic modifications in existing urban spaces invites the utilization of a variety of systems, materials, and building geometries that would mitigate the extremes of local climate. If planned carefully, the urban climate that results will be as healthy and pleasant as the climate in the country, and this maximum comfort will be achieved at minimal cost.

CHAPTER 7

CLIMATE-SENSITIVE GUIDELINES FOR URBAN DESIGN

ENERGY INPUT

The aim of this chapter is to provide guidance to the urban planner. Thus, special attention has been paid to the design parameters of urban spaces to ensure the maximum possible year-round outdoor comfort at a minimum cost. The design process involves

1. ambient energy (that which surrounds us)
2. metabolistic energy (that which is consumed and digested to mitigate outside hostile forces)

An equilibrium between ambient and metabolistic forces is possible only with a steady input of energy. This concept of equilibrium (a balance between the external and internal environment) is applicable to a variety of different situations (for instance, the environment within and surrounding the human body as compared with the thermal environment within and surrounding a house) and must be understood accordingly.

Perfect equilibrium is an elusive but frantically pursued state, since the relationships involved are being continuously tipped off balance. A person walking on the street is exposed to ambient weather forces, which are received by a complex system of sensory perceptions. When ambient weather forces become too cool or too warm for human comfort, the internal metabolistic system is triggered to restore balance. This in turn might create a demand for an additional supply of energy for warming or cooling—an ice-cream cone or a bowl of soup may become irresistible. If such a counteroffensive fails, one may have to restore equilibrium by moving indoors, leaving urban spaces deserted. To prevent this occurrence, the designer must strive for equilibrium by optimizing energy input, which can be achieved by modifying the urban microclimate at street level. Although it is possible to use excess heat from large buildings to warm sidewalks in the winter or create warm air screens around entrances to buildings, the use of this surplus energy is rather limited at present. This chapter will therefore concentrate on the manipulation of ambient energy, which requires a close look at the Ecochart.

As mentioned in chapter 3, the composite impact can be read vertically on the Ecochart, but this can be a difficult task, requiring a great deal of professional experience and intuition. The phenological observations on the Ecochart (the monitoring of biological behavior in response to cumulative weather forces) appear to be the safest guide for estimating the quality of a given microclimatic situation or urban space.

Climatic elements, which are in fact energy carriers, transport energy by *radiation*—the process of emitting energy in the form of waves or particles, *conduction*—the use of a substance capable of transmitting heat, and *convection*—the transfer of heat by automatic circulation of fluids or gases. In order to maintain a balance in energy input and output, leading to the most desirable urban environment, the community planner and urban designer should take into consideration the following elements of the regional climate:

1. solar radiation
2. air temperature
3. air movement
4. precipitation

These factors have a profound effect on the livability of urban space and thus on the urban form in a particular geographic area. Since the major force behind the weather engine and its climatic elements is the sun, and since it is very important in shaping the urban form and urban spaces (urban masses and voids), an explanation of the generation of heliomorphic urban space follows.

DESIGN OF HELIOMORPHIC URBAN SPACE

In high-density urban areas, every building plays two important roles. First, it provides a protected internal environment—a stable thermal milieu within the narrow limits of the human comfort zone. Second, it is responsible, to a great extent, for the creation of the urban microclimate. Within the context of the first role, the design of buildings for human occupancy is normally perceived as the general mandate of the architect and his or her group of consultants (mechanical, structural, and electrical). For decades the optimal milieu for a variety of human activities has been achieved mainly through energy input. A variety of energies such as electricity, gas, and oil were brought, sometimes over vast distances, into a building to light, heat, cool, supply, and condition the air, water, and so on. Relatively recently, it was discovered that it is possible to tap, tame, and store the wild and fluctuating energies around a building; most important, it was found that these energies can replace the more expensive conventional energies from a central distribution grid.

Although the general mandate of the architect has not changed, the strategy has. Now a building can be transformed into a quasi-biological system that sensitively responds to environmental variations, opening itself like a blossom, harnessing and absorbing ambient energies. On the other hand, when the outside environment is charged with energies that threaten to upset the building's internal balance or are dangerous to human life, the building shuts itself off like a fortress, rejecting and deflecting the pounding of the elements. The pursuit of this sophisticated interplay between a building and its environment has led to a renewed interest in understanding the climate and the behavior of the sun in particular. In order to maintain the internal equilibrium of a building without inflating the energy input, one must pay particular attention to overall shape, construction, outer shell design, selection of exposed materials, access to the sun, and the sun's penetration of interiors.

Most how-to books address the problems of a building built in isolation or in the context of a very low-density development. The issue has also been presented in the context of how a building can keep its internal environment within the human comfort zone and keep costs at a minimum. Little has been said or done, however, about keeping a building's external environment (urban public places) within the limits of the human comfort zone and at minimum cost.

This brings us to the second function of a building, particularly a structure that is constructed in dense, highly developed urban areas. Since all urban places are created by their surrounding buildings, each urban building is partially responsible for the creation of an urban space and its microclimate. Although the degree of responsibility varies, a single building, by virtue of its shape, height, exposed materials, and treatment at street level may be capable of destroying an urban space and eliminating pedestrian activities. The following sections will explore the ways a building can contribute to or improve an urban space and actions that can mitigate some of the negative effects of high-density development.

URBAN SURFACES

The extent of the livability of an urban space is effectively controlled by three factors:

1. intensity of solar radiation
2. exposure time
3. orientation

Intensity of Solar Radiation

Just prior to entering the earth's atmosphere, the value of direct solar radiation is 1353 W/m^2 (watts per square meter). This value, also called the *solar constant,* varies with changes in distance from the sun and with changes on the sun's surface. To our advantage, albeit a barely noticeable one, the earth is closer to the sun on December 21 (winter solstice) than on June 21 (summer solstice). Due to the "imperfect geometries" of the earth's path, the winter distance is about 96.5 % of the summer distance.

As a result of the cumulative effect of the atmosphere's thickness, the angle of sunbeam incidence, and atmospheric composition and content (ozone, dust, water, vapor, carbon dioxide and other pollutants), the amount of direct solar radiation is reduced as it penetrates the earth's atmosphere. Clouds and atmospheric ingredients therefore reflect part of the incoming solar radiation, a certain amount is scattered by molecules in the atmosphere, and some is regained as diffuse radiation. Part of the radiation that reaches ground level is reflected by the earth's surface, but most of the energy is absorbed. It changes to heat and raises the temperature of the air, the ground, and that of surrounding objects (fig. 7-1). The intensity of radiation received on the earth's surface increases in direct relation to the height of the surface above sea level, because less is lost to the atmosphere. A commonly used scientific tool to measure solar radiation received is an *extinction coefficient,* which expresses the turbidity of atmosphere; it is determined by measuring site data and comparing it with the theoretical value of the sun's intensity. Empirical research results indicate that although atmospheric conditions (clouds in particular) modify the amount of energy that reaches a building,

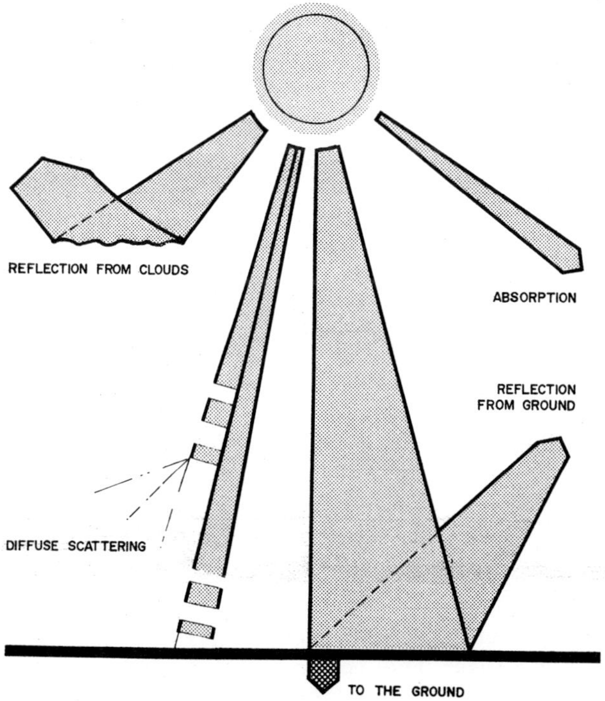

7-1. Solar radiation is reduced by the earth's atmosphere.

the cumulative effects of direct, diffuse, and reflected forms of solar radiation mitigate the negative conditions created by atmospheric obstructions and result in a roughly balanced situation for community planning.

Direct solar radiation is very important to an urban designer. Once the solar beam enters the vicinity of a building, it can be skillfully manipulated to improve the microclimate. A pedestrian can be exposed to direct, reflected, and diffuse solar radiation. These three varieties, like three basic colors, can be mixed to achieve certain desired results by manipulating the geometry of the built environment and the reflectivity or albedo of exposed surfaces. Again, this underlines the joint responsibility of the architect and the urban planner to maintain a high-quality urban environment.

Exposure Time

The regional Ecochart indicated the maximum possible sun hours in each month and the statistical averages of available sun. The differences are caused by cloud obstruction. For example, in the Hamilton area, a minimum of 69 hours of sunshine is to be expected in December; the annual maximum is in July, 287 hours. Although 69 hours a month of exposure to sunlight may not seem to be enough to maintain thermal comfort in our houses, it may be enough to improve work performance, health, and our dispositions. Thus, every effort should be made to let the winter sun into our living rooms, offices, and urban spaces.

The review of microclimatic conditions between buildings and in courtyards revealed that not only clouds, but buildings themselves, block the sun. Every object produces an impact area that is deprived of direct solar radiation. The architect and urban designer must therefore be able to determine the sizes of impact areas and evaluate every design or development proposal in terms of its positive or negative relationship to the surroundings. The construction of shadow-casting patterns is a fairly simple drafting exercise that is described in chapter 8.

Orientation

Objects are oriented, first, to maximize energy-harnessing capabilities (absorbtion and storage). Within the urban context, this means creating the optimum orientation of urban surfaces toward solar radiation. Second, objects are oriented to cast energy back to the environment. Within the urban context, this means creating the optimum orientation or urban surfaces to reflect the sun's radiation into a designated area.

In northern latitudes the incident angles of the sun's rays on horizontal and vertical surfaces should determine the orientation of buildings and the shape of public spaces. The interaction between the sun's beams and an exposed urban surface produces a variety of microclimatic conditions on the pedestrian level; it is also largely responsible for the urban heat island.

Thus, the successful designing or redesigning of urban spaces in northern latitudes is dependent upon a few basic relationships between solar rays and building forms.

1. Buildings with a high surface/volume ratio (S/V) are more sensitive to orientation than buildings with a low S/V.
2. High-rise buildings are more sensitive to orientation than low-rise buildings.

These relationships must be manipulated to create built forms that respond to environmental variations (see chapter 4).

The subject of building orientation has been widely researched, publicized, and discussed. It is usually recommended that a heavily fenestrated wall be given a southern exposure; the wall can deviate significantly (up to 30°) from due solar south. With the present state-of-the-art technologies, a passive-solar, single-family residence is still a rather crude device to harness solar energy in comparison with a photovoltaic collector. Therefore, successful designers of northern houses that incur no heating costs rely on the use of metabolistic heat (the by-product of human activities) by superinsulating homes.

Figure 7-2, for example, shows the top view of a transparent Skydome (see chapter 4). The black lines indicate sun paths as seen from the center of the dome from sunrise to sunset on June 21, March 21, September 21, and December 21. Seven points were selected along these three sun paths: numbers 1, 3, 5 are due south; numbers 2, 4, 6 are due southwest (or southeast); number 7 is due west (or east).

In the next step, the observer in the center of the dome is replaced by a flat, rectangular plate on a turntable. This rectangular plate or generic solar collector can be tilted to any desirable angle.

Figure 7-3, point 1 on the Skydome, shows the square plate oriented due south. The azimuth indicates the time of the experiment—winter solstice, December 21 at 12:00 hours. It also specifies the location in terms of latitude, 44°N; the sun altitude is 23.6°.

When the plate begins to change its tilt, the sunbeam hits the plate at a different angle. The change in angle of incidence relates to the amount of energy received:

0°—minimum or zero energy received
90°—maximum possible amount received.

This flat, square revolving plate may be used for a variety of representations. It can serve as an illustration or explanation of topography and its uneven attainment of energy, and it can assist in determining the most efficient or optimal position for solar collectors and the orientation of buildings. It can also aid in evaluating the solar energy gain of urban surfaces. Since the urban form is composed primarily of vertical and horizontal surfaces (roofs, walls, sidewalks), these two positions are especially important for architects and urban designers. For example, in the case shown in figure 7-3—point 1 on the Skydome—the sun altitude is 23.6°. The tilt of the plate for maximum energy receivership for this particular orientation (due south or 12:00 hours), and for this particular time of year (December 21), would lie between a 45° sloped surface (roofs, collectors, or a 100-percent sloped terrain) and a 90° vertical surface (walls or sheer cliffs in the natural environment).

In figure 7-4—point 2 on the Skydome—it is winter solstice, December 21, and the orientation is southeast and southwest; the latitude is 44°N. The sun altitude is 9°, a very low winter sun; thus, maximum energy can be obtained from surfaces close to the vertical position.

In figure 7-5—point 3 on the Skydome—the plate is oriented due south during equinox, March 21 and September 21, at 44°N latitude. The sun altitude is 47° and the time is 12:00 hours. The maximum sun energy is now received by 100-percent slopes or by 45° roofs and collectors; flat horizontal surfaces (roofs, sidewalks, parking lots) and vertical surfaces (walls) receive considerably less energy, but they are about equal in their receivership. This means that if it is not possible to utilize a 45° slope, there are two other options within the orthogonal system, either a horizontal or a vertical surface.

Figure 7-6—point 4 on the Skydome—shows a southeastern and southwestern orientation of the plate during equinox at 44°N latitude. The time is 09:00 hours or 14:20 hours and the sun altitude is 36.6°. This places the point of maximum energy receivership approximately halfway between the horizontal and vertical position, leaning more toward the vertical one. In the orthogonal system, it would therefore be somewhat advantageous to select a vertical surface.

In figure 7-7—point 5 on the Skydome—the plate is oriented south during summer solstice, June 21. The time is 12:00 hours and the latitude is 44°N. The high altitude of the sun, 70.4°, makes the horizontal surface at nearly the best angle for maximum energy receivership.

Figure 7-8—point 6 on the Skydome—shows a

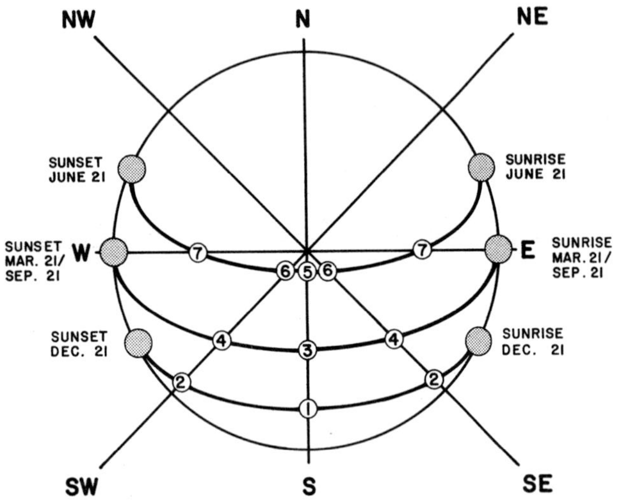

7-2. Skydome as viewed from above.

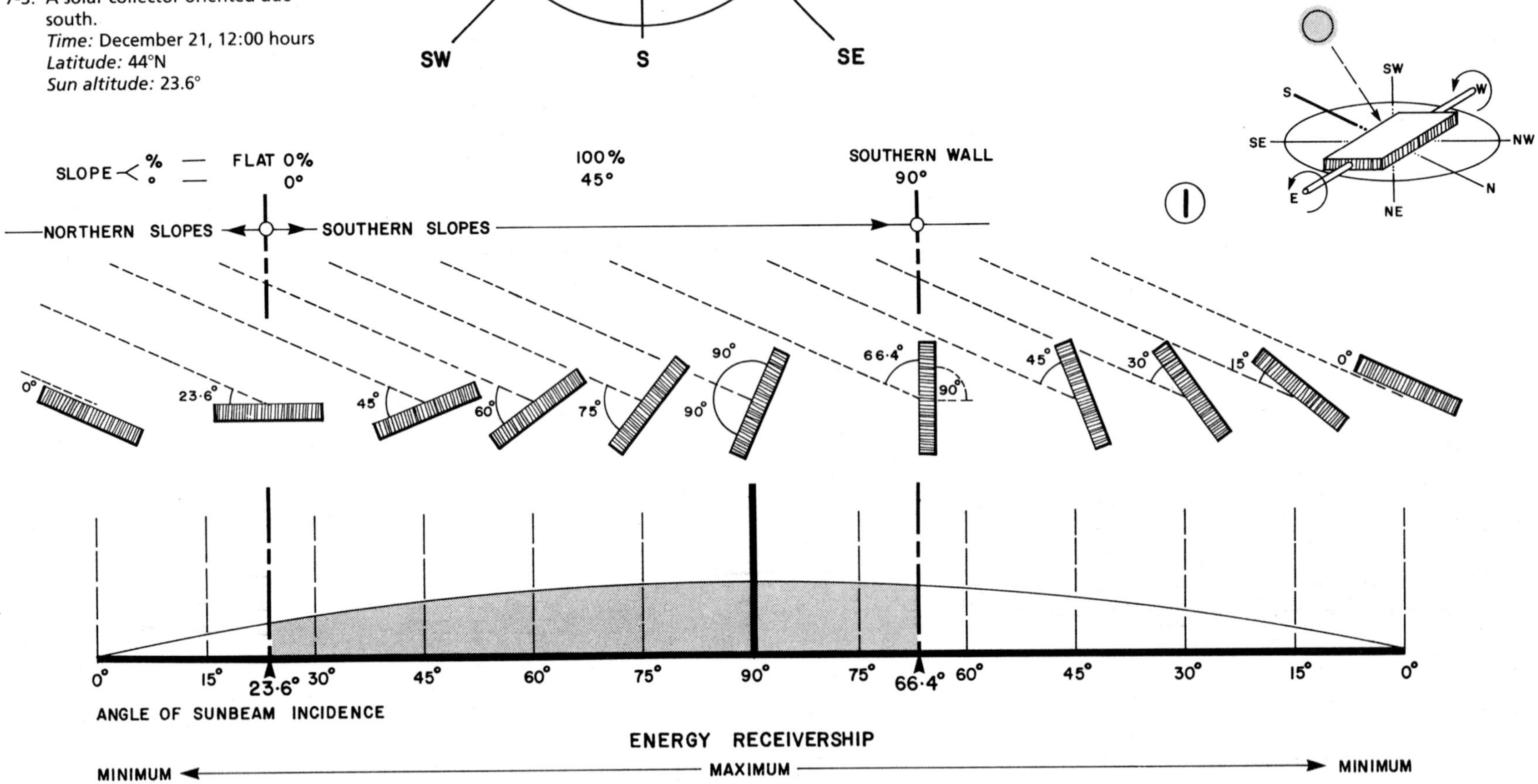

7-3. A solar collector oriented due south.
Time: December 21, 12:00 hours
Latitude: 44°N
Sun altitude: 23.6°

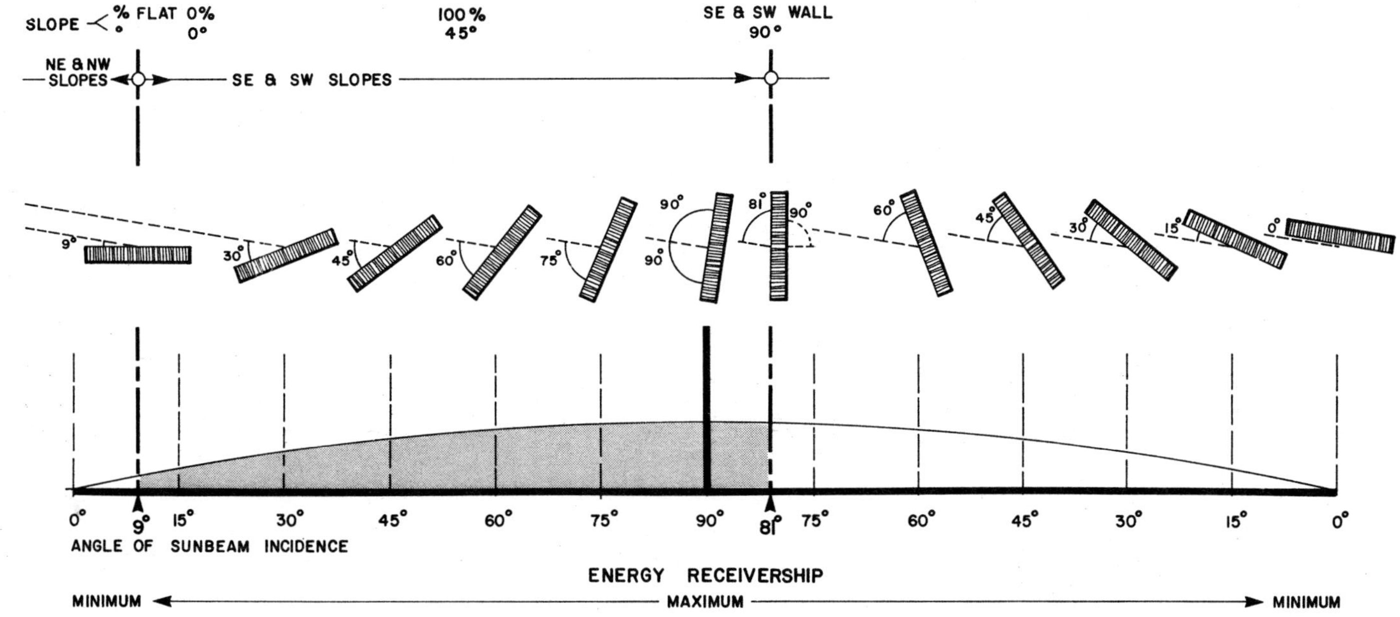

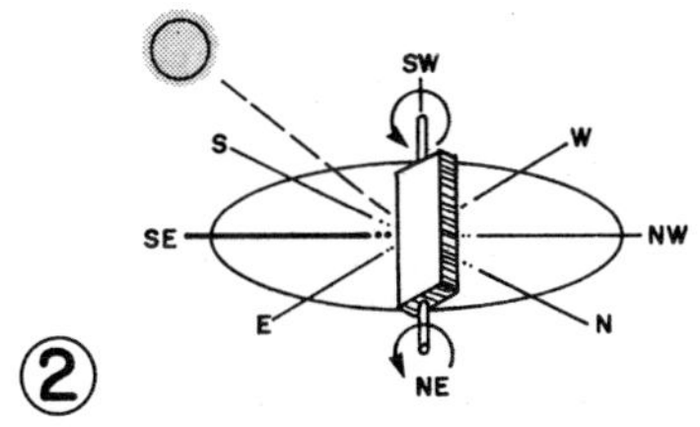

7-4. A solar collector oriented due southeast (or southwest).
Time: December 21, 08:45 hours (15:15 hrs)
Latitude: 44°N
Sun altitude: 9°

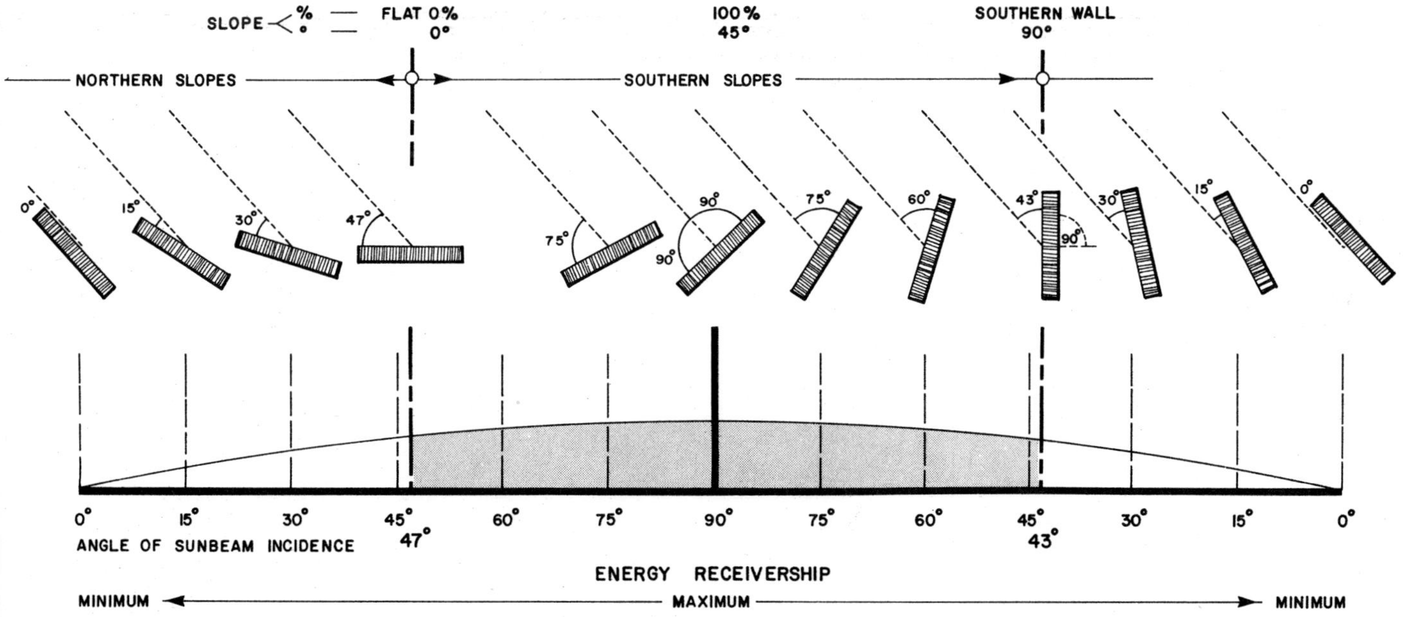

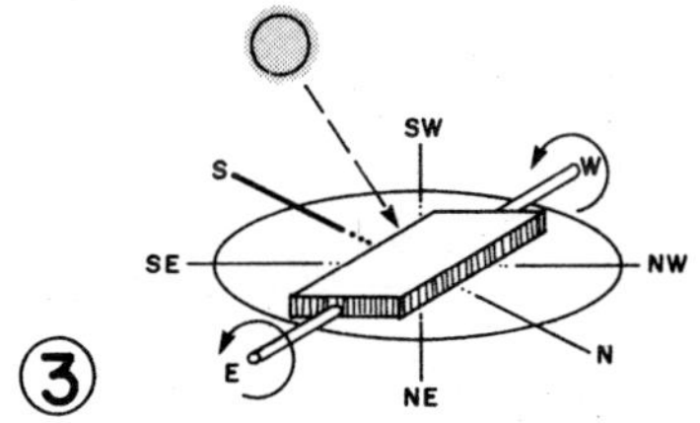

7-5. A solar collector oriented due south.
Time: March 21 and September 21, 12:00 hours
Latitude: 44°N
Sun altitude: 47°

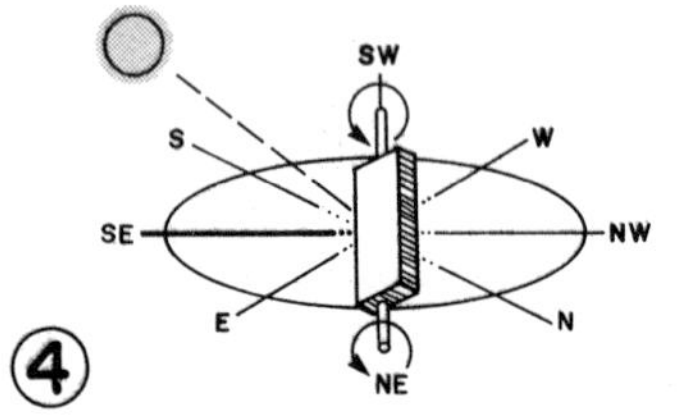

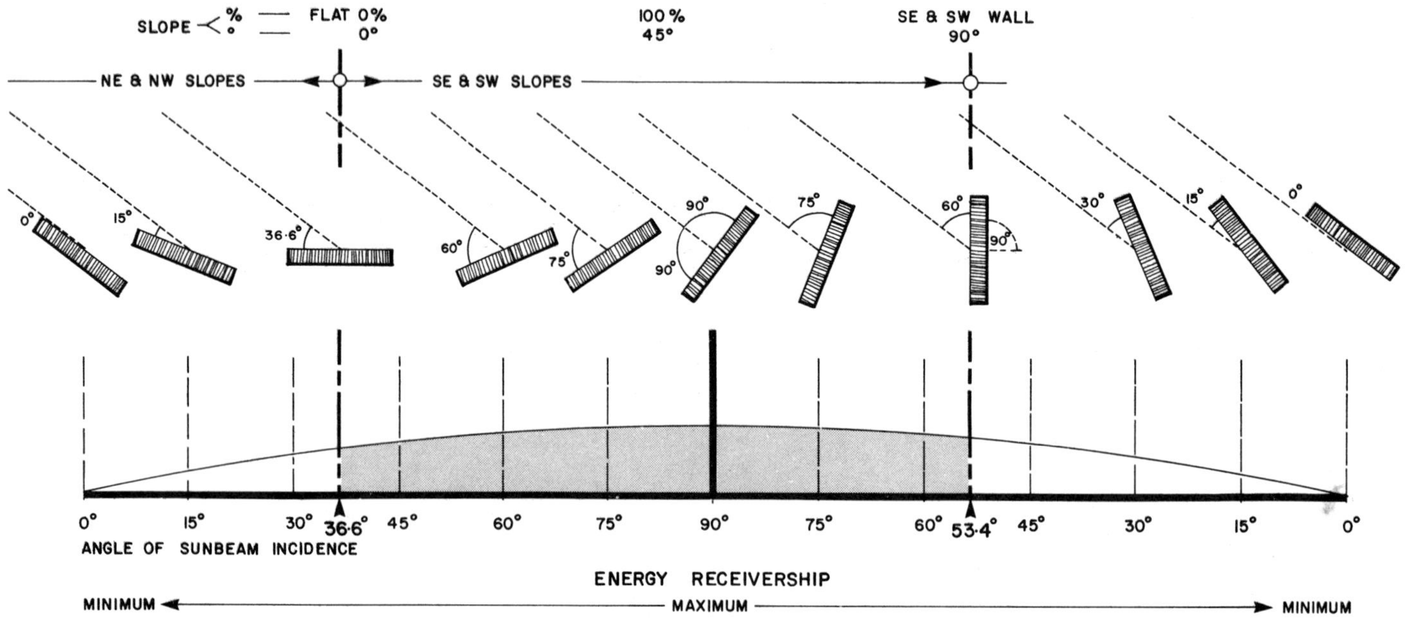

7-6. A solar collector oriented due southeast (or southwest).
Time: March 21 and September 21, 09:40 hours (or 14:20 hrs)
Latitude: 44°N
Sun altitude: 36.6°

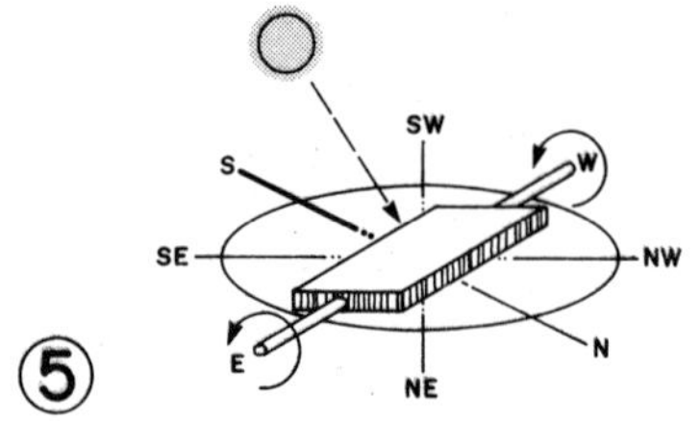

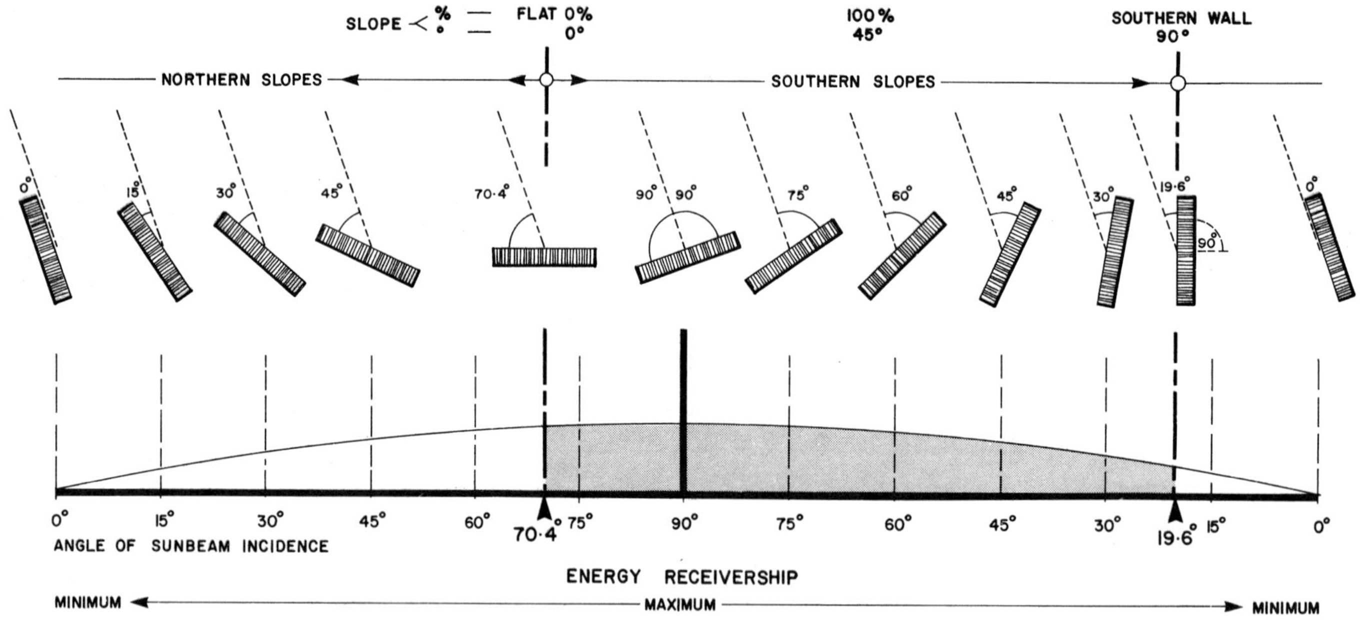

7-7. A solar collector oriented due south.
Time: June 21, 12:00 hours
Latitude: 44°N
Sun altitude: 70.4°

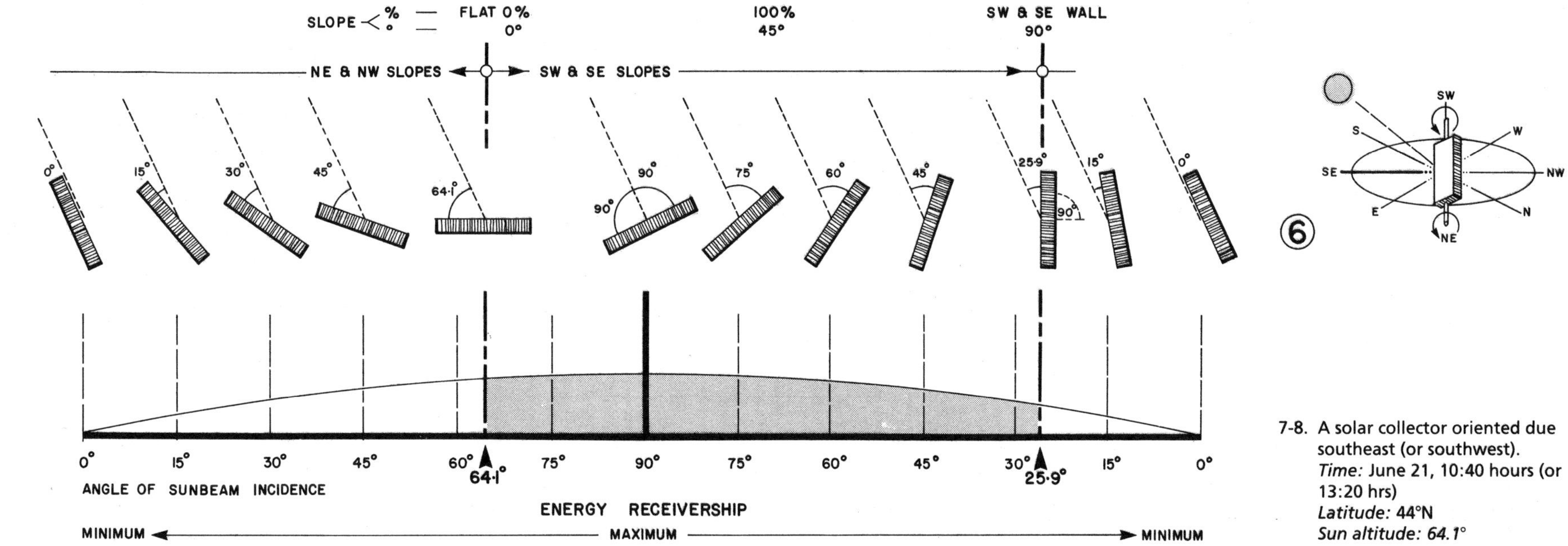

7-8. A solar collector oriented due southeast (or southwest).
Time: June 21, 10:40 hours (or 13:20 hrs)
Latitude: 44°N
Sun altitude: 64.1°

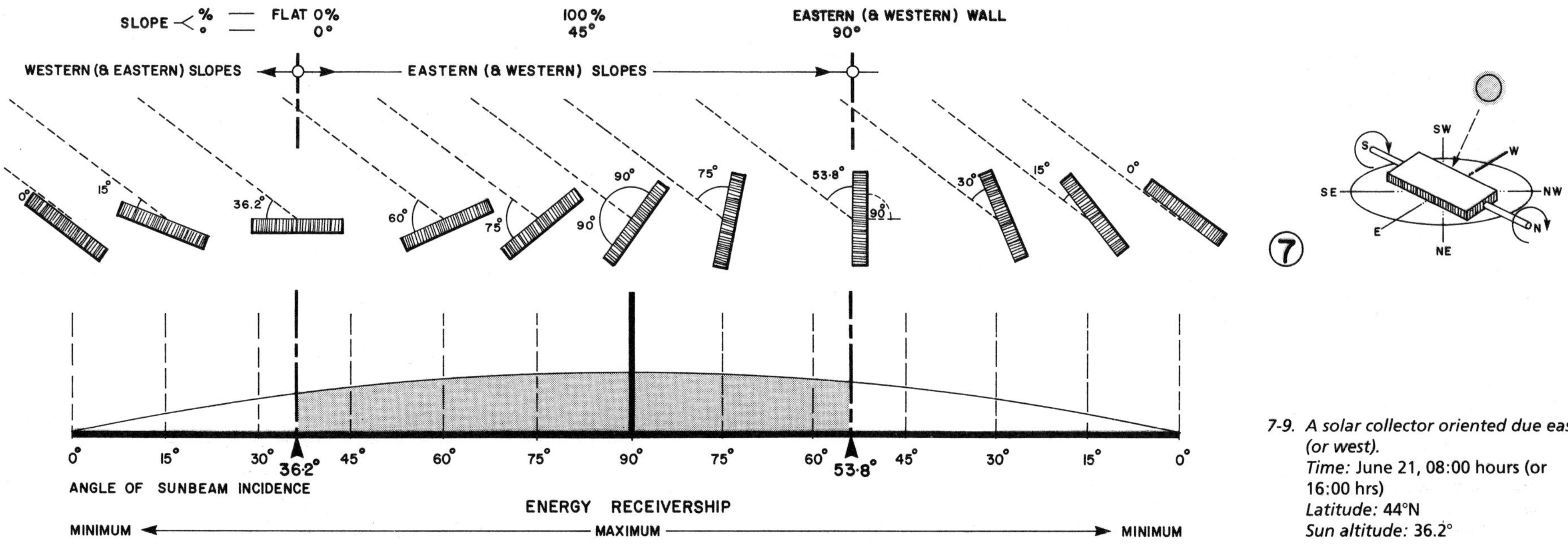

7-9. A solar collector oriented due east (or west).
Time: June 21, 08:00 hours (or 16:00 hrs)
Latitude: 44°N
Sun altitude: 36.2°

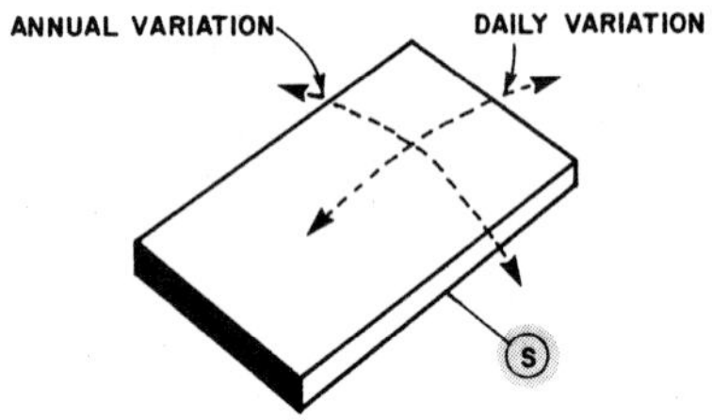

7-10. A tracking collector.

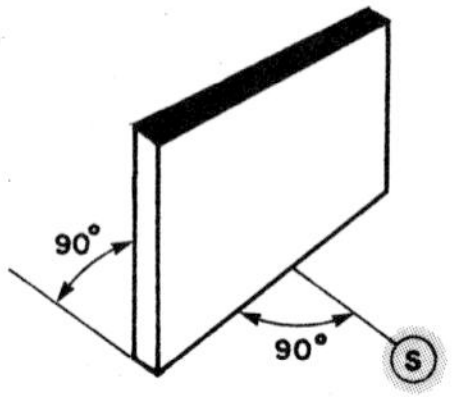

7-13. The collector is fixed due south, and the tilt angle is fixed at 90°.

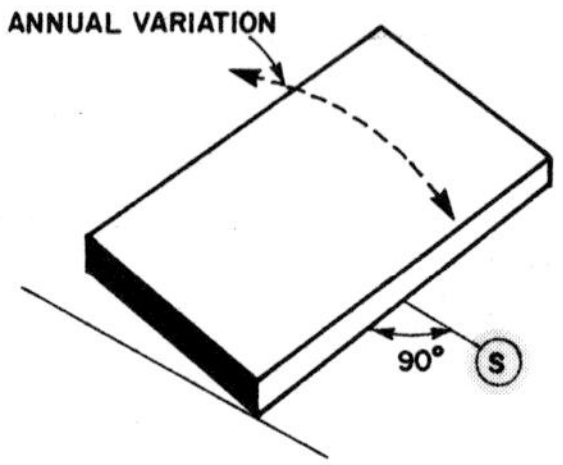

7-11. The collector is fixed due south, with an adjustable tilt angle.

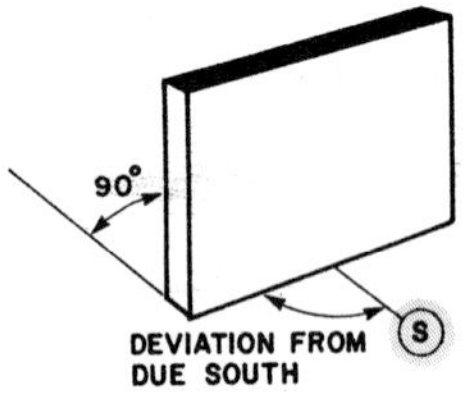

7-14. The collector's tilt angle is fixed at 90°, but orientation deviates from due south.

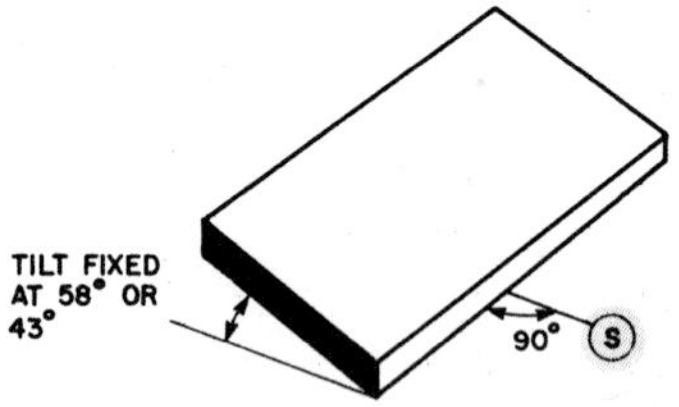

7-12. The collector is fixed due south, and the tilt angle is fixed to optimize annual or seasonal energy received.

southeastern and southwestern orientation of the plate during summer solstice on June 21, latitude 44°N. The times are 10:40 hours and 13:20 hours. The sun altitude is still very high at 64.1° but slightly lower than in point 5. This moves the point of maximum receivership marginally away from the horizontal surface. The horizontal and the 45°-inclined surfaces would receive about the same amount of energy, though this would be less than the maximum amount that is available. Within the orthogonal system, the horizontal surface is preferable.

Finally, in figure 7-9—point 7 on the Skydome—the plate is oriented west and east during summer solstice on June 21, latitude 44°N. The times are 08:00 and 16:00 hours. The relatively low solar altitude, 36.2°, places the point of maximum energy receivership approximately halfway between the horizontal and vertical surfaces, leaning slightly toward the vertical surface. The 45° slope would be close to the point of maximum receivership. In the orthogonal system, the vertical surface would be marginally preferable over the horizontal surface.

The previous exercise clearly illustrates that the best orientation for maximum solar gain depends on two variables, the orientation and the tilt. The more the flat plate deviates from the ideal position, the less solar energy received. A few guidelines can summarize this lesson for a design practitioner.

1. A tracking device is the most efficient arrangement; a collector follows the sun's path and thus the sunbeam's interception angle is always 90° (fig. 7-10).
2. The collector is fixed due south and the angle of tilt is adjusted to optimize seasonal variations—the different sun altitudes between June 21 and December 21 (fig. 7-11).
3. The collector is fixed due south and the angle of tilt is fixed to receive optimal sunlight during the winter season. The conventional formula is latitude + 15; for example, in Hamilton it is 43 + 15 = 58° (fig. 7-12).
4. The collector is fixed due south and the tilt angle is fixed at 90° (in a vertical position) (fig. 7-13). (To optimize sun energy receivership in a fixed orthogonal framework, the tilt angle in northern latitudes is 90° and in southern latitudes, 0°.)
5. The collector is fixed in a vertical position, but it deviates from due south (fig. 7-14).

If, in the fifth case, the collector should represent the southern wall of a passive solar residence, then according to empirical testing, a deviation of + or − 20° from due south would have a negligible effect on the thermal milieu inside the house.

The relaxed attitude toward orientation, and sometimes even toward overshadowing, is usually justified because current construction techniques are often crude, many basic design concepts are archaic, and an immediate return on investment governs the planning and development of most human settlements. Indeed, a single-family residence taken from a current production line is a poorly functioning energy-converting machine.

Reflected radiation and urban form is a topic rarely investigated in literature, yet, for the northern urban climate, it is extremely important. The low winter sun casts long shadows and thus creates large impact areas. In a low-density setting, these areas of sun deprivation can be minimized by using natural topographical variations wisely.

In densely developed areas, although full exposure to direct radiation is preferred, it is sometimes difficult or impossible to achieve this relationship with the sun. A certain amount of overshadowing inevitably takes place, and some public spaces will occasionally, or even permanently, be deprived of sun. Figure 7-15 shows shadow patterns on typical street grids. Jeffersonian and Spanish grids are often used as examples of typical

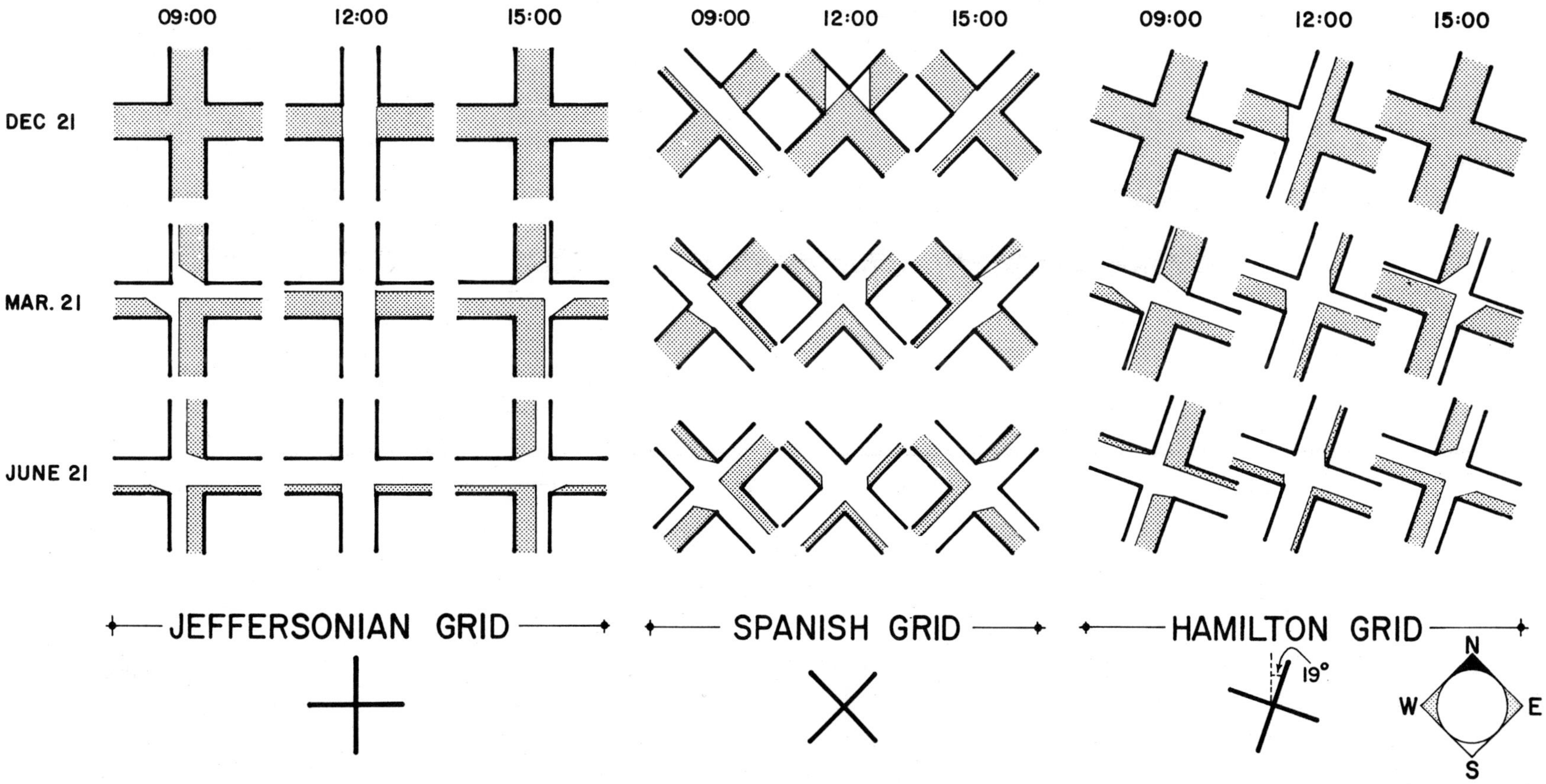

7-15. The relationship between shadows and insolation at street level in different street grid layouts.

street grids. However, not all cities have rectangular grids, and if they do, their alignment is unique, as is the case in Hamilton (19° off due south). The other variables are latitude and the profile of the street corridor. Any assessment must therefore begin with a description of the shadows on a street in a given area at a given time (fig. 7-16).

The low angles of the winter sun can be used to compensate for a lack of direct solar radiation; reflected sunlight brightens the environment around buildings and thus softens the effects of an oppressive mass of urban structures, both in terms of the amount of net radiation and the quality of the urban environment.

Direct and reflected radiation are affected by relationships among the following: building geometry, performance characteristics of horizontal surfaces and facades (the ratio of reflected to incident radiation), and sun angles. It should be stressed again that many recommendations found in energy conservation publications apply only to buildings located in relative isolation.

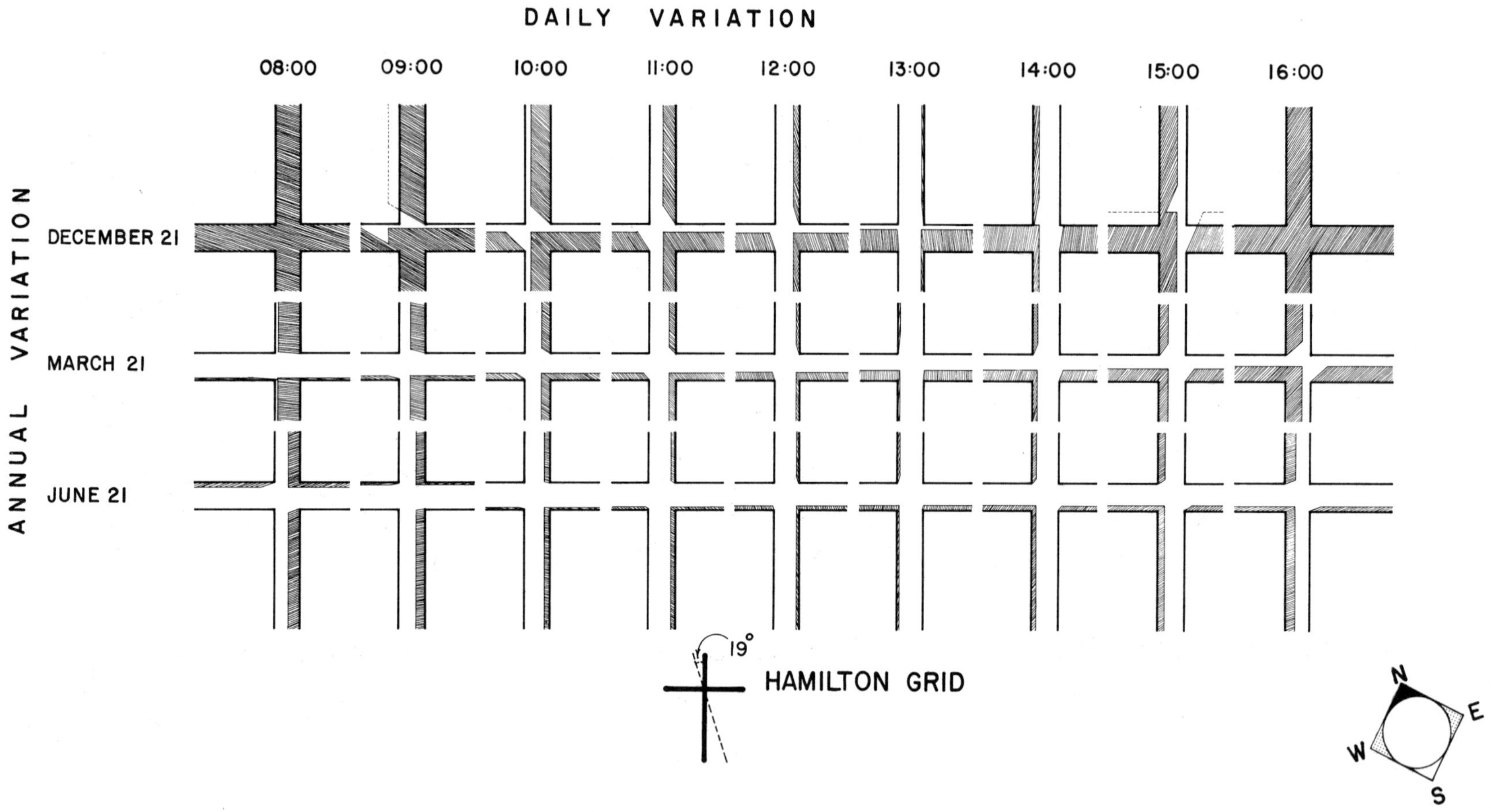

7-16. The relationship between shadows and insolation at street level in a typical Hamilton street grid layout.

Many recommended relationships between the shape, orientation, and irradiation of buildings may therefore be ineffective if the buildings are grouped in tight formation.

The dictum of collective responsibility throughout the design process is reinforced. The architect and urban designer must understand and together take responsibility for the relationship between a proposed building and any resulting loss of direct sun access due to overshadowing. A solution must satisfy private interests as well as those of the community.

Larger buildings, generally located within an urban heat island and with a low surface/volume ratio, are less susceptible to environmental variations. In such cases, strategies for improving or economizing on the internal thermal environment should be directed toward preventing heat loss and toward the efficient storage and utilization of metabolistic heat. Since large buildings in tight formation are typical of building forms in highly developed urban areas, these buildings must allow sunlight to reach urban public spaces.

Reflected radiation is an exciting design concept that architects and designers have explored with relish in the past. The baroque period, for example, used light to create dramatic designs; mirrored walls and reflecting pools were developed to perfection. Not only the geometries of a bouncing beam are of importance to designers but also the overall potential to create a desirable effect within the constraints of a given situation. In northern latitudes, the urban designer should recognize that the most important reflective surface, particularly in the winter season, is the vertical one (fig. 7-17).

Table 7-1 summarizes the relationship between sun altitudes and the position of a reflective surface for maximum reflection.

The guidelines in Table 7-1 can be modified to achieve the desired results by tilting or curving the horizontal and vertical surfaces. In addition, the designer must take into consideration sun azimuth angles. Thus, in plan, reflections would follow the pattern shown in figure 7-18.

TABLE 7-1
Sun Altitudes and Position of a
Reflective Surface for Maximum Reflection

	ALTITUDE OF SUN	
	HIGH	LOW
Reflective surface	horizontal	vertical
Geographic location	southern latitudes	northern latitudes
Season	certain northern latitudes during the summer period	certain southern latitudes during the winter period
Time of day	certain northern latitudes during the summer period at noon	certain southern latitudes during the winter period in the morning and evening

The daily sun path provides a designer with a multitude of opportunities to direct light into desired areas of an urban space. Even a cursory look at the diagrams herein reveals that maximum values of reflected radiation in nothern latitudes can be achieved in medium- and high-density urban forms.

A number of recent experimental studies, for example, by Anis Siddiqi, indicate that the reflected radiation component is an important parameter for siting buildings, establishing their geometries, and locating fenestration on their facades. A building's shape, its relationship to adjacent buildings, the size and shape of its impact areas, the orientation and albedo of exposed vertical surfaces, and the albedo of horizontal urban surfaces all contribute to the amount and quality of reflected radiation in an urban environment as perceived by the individual users of urban spaces.

Table 7-2 lists selected examples of various surfaces and the percentage of solar radiation that they reflect.

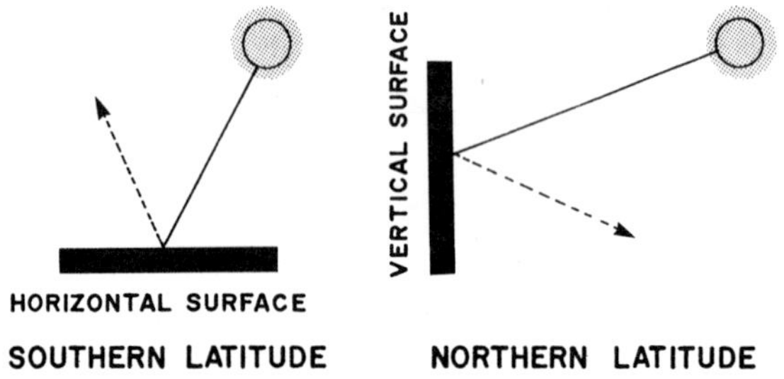

7-17. Reflecting surfaces.

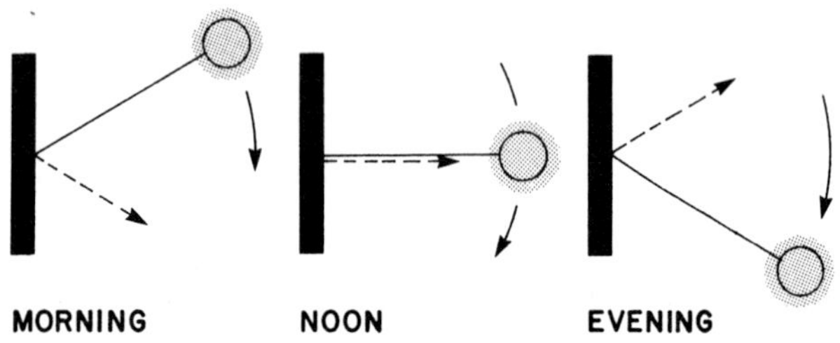

7-18. The daily variation of reflecting surfaces.

TABLE 7-2
Percentage of Reflection of Various Surfaces

NATURE OF SURFACE	PERCENTAGE OF REFLECTION
Ground	
dry	10–25
wet	8–9
Sand	
dry	18–30
wet	9–18
Rock	12–15
Glass	32
Green grass	3–15
Green leaves	25–32
Bricks	23–48
Asphalt	15
Snow	up to 90
Water pool	up to 90

Further research into the development of simple, quantitative evaluation techniques is urgently needed, so that a predesign appraisal of the most important variables and an assessment of the resulting direct and reflected radiation regime could be made. This in turn would be a most useful designer's tool for highlighting the efforts of specific plan arrangements and for the final evaluation of the thermal microclimate of a given urban area.

The geometries of a reflecting surface can be utilized and manipulated like that of a solar collector: tracking and curved mirrors can bounce sunlight to any chosen area. The various albedo values of exposed urban surface materials can enhance the distribution and quality of reflected sunlight. In a super-high-density environment (like Manhattan), sunlight can be delivered through building canyons to pedestrians by bouncing it between facades in a manner similar to that of light traveling within a glass fiber.

Reflected radiation holds enormous potential for enriching the urban environment, particularly during the cold season. The reflected sun can illuminate and warm areas and large tracts of the urban streetscape that would otherwise remain dark and cold for most of the year. Consciously designing with reflected sunlight opens the door to new ideas and imaginative solutions and helps to point out the problems encountered in planning practice that have thus far been barely acknowledged. For instance, mirrored facades might be introduced to brighten streets and public spaces that would otherwise remain dark for long periods. These facades might also be used to help thaw snow or enhance landscaping design. Curved surfaces can also amplify the light and heat of the sun and deliver them to a public pool or a playground. Direct and reflected sunlight can be used to create an entirely new approach to urban design.

STREETS AND PUBLIC SPACES

Figure 7-19 illustrates the nature of reflected sun geometries in an urban environment. An observer facing north on an east/west street sees a blast of sunlight bounced from the reflective surface of a south-facing facade. The latitude is 43°N; the time is 10:00 hours on March 21; the sun altitude is 39.3°, and the sun azimuth is 139.8°.

Reflected images continuously change with respect to the apparent movement of the sun across the sky. Although it may appear complex, the construction of the reflected sun path is not a difficult exercise. By introducing the "phantom world" concept (the illusion of a three-dimensional world as seen in a mirror), the drafing task is made as easy as shadow casting (see chapter 8).

Figure 7-20 illustrates, in simple terms, the temporal characteristics of the relationship between the sun and street. In this case, the observer is again standing on an east/west street facing north at 43°N latitude, during the summer, on June 21. At 07:00 hours, the observer

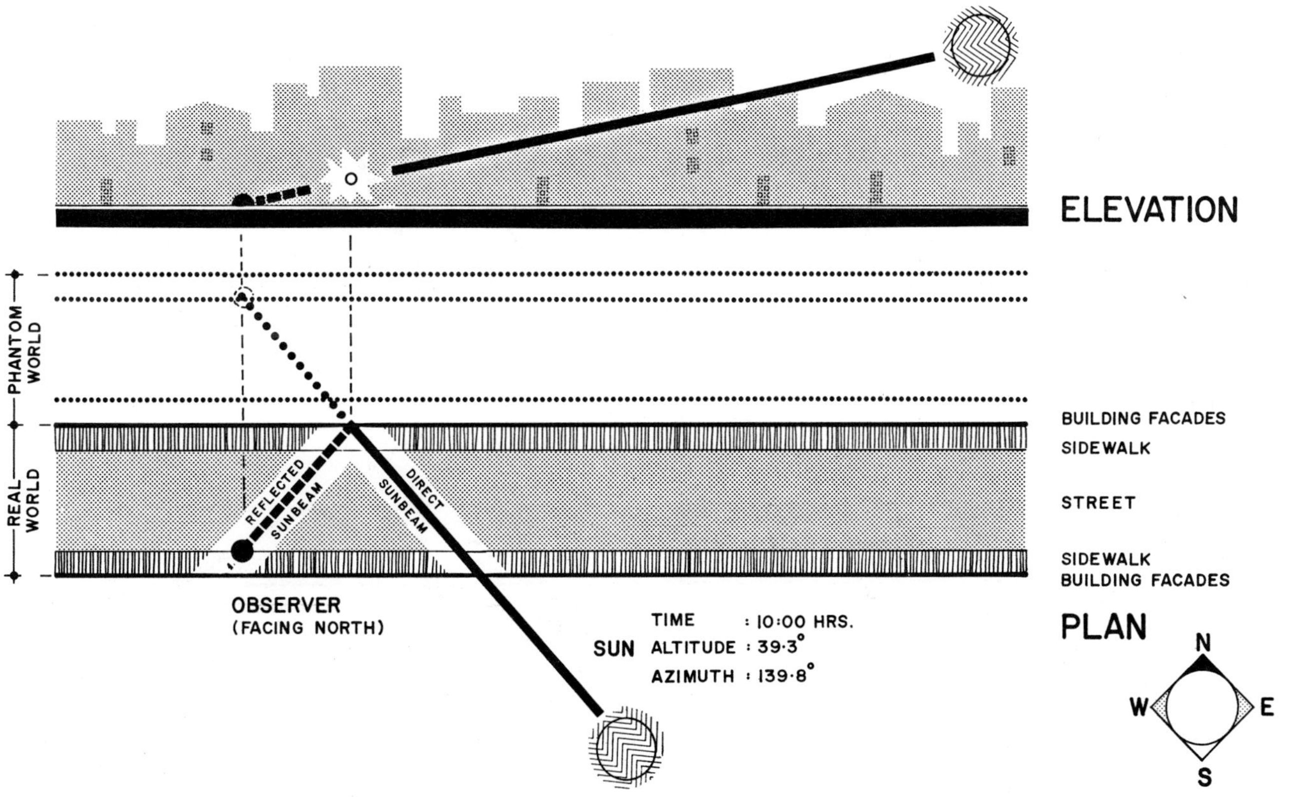

7-19. The reflected image of the sun in an urban street situation at a given time.

is exposed to the low early morning sun. One hour later, the sun's rays are parallel with the street axis, exactly east/west. The sun then moves behind the observer, illuminating objects in front of him or her and thus intercepts the mirror. After 08:00 hours, the observer is exposed to sun reflected from the mirror. If the geometries of the surrounding buildings are favorable, the observer may be exposed to the sun's direct *and* reflected rays (a great asset in cold, northern urban spaces). In this case, his or her body would cast two shadows (fig. 7-21). The intensity (darkness) of the second shadow indicates the reflective quality of the mirror. Two, three, or more shadows may be constructed in an urban environment (indicating that several reflective surfaces are in place) by manipulation of the geometry of surrounding surfaces (fig. 7-22).

Reflected radiation is generated by both horizontal and vertical surfaces (fig. 7-23). At a given time, a point on a horizontal surface and one on a vertical surface simultaneously receive and reflect direct solar radia-

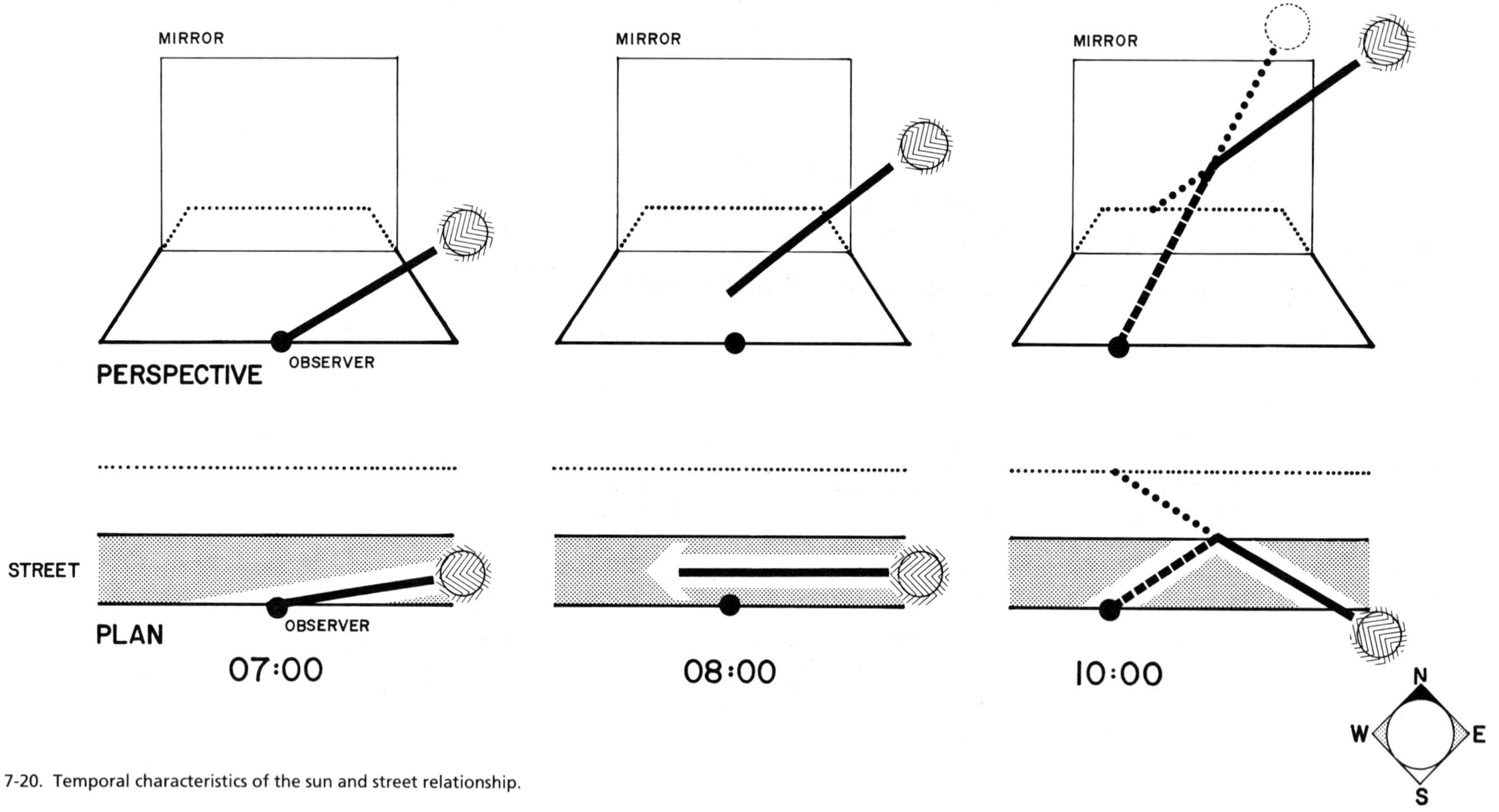

7-20. Temporal characteristics of the sun and street relationship.

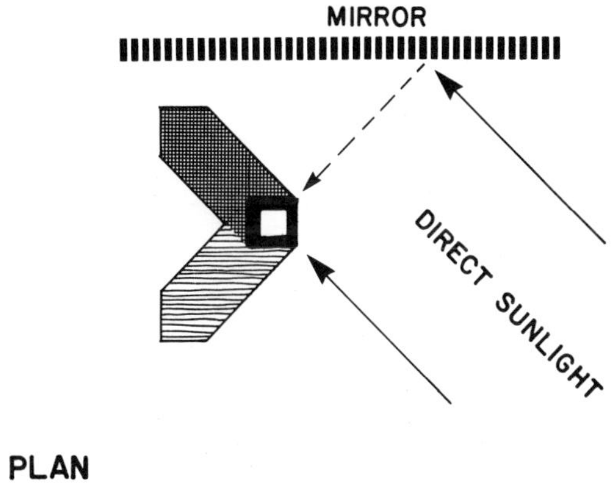

7-21. Shadows generated by direct and reflected sunbeams.

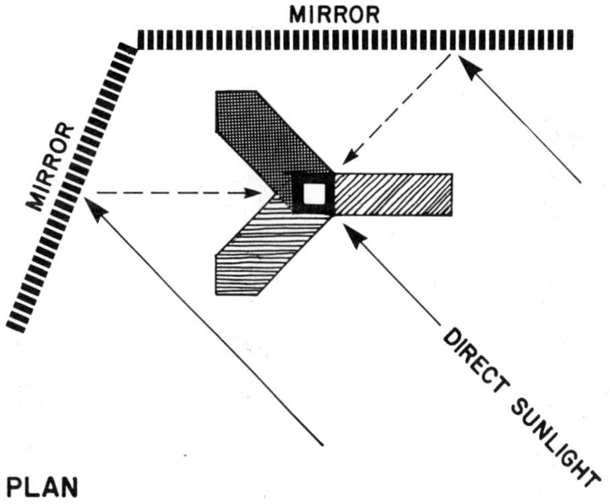

7-22. Shadows generated by direct and several reflected sunbeams.

tion. Under special circumstances, these two streams of reflected radiation may be equal, but usually they are not. Color, surface, texture, reflectivity, and the angle of incidence can make one stream of reflected energy stronger than the other. The designer must be aware of these variables in order to deliver sufficient light and warmth to the street, especially in areas that can be reached only by reflected sunlight (fig. 7-24).

For an urban designer committed to an active role in shaping the urban form with the use of direct and reflected sunlight, it is useful to know that reflecting and impact points can exchange functions if the direct sun can reach both points; a point on a horizontal surface (a white stone pavement, or a reflecting pool of water or ice), called the *impact point,* can become a reflecting point (see fig. 7-23). Any object placed between these two points would receive light from three sources, one direct and two reflected (fig. 7-25).

With the application of these concepts, the urban form, energized by the sun, can emerge as a complex

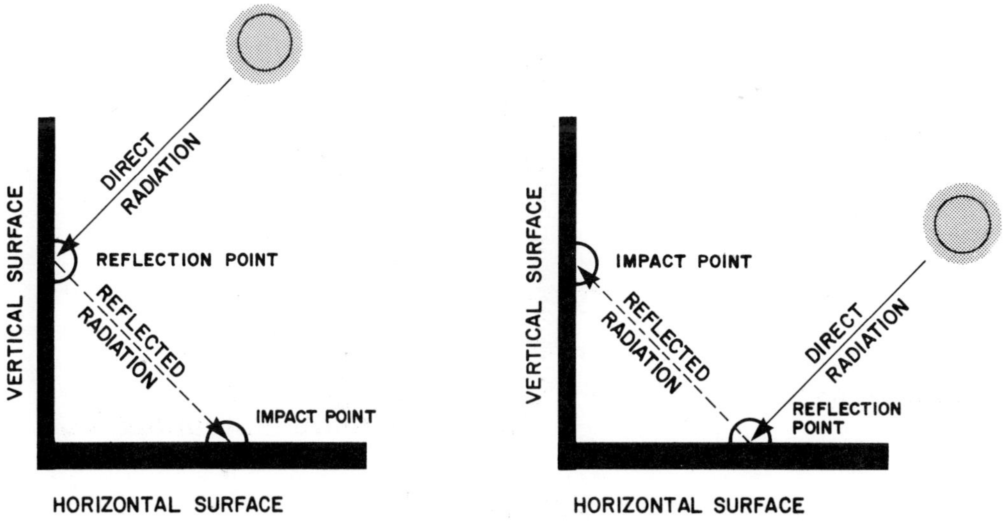

7-23. Reflected radiation from horizontal and vertical surfaces.

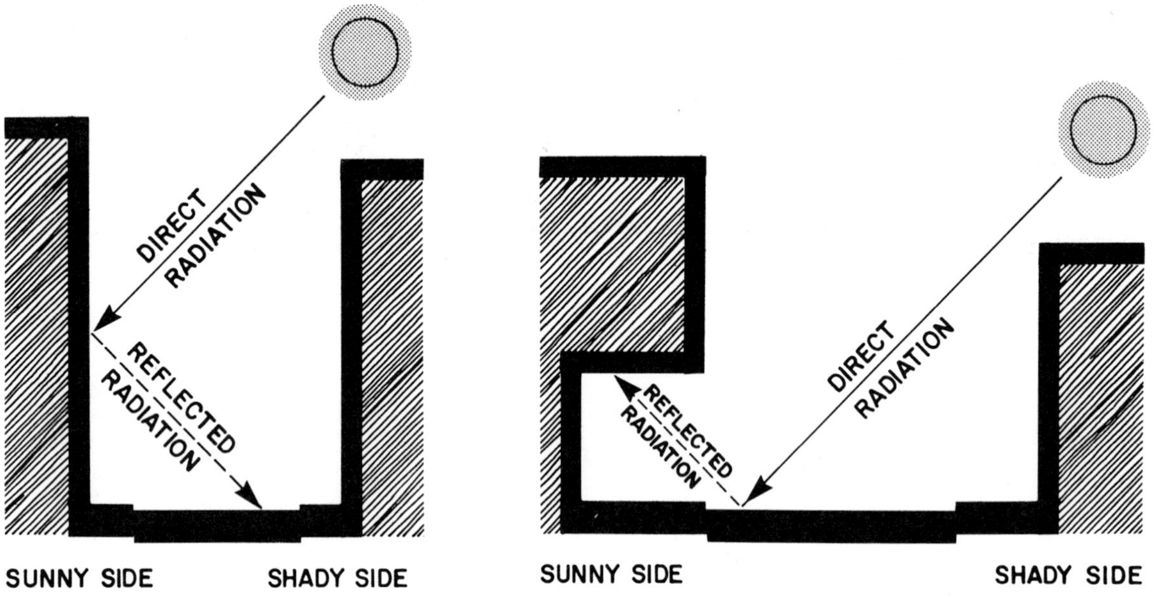

7-24. Certain urban surfaces can be illuminated only by reflected radiation.

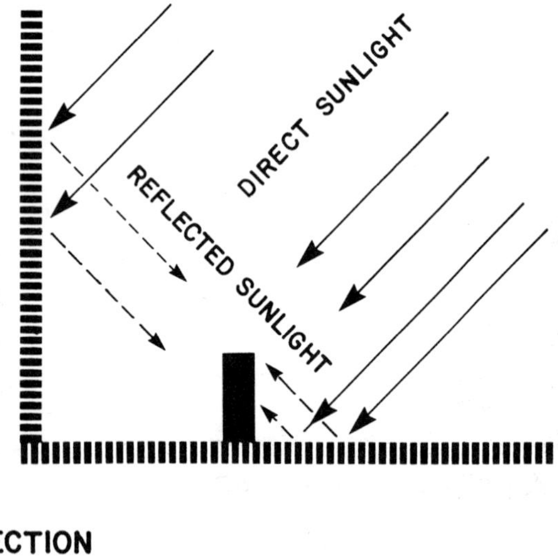

7-25. An object illuminated by direct and reflected sunlight.

optical instrument and as an engineering and artistic statement capable of making outdoor spaces comfortable year-round. This space would radiate warmth as if it possessed an internal source of energy whenever the sun is visible.

To determine which technique to use, how to use it, and to what extent to use it, an urban designer must carefully scrutinize the mesoclimatic data. Furthermore, the designer must be fluent in direct and reflected sun geometries. Figure 7-26 shows examples of dispersed reflected light (DRL) on an east/west street during the winter solstice, December 21. A narrow reflecting strip located on a facade bounces sunbeams in different directions at different times. An observer has to move in order to see, or be exposed to, the reflected image of the sun during the different hours of the day.

Figure 7-27 shows focused reflected light (FRL) on an east/west street. In this case, a stationary observer would be exposed to reflected sunlight from sunrise to sunset if a sufficiently large area of reflecting facade is provided.

An annual pattern of reflected sun images (reflecting points) is shown in figure 7-28. It outlines an area within which the reflected sun would appear on a south-facing facade between 09:00 and 15:00 hours daily throughout the year. The shown pattern is in fact a projection of the Skywindow on a vertical surface.

In northern latitudes, the two sides of an east/west street possess different characteristics and different light and thermal environments. A sufficiently wide east/west street has one side exposed to the sun throughout the year. The other side, with the exception of the early morning and late-afternoon hours in the summer, is practically deprived of direct sun radiation (fig. 7-29). Each latitude requires a different ratio between the width of the street and the height of the buildings to ensure that the sunny side is indeed sunny.

In developed areas of the city, this ratio is usually lost, and the sunny area is often overshadowed. The use

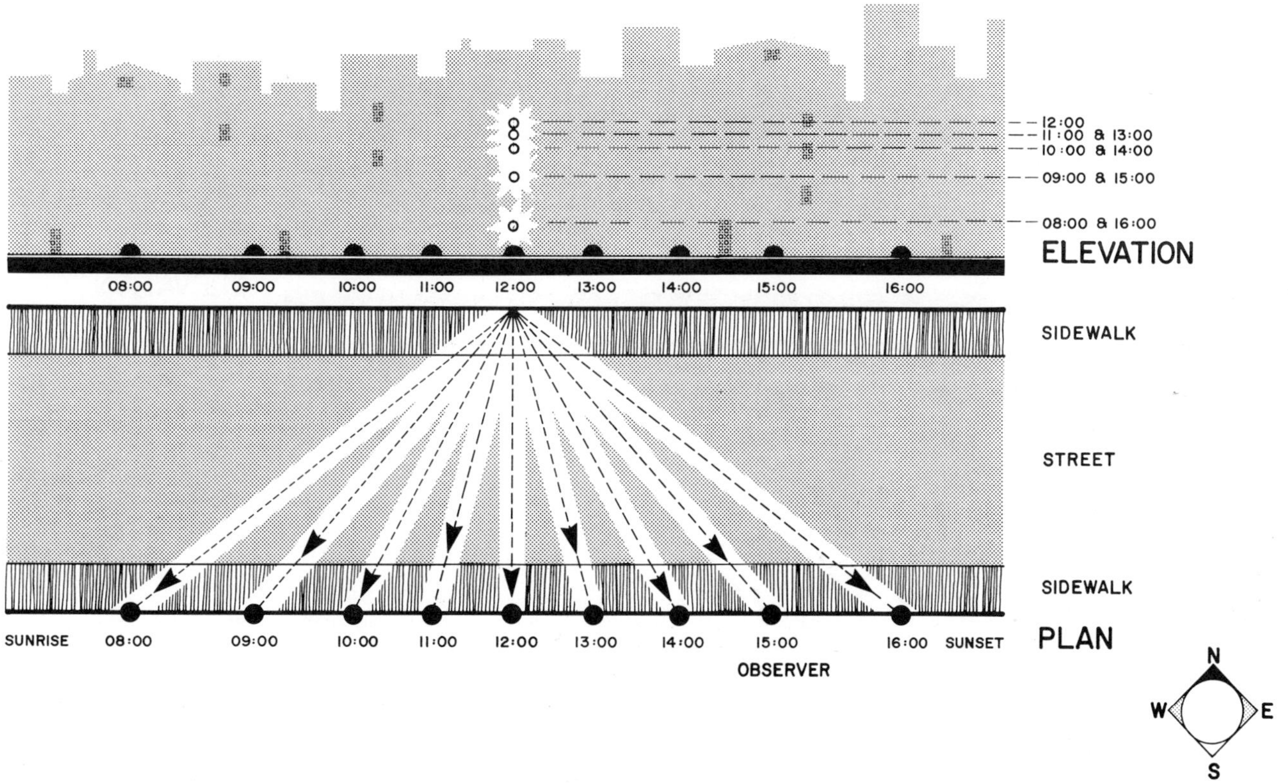

7-26. An example of dispersed reflected light (DRL) on an east/west street during winter solstice. An observer must move to see the reflected image of the sun.

of highly reflective facades on the sunny side of the street reverse the warm and cold environments; the sunny side becomes overshadowed and the dark side becomes illuminated (fig. 7-30). The overall pattern of direct and reflected sunbeams on an east/west street is shown in figure 7-31.

A north/south street has a different set of opportunities and constraints. The side exposed to the sun before noon is called the *morning side*. The side opposite is the afternoon side. At noon, 12:00 hours solar time, the sun runs parallel with the street axis and begins to cast light on the afternoon side. Thus, the north/south street is symmetrically divided (fig. 7-32). It should be noted, however, that because of the general daily temperature profile, the afternoon side is usually much warmer. As in the previous case, higher buildings with reflective facades may reverse the light and dark cycle (figs. 7-33, 7-34).

An example of DRL on a north/south street during the morning hours of the winter solstice is shown in

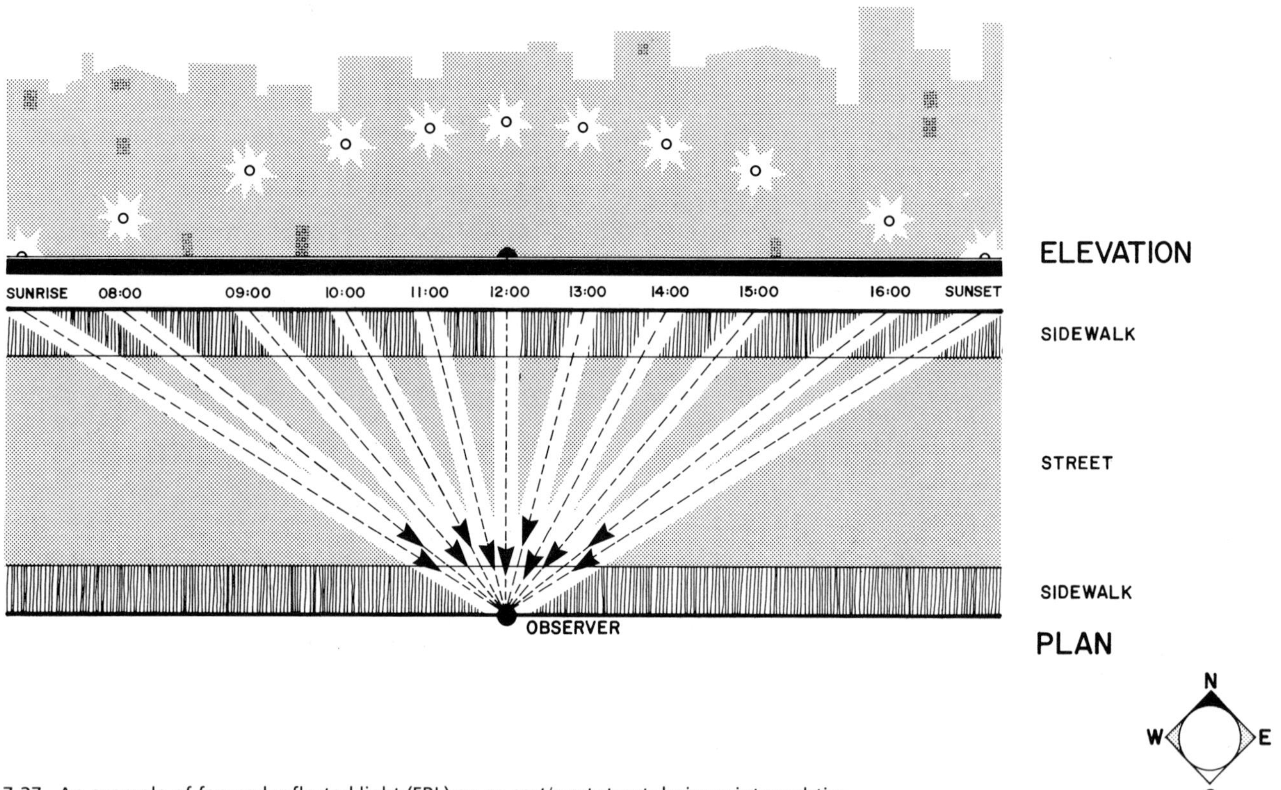

7-27. An example of focused reflected light (FRL) on an east/west street during winter solstice.

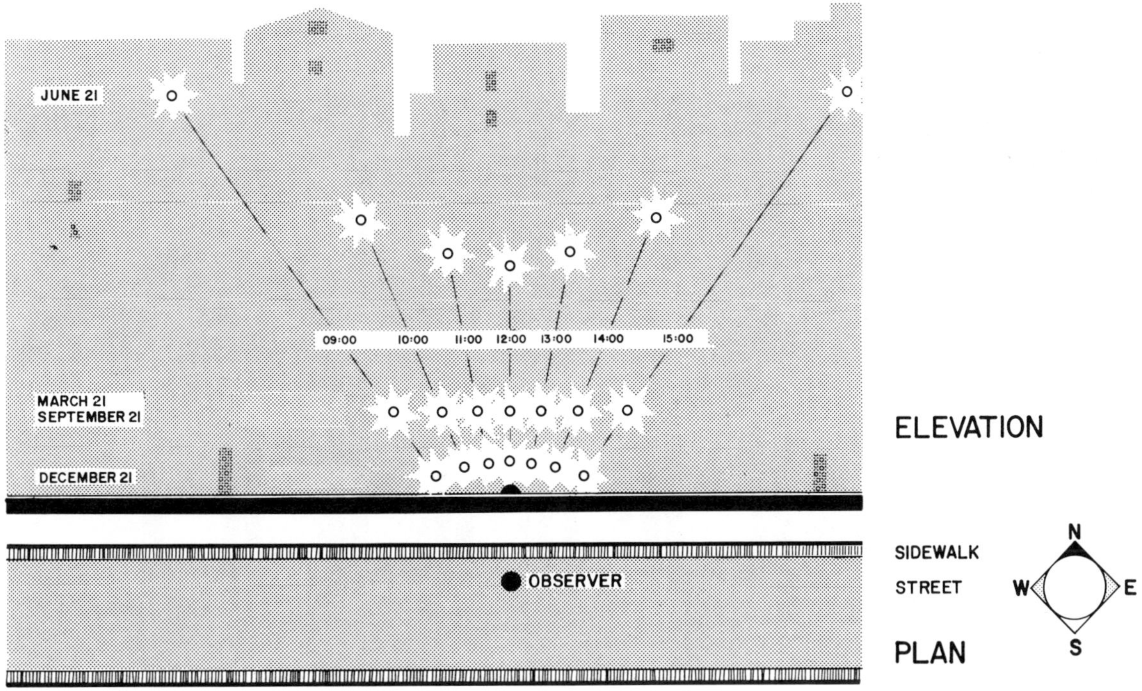

7-28. An annual pattern of reflected sun images.

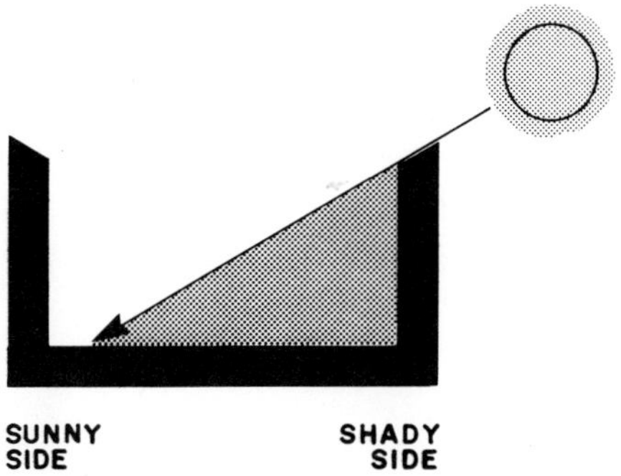

7-29. Two sides of an east/west street (an example of a low-rise development).

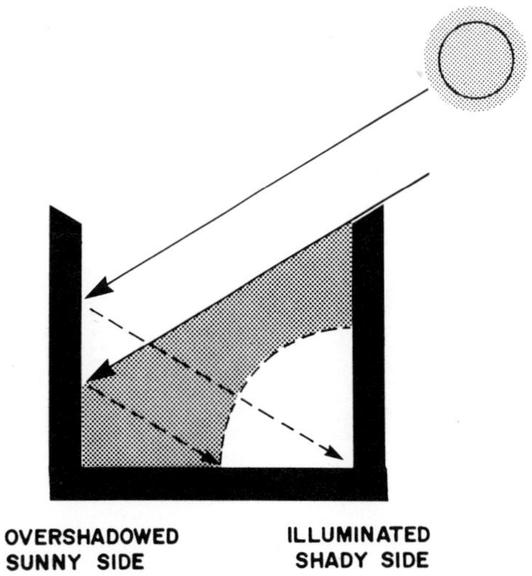

7-30. Two sides of an east/west street (an example of a high-rise development).

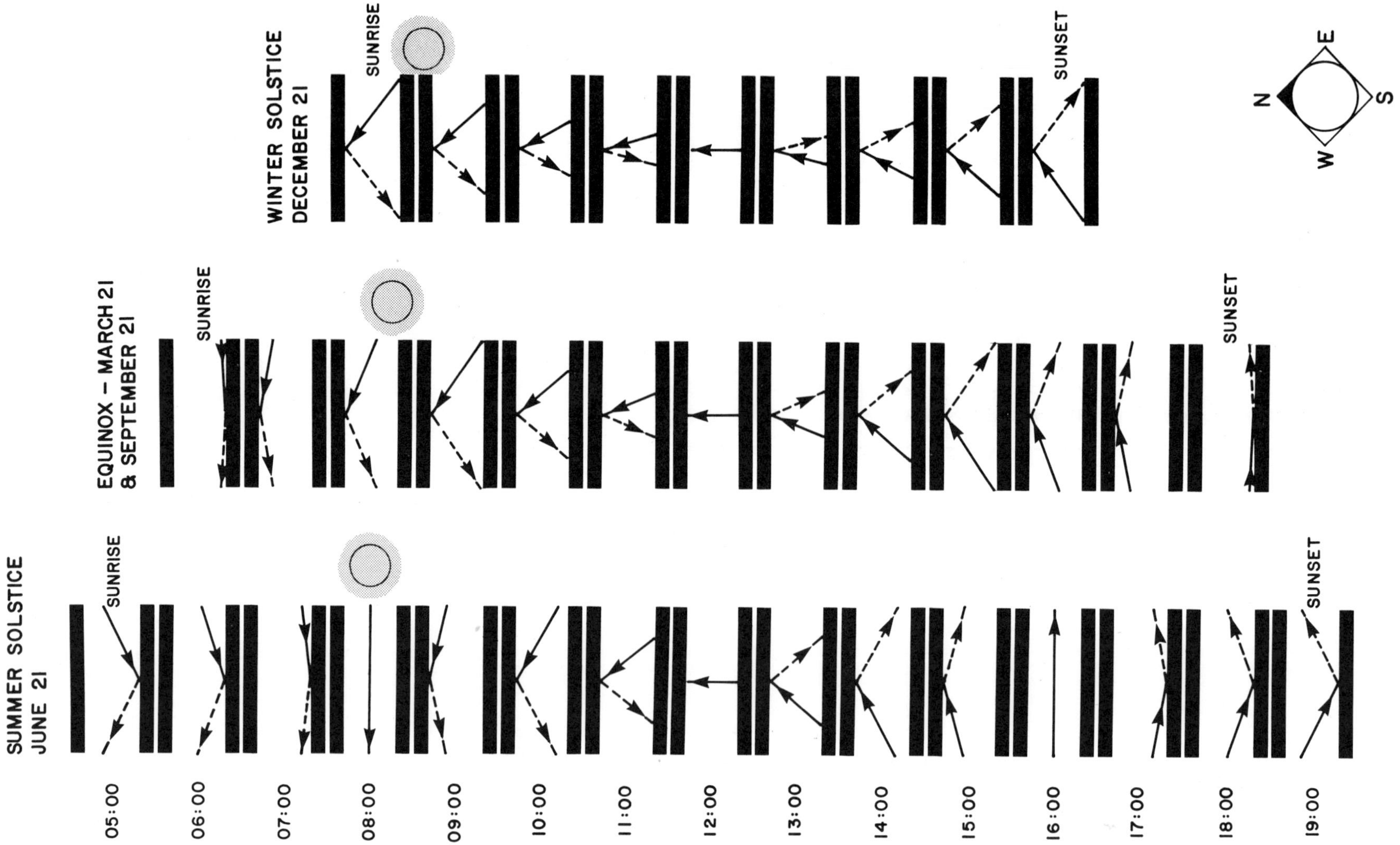

7-31. Behavioral characteristics of direct and reflected sunbeams on an east/west street during summer solstice, equinox, and winter solstice.

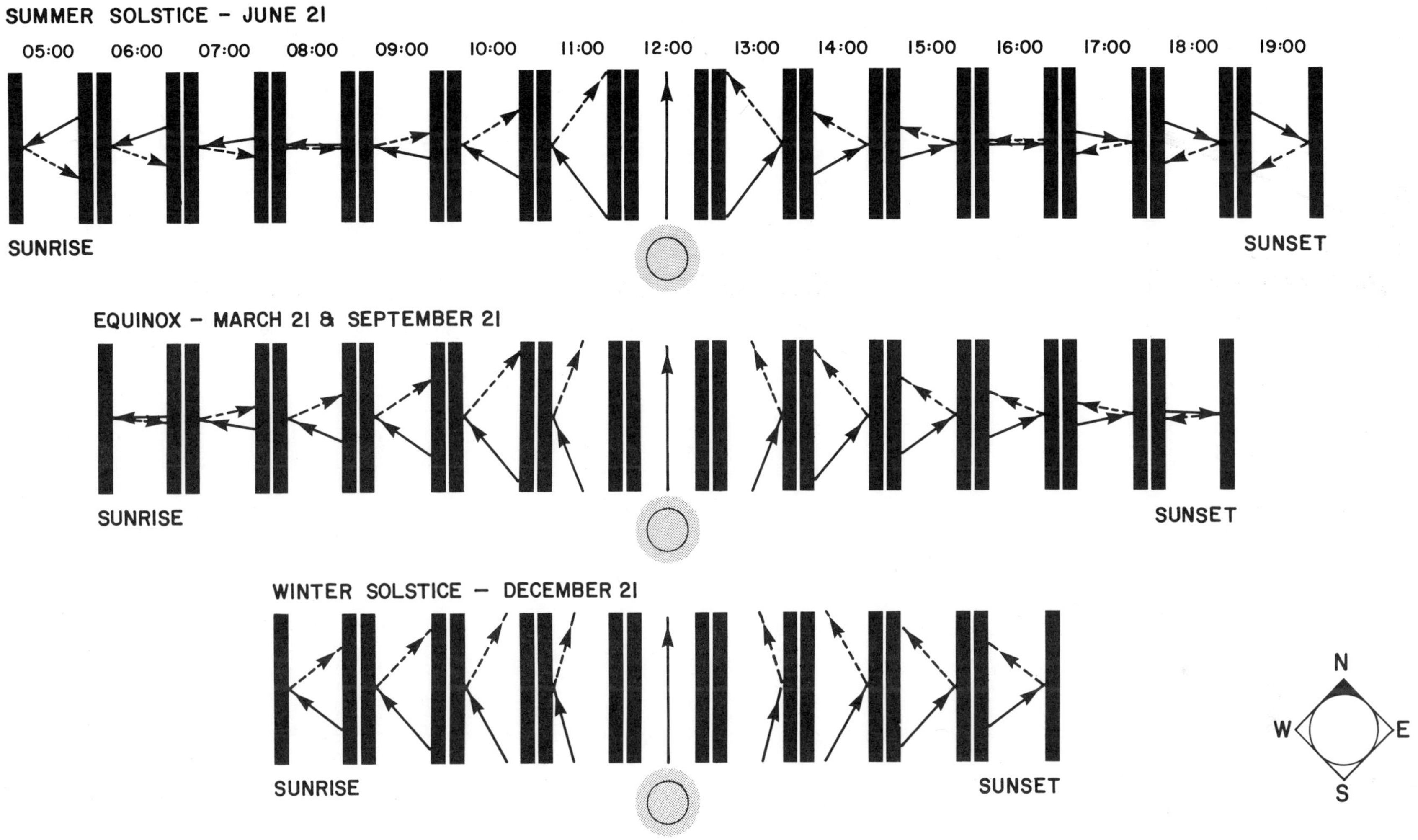

7-32. The behavioral characteristics of direct and reflected sunbeams on a north/south street during summer solstice, equinox, and winter solstice.

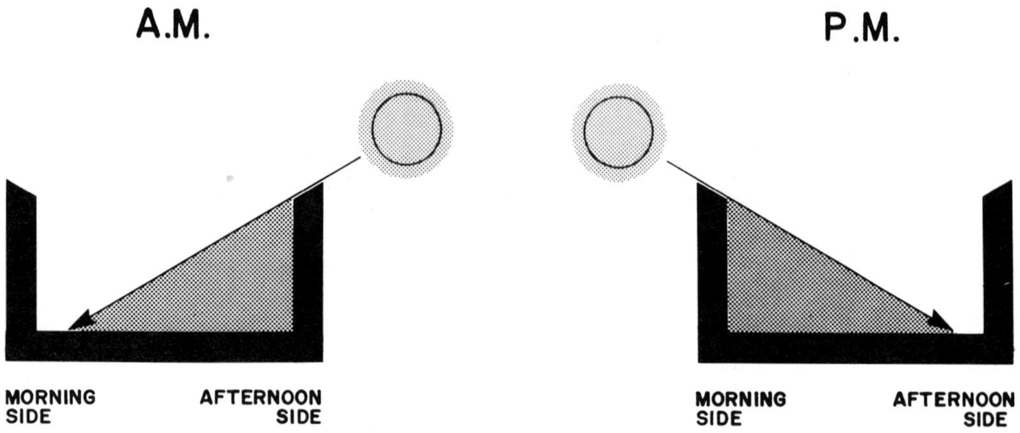

7-33. A low-rise development on two sides of a north/south street.

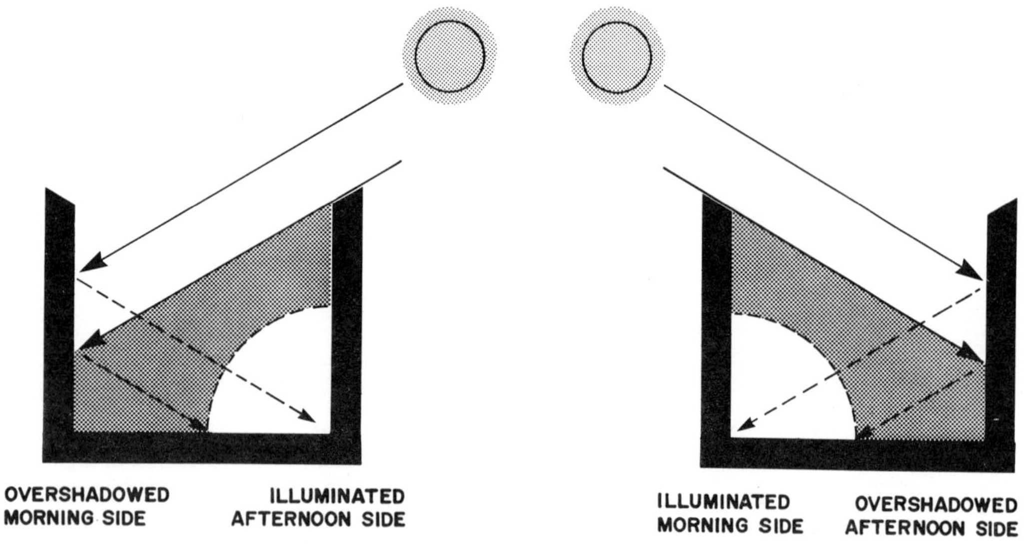

7-34. A high-rise development on two sides of a north/south street.

figure 7-35. A narrow, reflecting strip on the western side of the street (the morning side) bounces the sunbeams in different directions between sunrise and 12:00 hours.

Figure 7-36 shows FRL on a north/south street. A stationary observer can be exposed to reflected sunlight from sunrise to 12:00 hours if a sufficiently large area of reflective facade is provided. The same principles apply to the other side of the street. Figure 7-37 shows the DRL pattern in the afternoon hours, and figure 7-38 shows the FRL pattern. Unlike figure 7-28, the projection of the annual pattern (the Skywindow) on a north/south street must be divided into two symmetrical segments before and after noon, 12:00 hours solar time (figs. 7-39, 7-40).

Truly focused sunlight can be achieved with the use of curved mirrored surfaces. Within the focus of such an instrument, blindingly intense light can be created and the resultant temperatures could be high enough to melt iron ore. Although no urban designer would want to achieve extremely high temperatures whenever the sun came out, using the calculated curvatures of mirrored facades, a springlike and perhaps even a summerlike microclimate may be produced in a designated area for all the winter sun hours recorded in the Ecochart.

A model shown in figures 7-41 and 7-42 illustrates opportunities for climate modification with solar energy. The target area—a solar pit—is illuminated with direct and reflected sunlight between 09:00 and 15:00 hours throughout the year. In reality, this enhancement with reflected light would be desirable only during the cold period of the year; this varies from region to region, but in most northern localities, it would not exceed six months. Thus, the size of the reflective facade can be cut by half.

Figure 7-43 explains the mechanics of direct and reflected insolation at different hours of the day and during different seasons of the year. It also indicates the extent of permissible obstruction of the sun. This

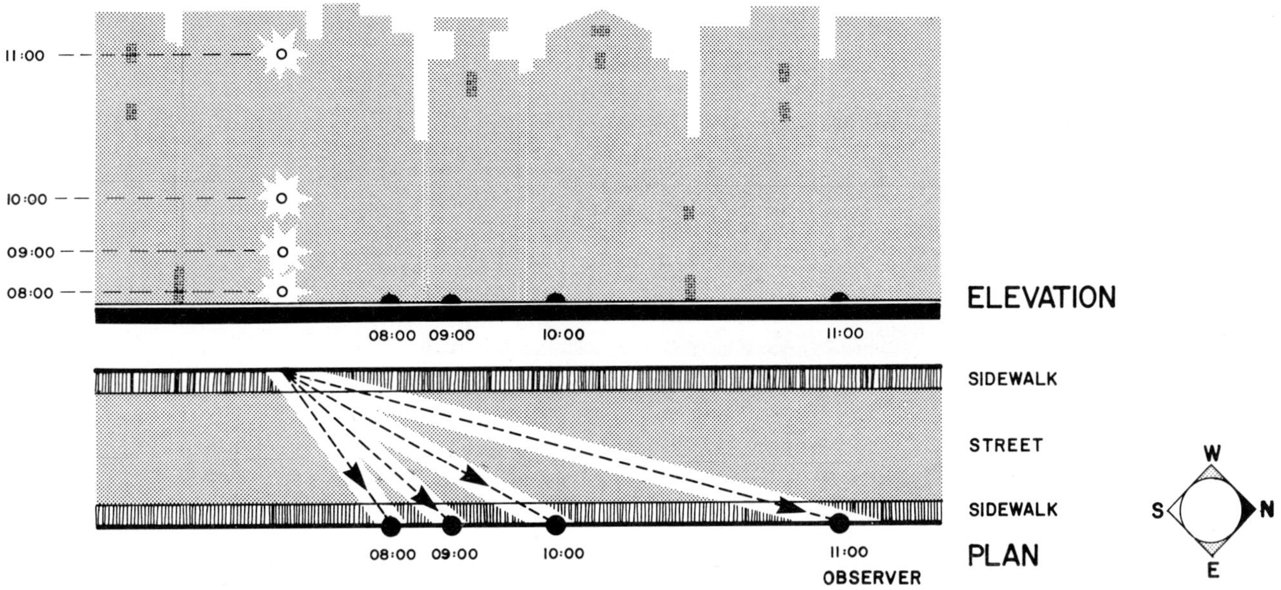

7-35. An example of DRL on a north/south street during winter solstice before noon (an observer must move to see the reflected image of the sun).

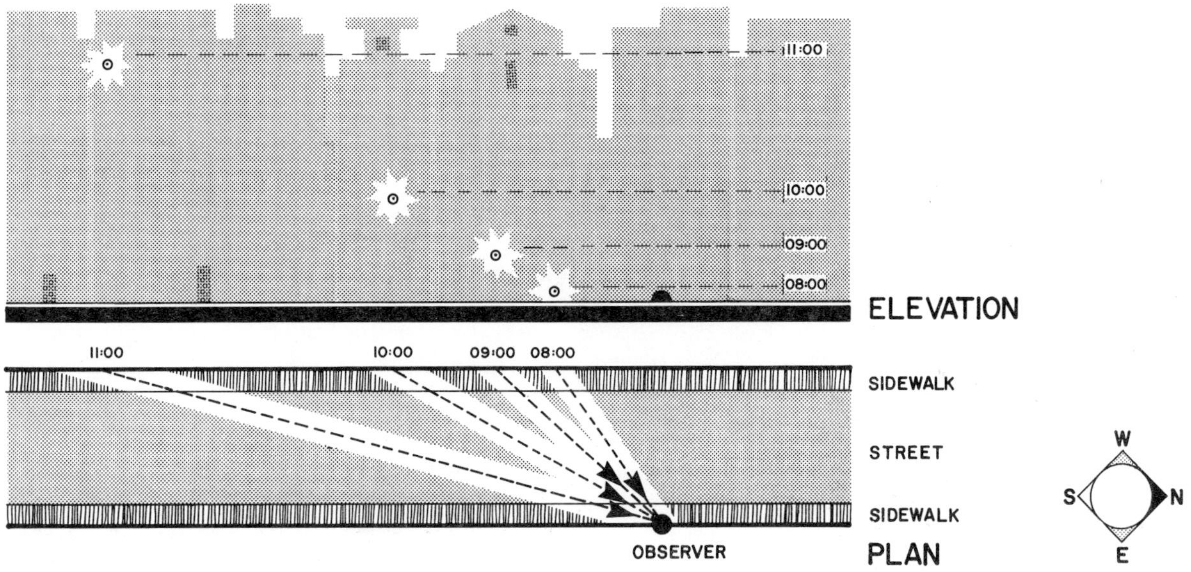

7-36. An example of FRL on a north/south street during winter solstice. An observer can see reflected sunlight from sunrise to 12:00 hours if a sufficiently large area of reflective facade is provided on the western side of the street.

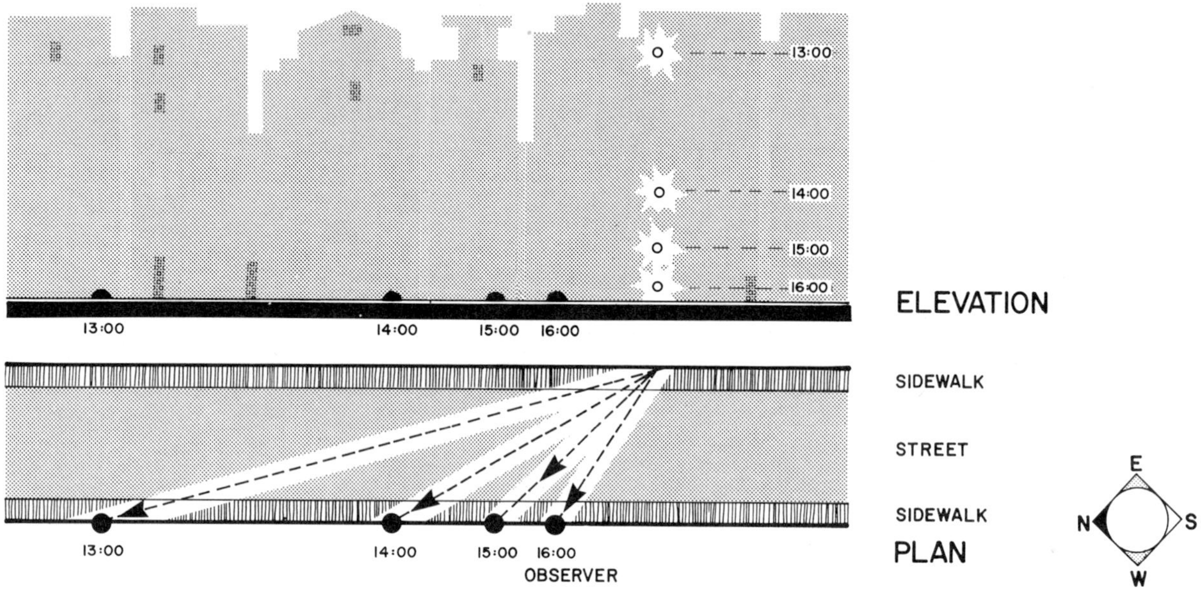

7-37. An example of DRL on a north/south street during winter solstice. A narrow, reflecting strip on the eastern facade bounces the sun's beams in different directions between 12:00 hours and sunset. An observer must move to see the reflected image of the sun.

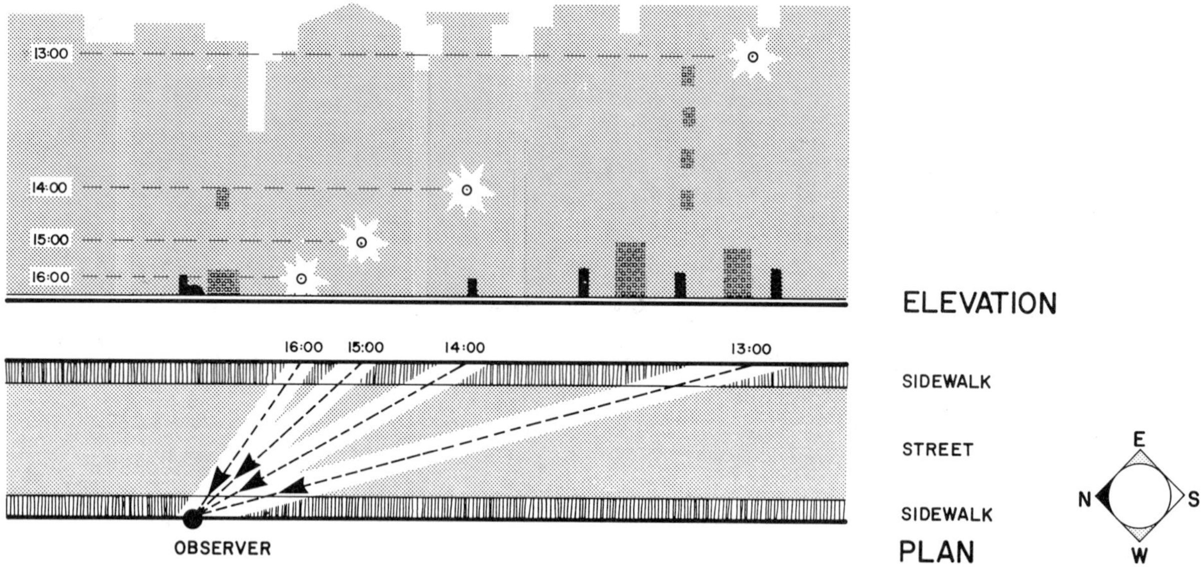

7-38. An example of FRL on a north/south street during winter solstice. An observer can see reflected sunlight from 12:00 hours to sunset if a sufficiently large area of reflective facade is provided on the eastern side of the street.

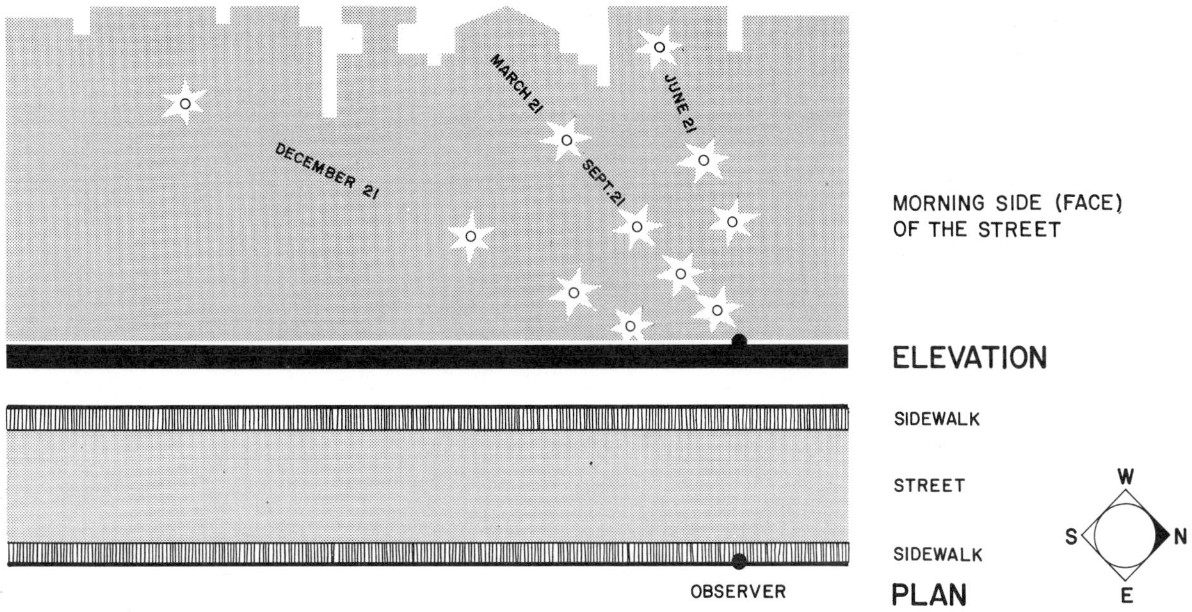

7-39. Annual and diurnal variations of FRL on a north/south street before noon (A.M.).

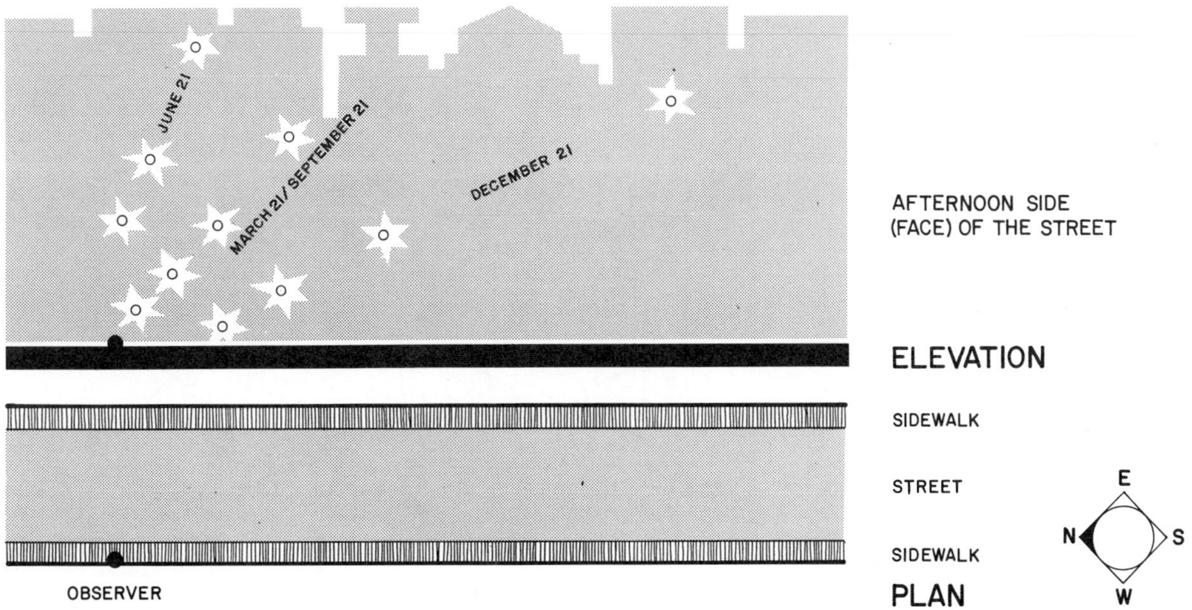

7-40. Annual and diurnal variations of FRL on a north/south street after noon (P.M.).

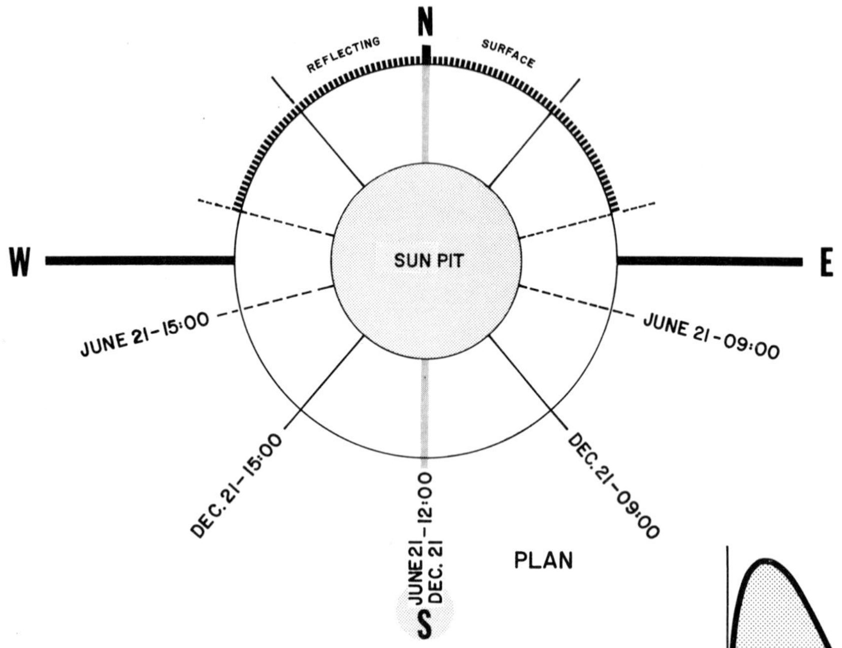

7-41. This model demonstrates the spatial geometry that permits a target area—the sun pit—to be exposed to direct and reflected sun radiation during 09:00 to 15:00 hours throughout the year. From this spatial arrangement, guidelines for maximum utilization of sun radiation in an architectural and urban environment can be derived.

7-42. These perspectives illustrate the sun pit model. Such a spatial arrangement can inspire a large number of architectural and urban design solutions, such as public or private swimming pools, playgrounds, neighborhood parks, and public spaces.

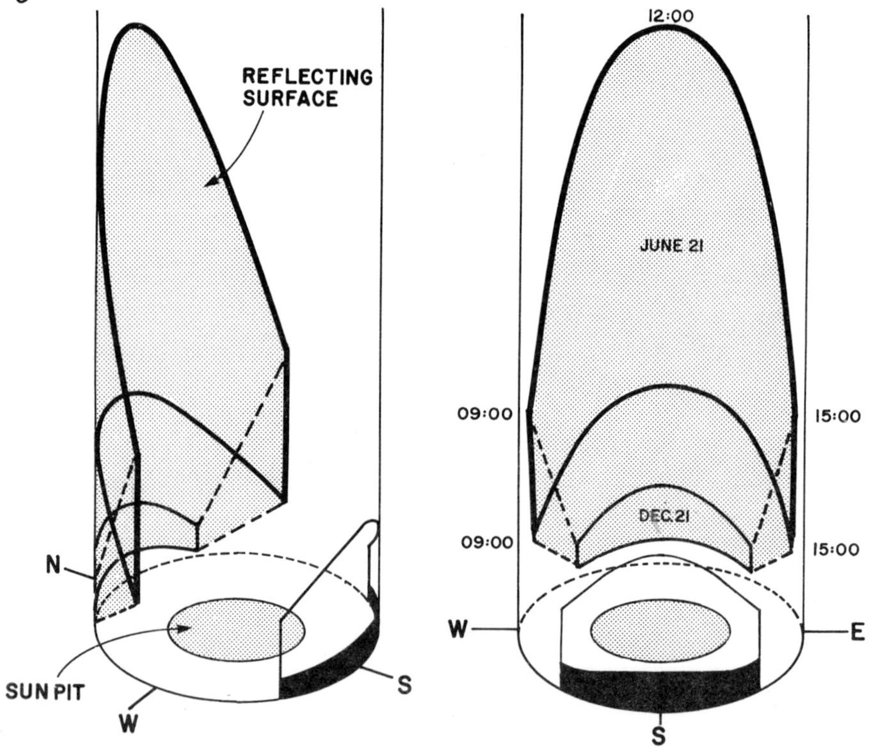

exercise can be directly utilized in determining the geometries of street corridors. Figure 7-44 shows how to derive height restrictions on east/west streets. Figure 7-45 shows how to ensure that a facade on the dark side of an east/west street receives illumination via reflected sunlight. Figure 7-46 illustrates the derivation of optimal geometries on north/south streets.

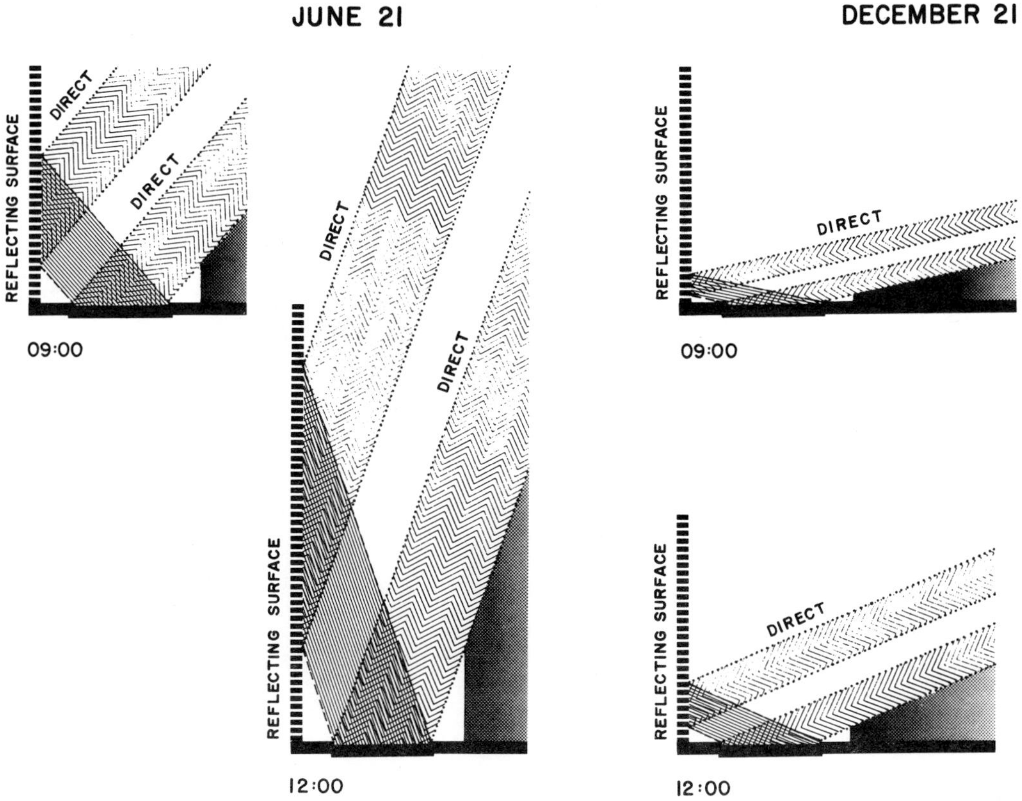

7-43. Sections of the sun pit model for summer and winter solstice indicate the vertical dimensions of the required reflecting surface and the permissible extent of obstruction.

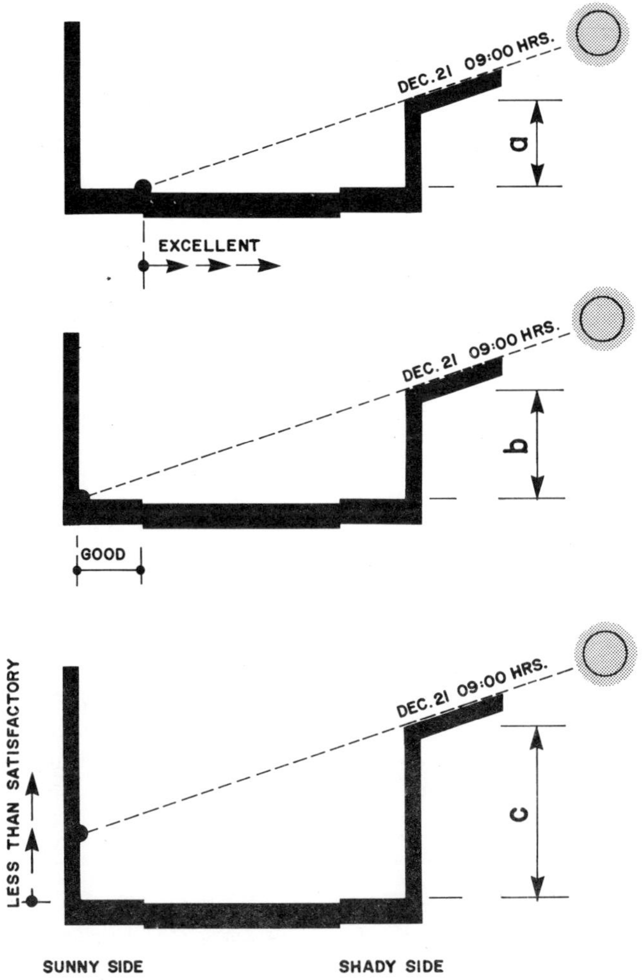

7-44. Height restrictions on an east/west street. A building height that is equal to or less than "a" is rated excellent. A building higher than "b" would be considered less than satisfactory.

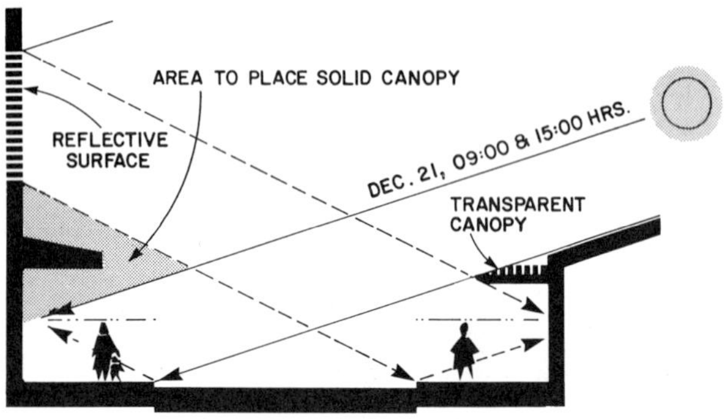

7-45. The geometry of an east/west street to ensure direct and reflected illumination of both sides of the street during winter solstice, 09:00 and 15:00 hours (note that the canopy on the shady side is transparent to allow reflected light from the sky to penetrate).

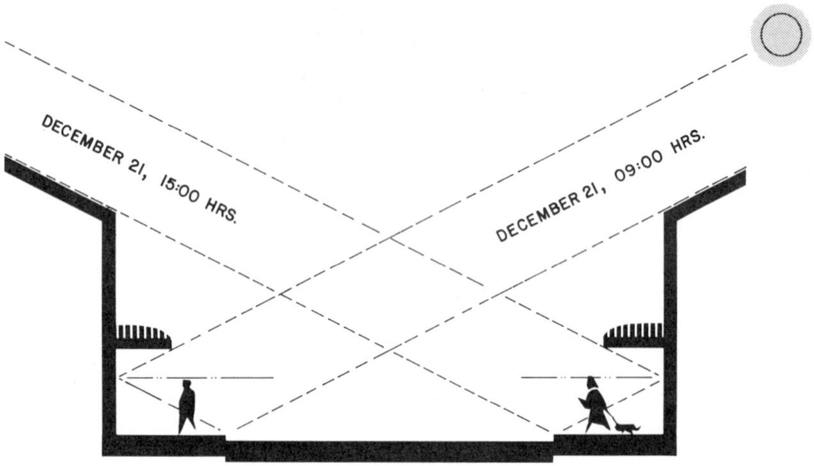

7-46. The geometry of a north/south street to ensure direct illumination of both sides of the street during winter solstice at 09:00 and 15:00 hours (if canopies are constructed, they should be transparent to allow penetration of light reflected from the sky).

AIR MOVEMENT

Until recently, the problems associated with wind were primarily related to the structural integrity of buildings and the development of design details for building envelopes that relate to air tightness and the prevention of heat loss caused by wind. But as the city began to pierce clouds, inhospitable, even dangerous, conditions were created at street level. Winds, for example, were reported to blow people through the city canyons like autumn leaves. The wind in Toronto, Buffalo, and Seattle plazas has been so overpowering at times that safety lines had to be installed to guide pedestrian movement. Paradoxically, this negative impact resulted from architectural and engineering achievements, the availability of new materials and construction techniques, and a prosperous economy. The danger is amplified during the winter, since cold air is more dense (for example, a unit volume of cold air contains more molecules than warm air). The greater the density, the greater the turbulent energy.

Consequently, the impact of wind on public urban spaces has become a major design consideration, particularly in northern latitudes, which means that the designer or design team should be able to predict the magnitude of wind speeds at street level. Such a task involves an understanding of building aerodynamics as well as an intimate knowledge of regional wind climatology (information available from the Ecochart). Although urban and architectural designers should have a good working knowledge of aerodynamic principles, an experienced aerodynamics professional must be involved in the early design stages of every large project.

At present, three methods can be used, with various degrees of efficacy, to incorporate aerodynamics into the design process and assess the impact of a proposed development.

1. MATHEMATICAL MODELS. A few mathematical methods allow computer-graphic modeling of air-

flow patterns around a variety of objects. These tools are still in the experimental stages and are used primarily in the aerospace and automotive industries.

2. PROFESSIONAL EXPERIENCE AND INTUITION. An experienced aerodynamics professional should be able to assess or predict the basic pattern of air movement among buildings, but only probable results can be achieved.
3. WIND SIMULATION. Given the limitations of the first two avenues, wind tunnel studies and water tank simulation techniques are currently the only means that allow a quantitative prediction of wind speed and movement among buildings.

Thus, a brief review of the aerodynamic considerations associated with the urban environment should assist a designer in understanding certain basic principles. This understanding will help the designer to develop a sensitivity for complex wind patterns and to promote more efficient collaboration and communication with the aerodynamics consultant.

City Aerodynamics

Actual experience and wind tunnel experiments indicate that most undesirable effects are generated by structures taller than their surroundings: either a tall building or group of buildings in an open plaza (fig. 7-47); one tall building among low-rise surroundings (fig. 7-48); or a medium-size building in an open space (fig. 7-49).

Friction over natural or human-made objects causes wind speeds to differ according to altitude. The local weather data usually indicates the characteristics (speed, direction, and so on) of the gradient wind (wind at a certain altitude, called *gradient height*, that is no longer affected by ground friction). In the open country, the gradient height is about 274 m (900 ft), in suburbia it is 366 m (1200 ft), and in metropolitan areas it is 457 m (1500 ft) (fig. 7-50).

When wind reaches a tall building, the lower pressure near the ground results in a downstream, which bounces up at the sidewalk (fig. 7-51). Thus, the generated eddies and turbulences often produce a situation akin to Marylin Monroe's famous walk over the New York subway system's ventilation grids in the film, *The Seven Year Itch*. Coined the "Monroe phenomenon" by New York journalists, this situation is a mild example of an adverse spatial configuration; configurations exist that could accelerate the ground air movement to double that of the wind speed.

City aerodynamics can sometimes contradict common sense. For example, conditions on the windward side of a tall building can be improved by shielding the ground area with a small structure; however, depending on the geometries, a small building upwind may worsen the situation considerably. The stagnation point (point at which airflow meets an obstacle and is fragmented into many directions) is pushed toward the top of the tall building and the more powerful downstream, in combination with the negative pressure behind the small structure, generates a continuous, powerful vortex and turbulences around the corners of the building (fig. 7-52).

These critical relationships between two buildings were first studied and described by A. F. E. Wise; hence, they are called the *Wise effect*. Other generic aerodynamic occurrences include the Venturi effect, the row effect, the effect of cells, the effects of slots under buildings, the effects of corners, the channeling effect, and the pyramid effect.

Urban aerodynamics are complex. A single building generates a giant vortex, and a group of buildings generates additional problems with wind. A rush of air from high-pressure to low-pressure areas results in potentially dangerous conditions around buildings. Tunnel testing indicates that the air speed in front and around the sides of tall buildings may easily be twice as high as the ambient wind. Certain spatial configurations accelerate the wind speed up to three times. In open plazas adjacent to tall buildings, up to four times the

7-47. A tall building and a group of tall buildings in an open plaza.

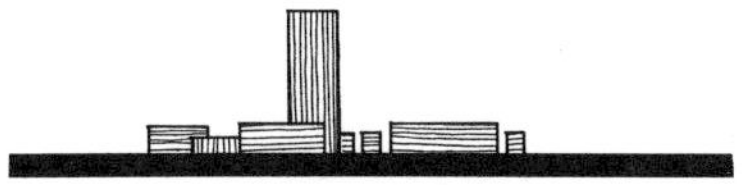

7-48. A tall building in low-rise surroundings.

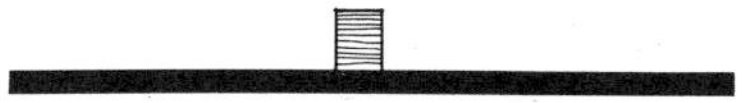

7-49. A medium-sized building in an open space.

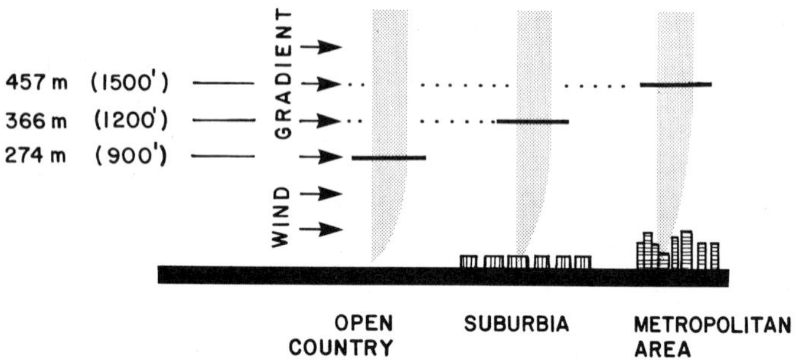

7-50. Wind gradient.

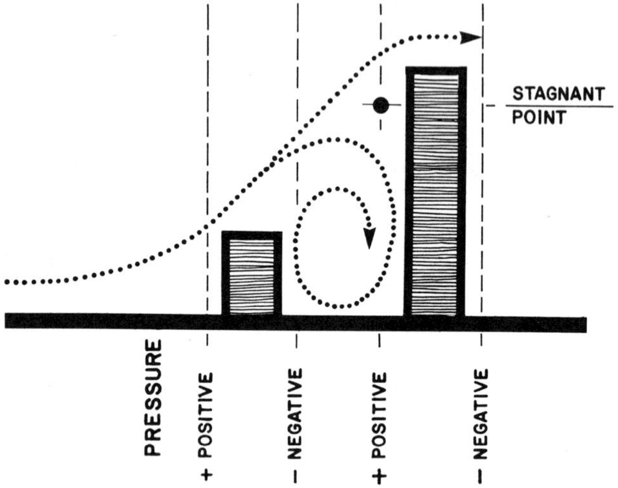

7-52. The Wise effect.

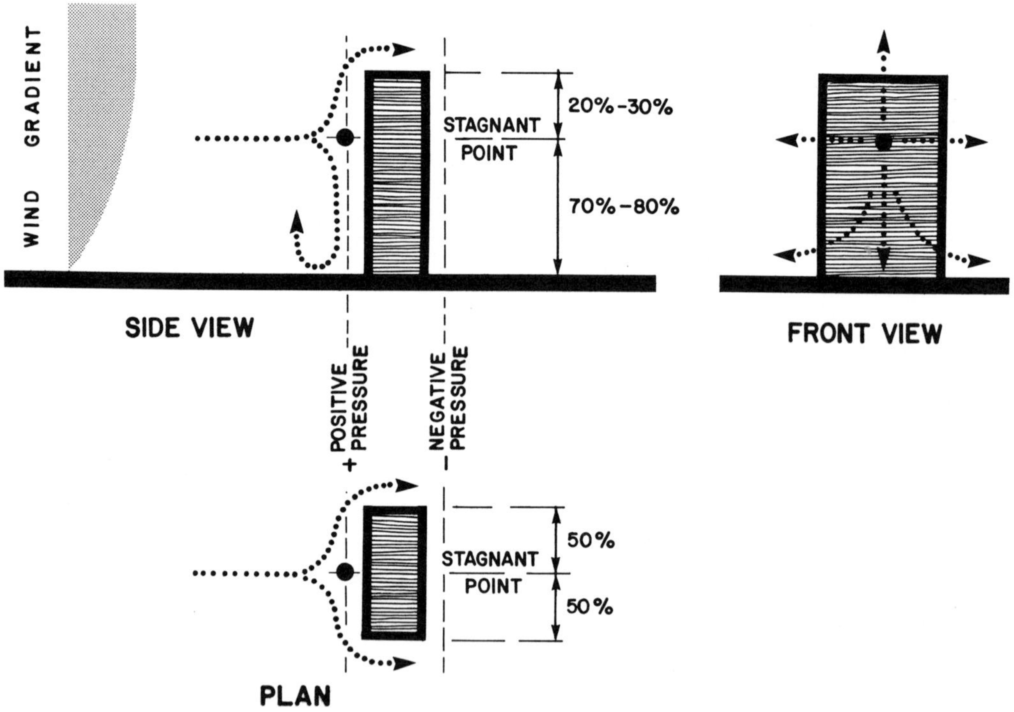

7-51. The aerodynamics of a tall building.

amplification can be anticipated. A typical example of this is Commerce Court in Toronto, Canada. Extremely high buildings, such as the World Trade Center in New York, may bring the amplification factor up to five in gusts and high-velocity winds. This potentially life-threatening situation is exacerbated by the presence of other elements, such as rain, low temperatures, ice, and snow; the air turbulence in a snowstorm can actually be seen. The movement of snowflakes, the nature of snow accumulation, the horseshoe patterns of snow around the base of buildings are tangible demonstrations of airflow turbulences.

Despite the temptation to question the rationality of certain architectural forms and urban spatial relationships, particularly if a high value is placed on the quality of the pedestrian street environment, designers should avoid sweeping conclusions. These wind conditions that hamper the use of public space are the result of the particular geometries of certain urban spaces and building configurations, of a particular azimuth range, and of the velocity and flow characteristics of air movement.

Human Comfort

In northern latitudes, designers must compensate for accelerated wind speeds, which result in increased heat loss, greater rain penetration into the building envelope, increased noise levels, and the accelerated aging of exposed materials. Urban planners must prevent flying garbage, sand, leaves, increased snowdrifts, spray from fountains, and damage to hanging signs, street lamps, furniture, plants, and trees. The wind's effect on the safety of individuals, impediments to their movement, and their comfort must be considered when designing cities in cold climates.

Open plazas are particularly hazardous during the winter when snow, ice, and slush impede walking. People fall because of slippery surface areas, their body weight and height (center of gravity), the type of shoes they wear (friction characteristics of soles), their physical fitness, their age, their posture, and the turbulence, speed, and direction of wind. The general consensus is that wind speeds in the neighborhood of 70 km/hr (20 m/sec), or approximately 45 miles/hr, make walking hazardous for the average person. Wind speeds below the threshold of physical danger are responsible for environmental situations related to different degrees of human comfort (fig. 7-53). Whereas in hot arid and hot humid areas air movement is usually desirable, in northern latitudes, particularly in the winter, protection against the wind is often mandatory.

In addition, it must be remembered that physical and thermal comfort are highly subjective and tolerance levels vary according to meteorological conditions, the physiological and psychological state of the individual, how active the individual is, and the warmth provided by an individual's clothing. For instance, children and the elderly are more prone to falls. The frequency of winds must also be considered, since an occasional gust may be tolerated, whereas steady air movement well below the danger threshold may discourage the use of outdoor cafes, lunches taken in a public park, or window shopping. Similarly, the same breeze that encourages outdoor activities and is welcomed in the summer may increase the chill factor in winter beyond one's tolerance level. Pedestrian comfort can be significantly enhanced during the winter season, and in cold climates in general, by providing the best possible shelters against the wind. The reverse, of course, applies to hot climates and hot seasons: induced air movement improves human comfort by eliminating heat through convection and through a higher rate of evaporation.

Design Strategies

During the design process, ten variables should be considered in the analysis of street-level wind conditions around tall buildings:

1. general wind characteristics of the region
2. topography
3. roughness of the surrounding terrain
4. density of buildings in the immediate area
5. relative height of other buildings and their proximity
6. size and shape of the open space and street
7. height and shape of the individual building
8. orientation of the building and the azimuth of the prevailing winds
9. presence of narrow passageways
10. conditions at the base of the building (podium, pilots, cutback, openings)

These are extremely complex considerations and only an aerodynamics professional can attempt to address them. A laboratory verification is always mandatory. Wind tunnel and water tank simulations are powerful verification tools for examining the worst effects as well as for providing a complete set of information data for a particular spatial setting.

In urban design projects, an assessment of the wind

WIND COMFORT CHART

EFFECT CAUSED BY THE WIND: ON LAND	EFFECT CAUSED BY THE WIND: ON SEA	SPEED km/h	BEAUFORT NO.	SUMMARY OF WIND EFFECTS ON PEOPLE, BASED ON BEAUFORT SCALE: AIR MOVEMENT	GENERAL EFFECT ON PEOPLE	TENTATIVE COMFORT CRITERIA FOR WINDINESS TEMP. 10°C+: HILL SLOPES DEMANDING THE SAME MUSCULAR POWER %	∡	AREAS: SIDEWALKS / ACTIVITY: WALKING	PARKS - SKATING RINKS / STROLLING SKATING	PARKS PLAZAS / STANDING SITTING SHORT - EXPOSURE	OUTDOOR - RESTAURANTS BANDSHELLS STADIA / STANDING SITTING LONG - EXPOSURE
Still smoke rises vertically	surface mirror-like	0 - 2	0	Calm	Not noticeable	-	-	-	-	-	-
Smoke drifts but vanes remain still	Only ripples form	3 - 5	1	Light Air	Not noticeable	-	-	-	-	-	-
Wind felt on face, leaves rustle, vane moves	Small, short wavelets, distinct but not breaking	6 - 11	2	Light breeze	Wind felt on face	-	-	○	○	○	○
Leaves and small twigs move constantly, streamer or pennant extended	Large wavelets beginning to break, glassy foam, perhaps scattered white horses	12 - 19	3	Gentle breeze	Wind extends, light play; hair is disturbed, clothing flaps	-	-	○	○	○	◔
Raises dust and loose paper, moves twigs and thin branches	Small longer waves fairly frequent white horses.	20 - 29	4	Moderate breeze	Raises dust, dry soil and loose paper, hair disarranged	-	-	○	○	◔	◕
Small trees in leaf begin to sway	Moderate waves, distinctly elongated, many white horses, perhaps isolated spray	30 - 39	5	Fresh breeze	Force of wind felt on body; drifting snow becomes airborne; limit of agreeable wind on land.	5%	3°	○	◔	◕	◕
Large branches move, telegraph wires whistle, umbrellas hard to control	Large waves begin with extensive white foam crests breaking; spray probable	40 - 50	6	Strong wind	Umbrellas used with difficulty; hair blown straight; difficult to walk steadily; wind noise unpleasant; windborne snow above head height (blizzard).	10%	6°	◔	◕	◕	◕
Whole trees move; offers some resistance to walkers	Sea heaps up, lines of white foam begin to be blown downward	51 - 61	7	Near gale	Inconvenience felt when walking	14%	8°	◕	◕	◕	●
Breaks twigs off trees, impedes progress	Moderately high waves with crests of considerable length; foam blown in well-marked streaks, spray blown from crests	62 - 74	8	Gale stormy wind	Generally impedes progress; great difficulty with balance in gusts	20%	11°	●	●	●	●
Blows off roof tiles and chimney pots	High waves, rolling sea, dense streaks of foam; spray may already reduce visibility	75 - 87	9	Strong (storm) gale	People blown over by gusts.	33%	18°	●	●	●	●
Trees uprooted, much structural damage	Heavy rolling sea, white with great foam patches and dense streaks, very high waves with overhanging crests; much spray reduces visibility	88 - 101	10	Heavy storm or whole gale	DEATH ZONE						
Widespread damage	Extraordinarily high waves, spray impedes visibility	102 - 116	11	Hurricanelike storm	DEATH ZONE						
Widespread damage	Air full of foam and spray, sea entirely white	117 - 132	12	Hurricane	DEATH ZONE						

SOURCES:

- Penwarden, A.D. and Wise, A.F.E., 1975: wind environment around buildings, Chapter 5 - Comfort and Safety Conditions, Dept. of Environment, B.R.E., HMSO, London, pp.40-42.
- Davenport, A.G., 1972: An Approach to Human Comfort Criteria for Environmental Wind Conditions; Paper No. 8. - Conf. Teaching the Teachers on Building Climatology, Stockholm, 1971. Statens Institut for Byggnads Forskning, Stockholm.
- Page, J.K., 1976: Application of Building Climatology to the Problems of Housing and Building for Human Settlements, Technical Note No. 150, WMO No. 441, World Meteorological Organization, Geneva, Switzerland

LEGEND

Relative comfort scale	Symbol	Acceptable if occurs not more than once
Perceptible	○	—
Tolerable	◔	A week
Unpleasant	◕	A month
Dangerous	●	A year

NOTE

At lower temperature, relative comfort level might be expected to be reduced by one Beaufort number for every 20°C reduction in temperature.

7-53. A wind comfort chart.

environment must be an integral part of the planning process. In order to establish the design parameters, two pieces of information are required:

1. the anticipated frequency as well as the worst possible wind conditions at street level
2. an evaluation of the acceptability of these conditions with respect to physical danger and pedestrian comfort

In cities located in northern latitudes or in regions with a long winter season, any air movement over 5 km/hr (maximum no. 1 on the Beaufort Scale) is potentially detrimental to the quality of the street-level environment and may discourage pedestrian use of public urban spaces. This introduces the need for a special design strategy, which could be based either on preventive measures or on remedial actions taken after the fact.

The first strategy would involve mathematical modeling and close work with the aerodynamics engineer and the testing laboratory during the conceptual design and design development stages. In fact, a number of major development projects in Canadian cities, including the McLean Hunter Building in College Park, the Tridel Building on Bay Street in Toronto, and other large redevelopment projects in downtown Calgary, Regina, Edmonton, and Montreal were modeled in laboratory water tanks. With their sleeves rolled up, architects sculpted rudimentary shapes in clay while the flow of water simulated wind turbulence and snowstorms.

If the ultimate goal were a laminar airflow among buildings and a windless street level, then architects' formal vocabulary would consist of oblong shapes, hemispheres, cylinders, curved surfaces, and round corners, since these do not present undue resistance to wind. In this case, the natural environment provides an important lesson: since elements in nature are continually seeking to restore a balance that has been upset by the energy of the sun (which affects the movement of air and water), there is a continuous struggle for equilibrium—the earth's surface and objects on it change in response to natural forces. The shapes of sand dunes, the shape of a hawk sailing through the sky, water-beaten boulders at the bottom of a waterfall, or a wind-shaped rock formation in the middle of a desert are all tangible results of natural forces altering the shapes of objects to offer minimum resistance to airflow and the flow of water.

Such a design strategy leads to the sleekness of a turbine blade, a form that would minimize airflow but would inevitably conflict with other objectives; buildings are constructed to shelter human activities and not to promote the sophisticated interplay of wind forces. Because buildings are created within engineering, economic, and cultural constraints, we are destined for an imperfect aerodynamic solution. Thus, there will always be downdrafts and turbulences. These will always require remedies.

There are a wide range of remedial actions available to modify the wind, including changes in the building form, the partial or total enclosure of pedestrian activity space, and the removal of affected pedestrian activity.

An undesirable situation at street level may be improved by revolving a tall building around on its vertical axis, so the narrow side of its corner faces the prevailing winds. Another possibility is to break the affected facade into a series of steps, which help to weaken the thrust of a downdraft (fig.7-54). In certain cases, turbulence can be eliminated by filing sharp corners into round ones (fig. 7-55).

The partial or total enclosure of pedestrian activity space is a remedial measure that can deflect, absorb, or dissipate the energy of moving air. This is usually achieved with a variety of solid or perforated barriers and blockages, which could be natural (berms, shrubs, trees) or human-made (fences, screens, grids, shelters, canopies). The efficiency of these wind screens vary with their permeability. Zero permeability works only

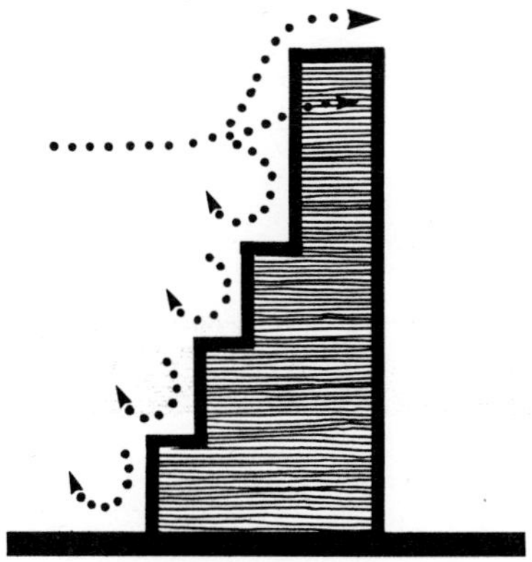

7-54. A stepped facade reduces the downdraft.

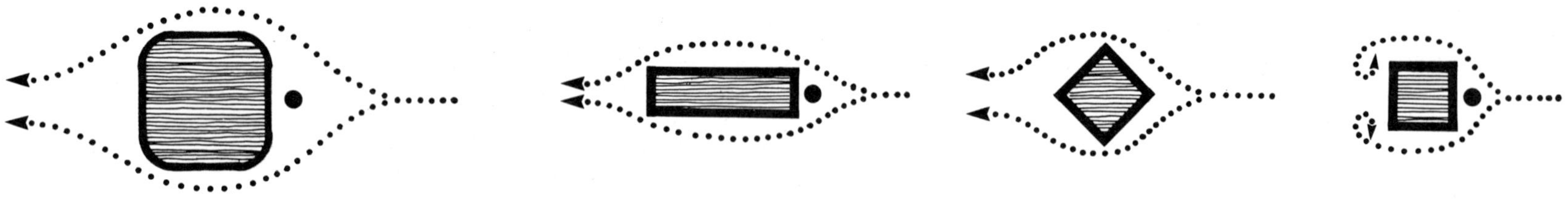

7-55. The airflow around different shapes.

under special circumstances; it is not necessarily the most desirable, because screens with zero permeability can accelerate the air speed around their edges by up to 100 percent, leaving devastating rotors and violent eddies behind. Observation and laboratory tests reveal that if a certain amount of air is allowed to penetrate a windscreen, the turmoil behind the windscreen is reduced. For example, 30 percent penetration does not create any eddies, and airflow is steady and laminar. However, when penetration climbs above 35 percent, there is a rapid decrease in efficiency. Any design of a livable street must therefore incorporate a whole range of horizontal and vertical screens with different degrees of permeability. This would include fences, walls, and shelters made from a variety of materials, such as concrete, ceramics, wood, plastic, and glass (fig. 7-56).

Trees and shrubs are natural, porous screens, or *wind sponges*. They are excellent in combing and streamlining turbulences, protecting against horizontal winds, diffusing downdrafts, and modifying the climate—a welcome contribution to the city streetscape. In northern cities, trees, shrubs, and other plants enhance an environment aesthetically, just as they filter and purify the air. Only coniferous trees, however, can be used to screen wind on an annual basis. In general, trees and shrubs can be used with great success provided that their growth stages and annual cycle are understood. Their high initial costs and costs of maintenance must also be taken into account.

Since a wind-flow pattern is unique to each building, remedial design intervention inevitably results in unique spatial solutions. This is an opportunity for a joint effort on the parts of the architect and urban designer to develop innovative solutions, to secure pedestrian comfort, and to create new aesthetics.

The removal of affected pedestrian activity is a regressive step if built into a design strategy. Pedestrian activity is rarely prohibited by decree simply because of wind created by building mass and urban void. An example of such an exception may be the pedestrian walkway at the +15 level, Toronto City Hall, where pedestrian activity was prohibited after a serious accident was caused by wind. In general, pedestrian activity disappears spontaneously whenever perceived discomfort or danger surpasses the tolerance level.

The Visual and Audio Perceptions of Wind

Wind is an energy carrier, visible to the human eye only if the air—its medium—is mixed with particles of water, solids, or certain gases. Thus, wind manifests itself only indirectly in the motion of objects such as passing clouds, rain, waves, sand, and snow, and in the destruction of objects. The quality of public spaces is seriously affected by waving trees, flying leaves and debris, the deflected smoke from chimneys, and by snowdrifts and snow piling, particularly in front of entrances and emergency exits.

Not all manifestations of wind are necessarily unpleasant. In fact, air movement can enrich the aesthetic qualities of a space; the physical motion of objects can be a part of artistic expression. This is deliberately exploited in kinetic art, where sculptures have dynamic components that are suspended, hinged, and pivoted. Such an elaborate three-dimensional system can be animated by conventional electrical energy, compressed air, steam, water, and by induced or natural air movement. By using the variable speeds of natural air movement, an unlimited number of spatial configurations and an endless steam of aesthetic experiences can result. Any air movement, wind, or gentle currents of hot or cold air, give motion and breathe life into these delicately balanced structures. They are ideally suited for open public spaces, parks, or pedestrian plazas.

On occasion, the kinetic energy of air movement produces sound; air in rapid motion around tall buildings can be heard. Certain architectural detailing and projections from a facade's plane may also be responsible for the generation of sound. Romantic literature in particular talks of certain moods created by the sounds of air penetrating and escaping buildings through cracks and blowing in or over open wells and chimneys.

Furthermore, wind can carry sounds to distant locations; the sound of trains or highway traffic can be heard miles away. In an urban environment, sound waves generated by traffic and human activities in a narrow street are amplified when they bounce between tall buildings; the wind then carries these sounds to other locations.

Throughout history, there have been a number of attempts to use the wind for artistic expression. The aeolian harp, for example, was an ancient musical instrument played without human participation. The divine musician was Aeolus, god of the winds. Whether blowing through strings or over the pipes of a syrinx, wind can produce eerie, mysterious sounds, creating a mood that can enrich the design and choreography of kinetic sculptures. Wind used to its best advantage can

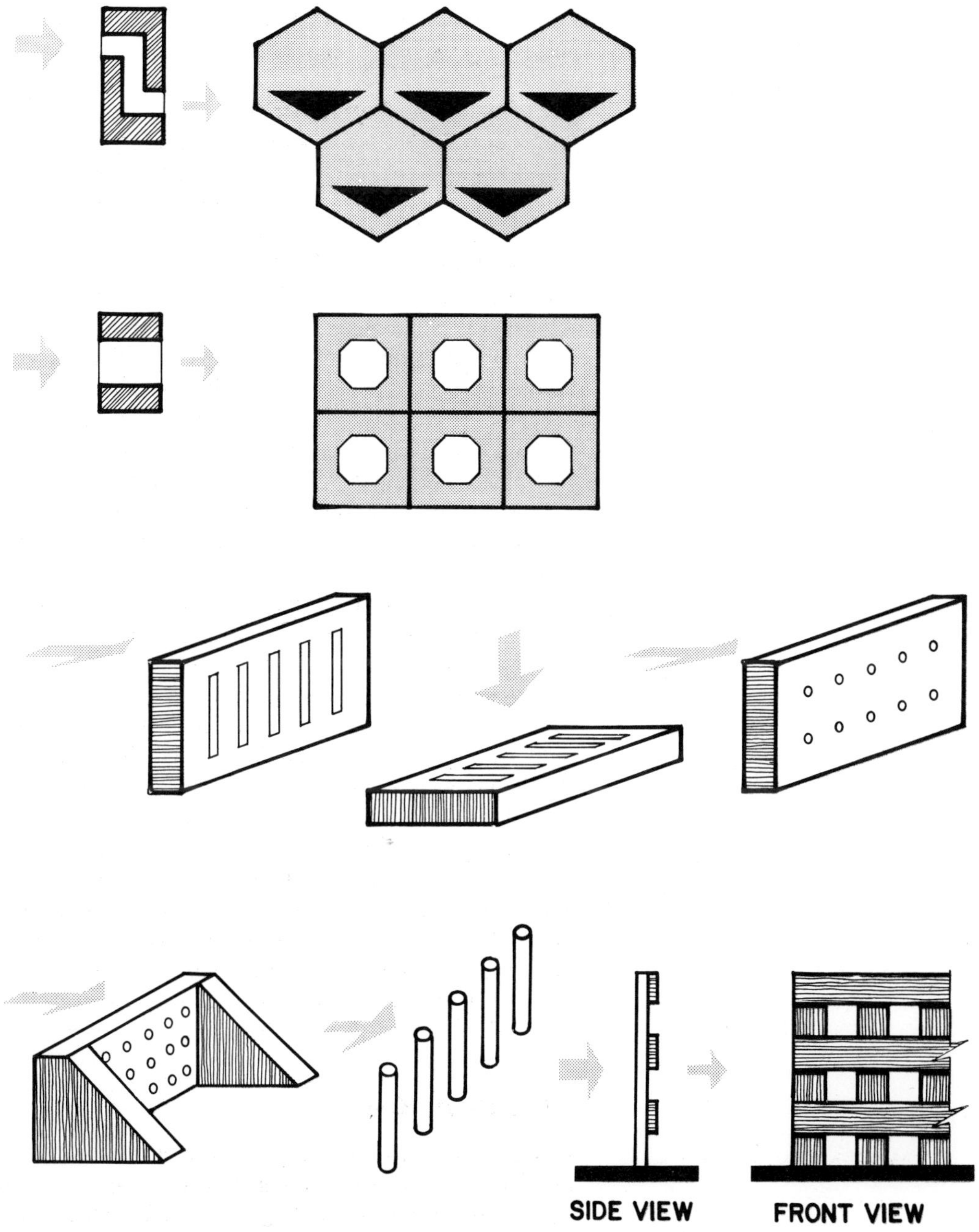

7-56. Vertical and horizontal screening devices.

beautify public places with a complex orchestration of motion, sound, color, and light.

Wind and the Generation of Urban Form—the Art of Compromise

The theory of the wind's effect on city planning and urban design has not changed much since the time of Vitruvius twenty centuries ago. Current literature often recycles time-honored principles: streets, road patterns, building form, and space should be designed to shield against the undesirable and be open to the desirable. Naturally, it would be wise to shield against prevailing stormy winds, especially during the winter season. On the other hand, a light breeze is welcome to carry away pollution and to cool through accelerated heat convection and evaporation. Unfortunately, even a moderate air speed, when combined with low temperatures, can result in a chill factor that can be withstood only with space gear.

Wind alone, however, cannot generate urban form. A workable compromise must be reached between many competing interests and mutually exclusive requirements. A sound planning and design strategy is derived from two fundamental requirements: exposure to the sun, both direct and reflected; and shelter against the wind, both natural and induced by buildings. In cold regions, air movement appears desirable only during the short, warm season. Consequently, the ideal city in a cold region should be windless. Such a pool of stagnant air, however, can be a health hazard if it is not clean. As a result, cold cities must be sunlit, windless, and pollution-free. This requirement for a clean environment places additional constraints on the planning, design, management, and administration of cold cities.

PRECIPITATION

Designers must be seriously concerned with precipitation. Northern latitudes must deal with both rain and snow. For example, as the Hamilton Ecochart indicates, that particular region is blessed with a considerable amount of rain, particularly during the warmer part of the year; the annual average is about 690 mm (27.6 in). The precipitation data shown in the Ecochart also indicates the monthly accumulation of snow. Snowfall can be anticipated in the six-month period between November and April; the annual accumulation averages about 170 cm (5.6 ft). However, this regional measurement must be adjusted to take into account the effect of the urban heat island (it should be reduced by about 14 percent). This adjustment represents a significant savings as compared with suburbia, where the municipality spends vast sums of money on removing snow from roads and sidewalks.

The amount of rain affects the type of roof construction and roofing material, the drainage system, the design and size of water leaders, downspouts, and the types of discard to the environment. The incidence of rain also determines a variety of construction details, particularly those related to the protrusion of elements through the roof, such as chimneys and vents, and those related to openings, such as windows and entrances. Precipitation may also influence the general layout. For example, a wise design response to the combination of rain and children is a mud room. But it is the mixture of rain and wind that demands ingenious design solutions from the architect and superb craftsmanship from the trades. An exciting high-rise project can turn into an architect's nightmare when rain moves horizontally and water defies gravity and travels upward in exactly the opposite direction from the one the designer may have originally envisioned.

Since the design of larger buildings must provide generous tolerances to allow for thermal expansion, submarine-style (water-tight) detailing is not usually possible. The architect must ultimately develop an ingenious system to control leaks, usually a double-shell envelope, which diverts the penetrating water back to the surface of the building by gravity or by negative pressure.

Rain poses a different set of problems for the urban designer. Urban dwellers often perceive rain as a nuisance that can significantly reduce their enjoyment of streets and urban spaces. While the urban designer is attempting to create a livable urban space for people, rain can halt movement. A certain degree of protection must be offered to the downtown shopper; hence, the recent proliferation of bus shelters and the use of a variety of time-tested architectural features, such as arcades. There is unlimited room for innovation. For instance, water can be retained in public spaces with a shallow pond receiving rainwater from surrounding roofs. Because it is shallow, it can be used with little risk of drowning. When dry, it might be converted to an arena for roller skating, skateboarding, or ice skating. A sophisticated urban design concept would attempt to capitalize on the rain not only in a purely functional sense but also by introducing new aesthetics and new art forms that are triggered by rain, or that use rain in movable, sound-producing sculptures, or in special light effects.

Temperatures below freezing, however, introduce a new set of rules for the urban design game. More than any other weather factor, precipitation in the form of snow can be the worst direct or indirect handicap to practically all human activities. In a spatially extensive yet highly organized, interdependent, and interrelated society, any interruption in the vital links can be a matter of life and death. Snowfall in the city may cause minor inconveniences or it could bring the city's movement to a complete standstill. Snowfall generally constitutes a hazardous situation since even a small accumulation tends to cause serious traffic accidents. A larger amount of snow may interrupt power and telephone cables. Snowfall may disrupt construction activities, industrial production, and trade. Even school attendance may be affected. Ultimately, the wide network of city services may be suspended.

The consequences in economic terms are severe: heavy financial losses due to reduced movement of people, goods, and information; acquired costs related to snow and ice removal; additional funds allocated and the operational and maintenance budget increased for schools, factories, and stores. Direct costs can result from damages and indirect costs from funds that are diverted from business expansion and social services.

But it should be emphasized that many of these problems are a result of structures in northern cities that were designed and built for a much warmer climate. The growing concern about the livability of northern cities must be addressed with attempts at fresh, genuinely northern urban-planning ideas.

CONCLUDING REMARKS

In analyzing and responding to local climatic conditions, urban design and planning must take into account not only the environmental stressors that directly affect human physical health and personality, but also the sociocultural environment. The design and use of spaces between buildings is an expression of cultural norms. Since every culture is, to a large extent, derived from the natural setting and from a ritualized response to the surrounding stressors, any public space design has hidden or explicit assumptions about climate, use of the space, and local life-style. When the reality and the assumptions do not coincide, public places are deserted.

These failures are painfully obvious in both extremely hot and cold regions. Public spaces in cold regions are routinely built to resemble such spaces in mild regions and are thus used only during the short summer period. Their desolation is accepted as a fact of life. However, even the prosperous economies of developed nations and their generous municipal budgets cannot indefinitely support the costs involved in constructing, maintaining, and upgrading urban spaces that are virtually empty.

PART 4

TOOLS

C H A P T E R 8

SOLAR TEMPLATES

Solar Templates allow a designer to perform a number of tasks quickly and accurately. These include:

1. determining shadows cast and the impact areas of solid objects
2. constructing Solar Envelopes in order to obtain the maximum heights and volumes permissible for development
3. performing the above tasks on flat, horizontal surfaces, inclined planes, and variable topography (all can be constructed from plans with no sections required)
4. determining spatial relationships between buildings and sun access to public spaces
5. determining solar access to the internal spaces of buildings
6. constructing devices that reflect sun from vertical and horizontal surfaces

CONSTRUCTION OF SOLAR TEMPLATES

The sun's position in the sky is defined by two angles (fig. 8-1). The first is the angle of elevation of the sun above the horizon, called *solar altitude*. This may range from 0°, when the sun is on the horizon (sunrise and sunset), to 90° in certain locations, when the sun is at its zenith (solar noon).

The second angle is determined by the position of the sun from east to west and is called the *solar azimuth*, which is measured clockwise from the north to the projection of the sun on the horizon. It equals 0° at north, 90° at east, 180° at south, 270° at west, and 360° at north, completing the circle.

Any shadow of an object is cast for a specific time of day and year. Figure 8-2 shows how shadows are cast by a pole at different times of the day. Once the position

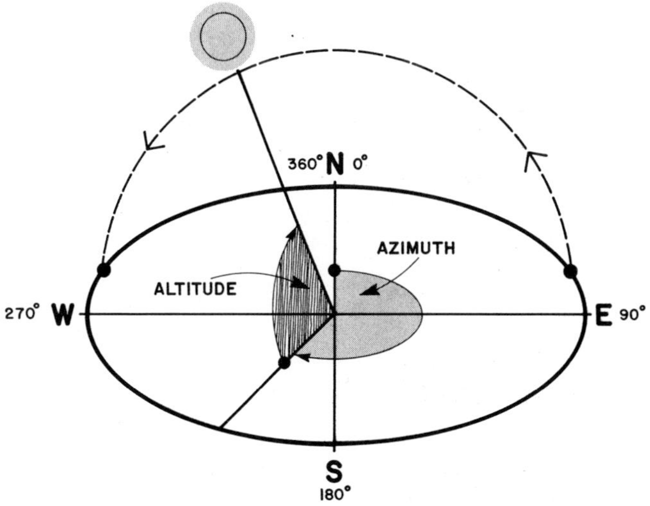

8-1. Solar altitude and solar azimuth.

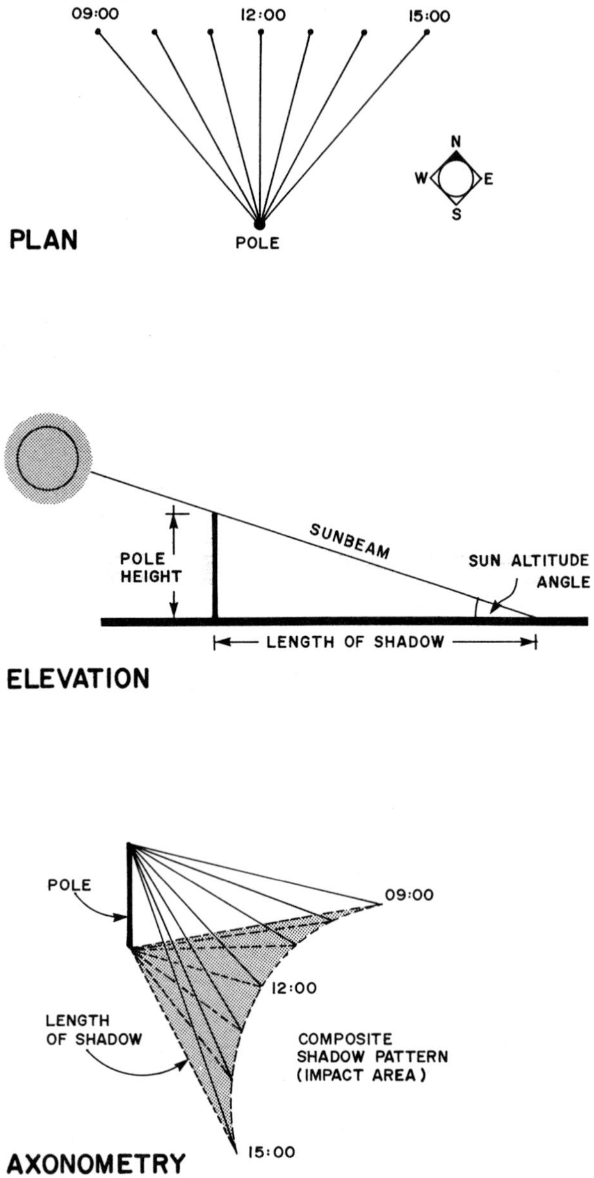

8-2. Directions of shadows cast by a pole are shown in plan. The axonometry shows direction and length.

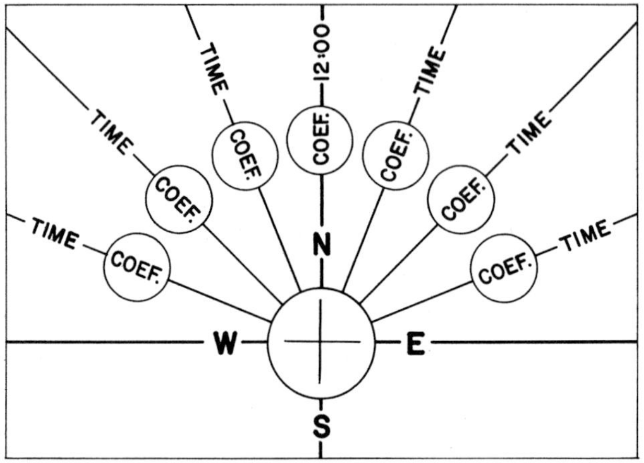

8-3. A generic Solar Template.

of the sun in the sky is known (solar azimuth and solar altitude), the following can be determined:

1. the direction of the shadow cast (determined by solar azimuth)
2. the length of the shadow cast (determined by solar altitude [shadow length = pole height × cotangent of solar altitude])
3. the height of the pole (determined by the length of the shadow [pole height = shadow length ÷ cotangent of solar altitude])

This can lead to the construction of a simple instrument to assist designers (fig. 8-3). The cross sections of the north/south and east/west directions indicate the position of a pole of unit height. Radial lines represent shadow directions for every hour from sunrise to sunset. Time is expressed in solar time in order to make the shadow at 12:00 hours point toward true north. The cotangent of solar altitude will be referred to as the coefficient in future. Coefficient values are computed from trigonometric equations (cotangent of solar altitude), which allow the designer to compute the length of a shadow (pole height × coefficient), or if the length of the shadow is known, to compute the height of the pole (shadow length ÷ coefficient). These geometries apply only to a given latitude and to a given day of the year.

Since the shortest and longest days of the year fall on December 21 and June 21 respectively, it is best to design all templates for the twenty-first. This would lead to a set of twelve templates, one for each month. For developing a design tool, the earth's path around the sun can be considered symmetrical along the June/December axis and therefore only seven templates are needed. A set of templates constructed for 43°N is shown in the following figures:

Figure 8-4, Template #1—December 21 (the shortest day of the year in the northern hemisphere)
Figure 8-5, Template #2—January 21/November 21
Figure 8-6, Template #3—February 21/October 21
Figure 8-7, Template #4—March 21/September 21 (spring and fall equinox)
Figure 8-8, Template #5—April 21/August 21
Figure 8-9, Template #6—May 21/July 21
Figure 8-10, Template #7—June 21 (the longest day of the year in the northern hemisphere)

A set of Solar Templates should be prepared for each degree of latitude. The requirements of design professionals, however, can be satisfied with Solar Templates constructed for every second latitude. For example, a 43° latitude set can be used within the 42° and 44° latitude strip (fig. 8-11). The same geometries apply to the southern hemisphere. Consequently, this set may be used within the corresponding strip between latitudes 42° and 44° south.

Since shadows in the southern hemisphere point south, it would be practical to construct a southern set of templates. Solar azimuth and altitude data are the same as are the hourly shadow direction lines and coef-

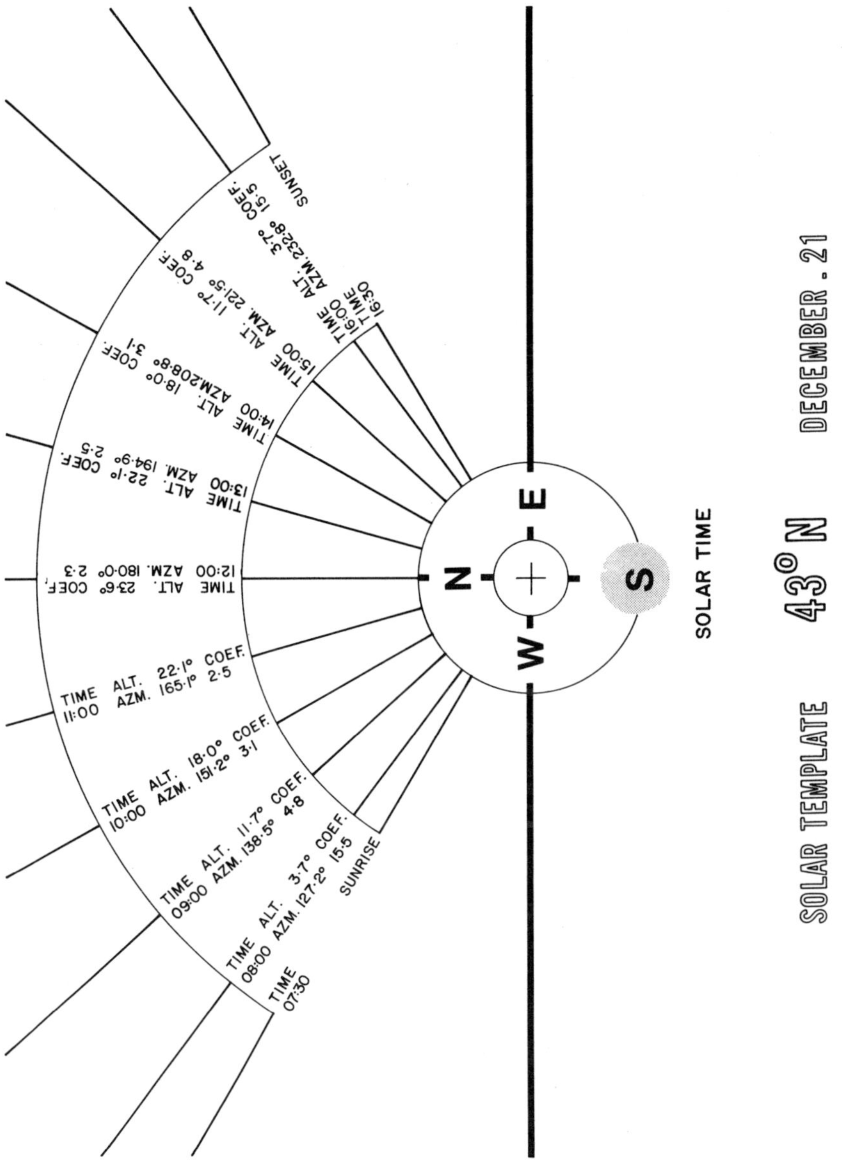

8-4. Solar Template #1—December 21 (the shortest day of the year in the northern hemisphere).

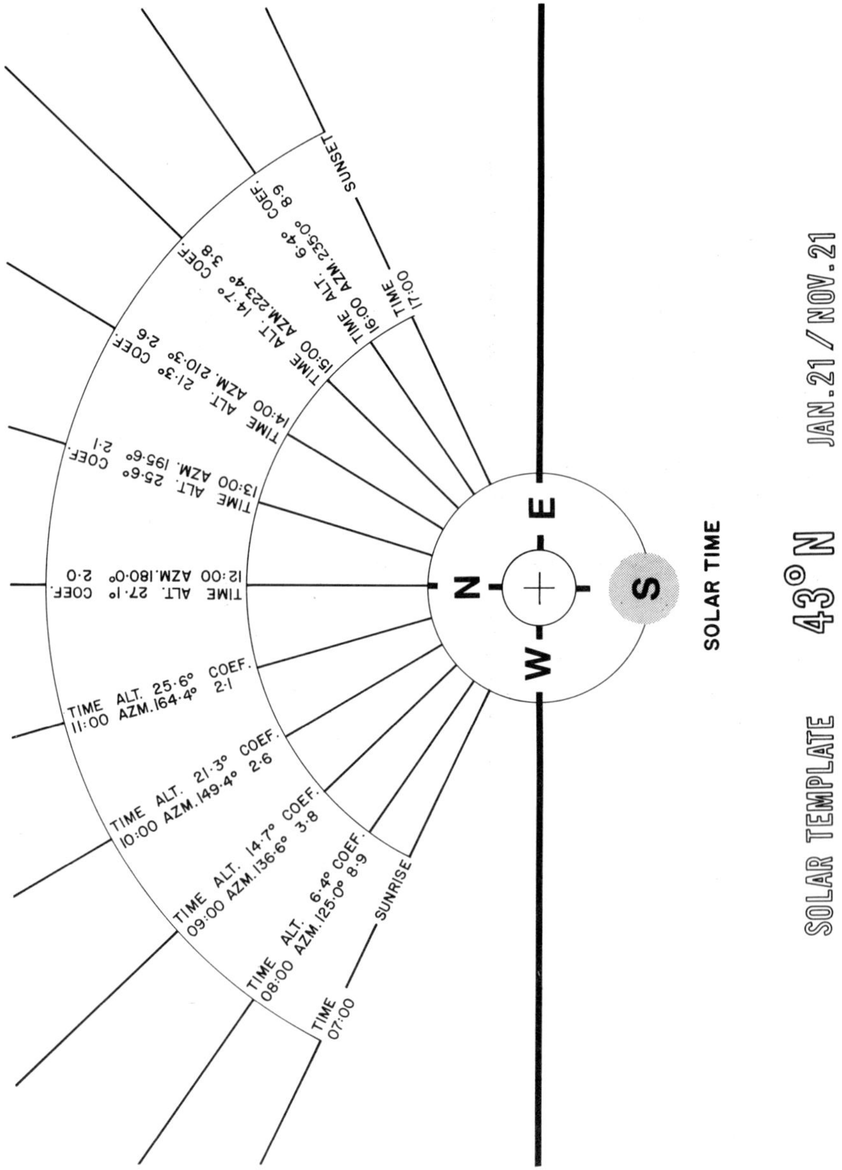

8-5. Solar Template #2—January 21/November 21.

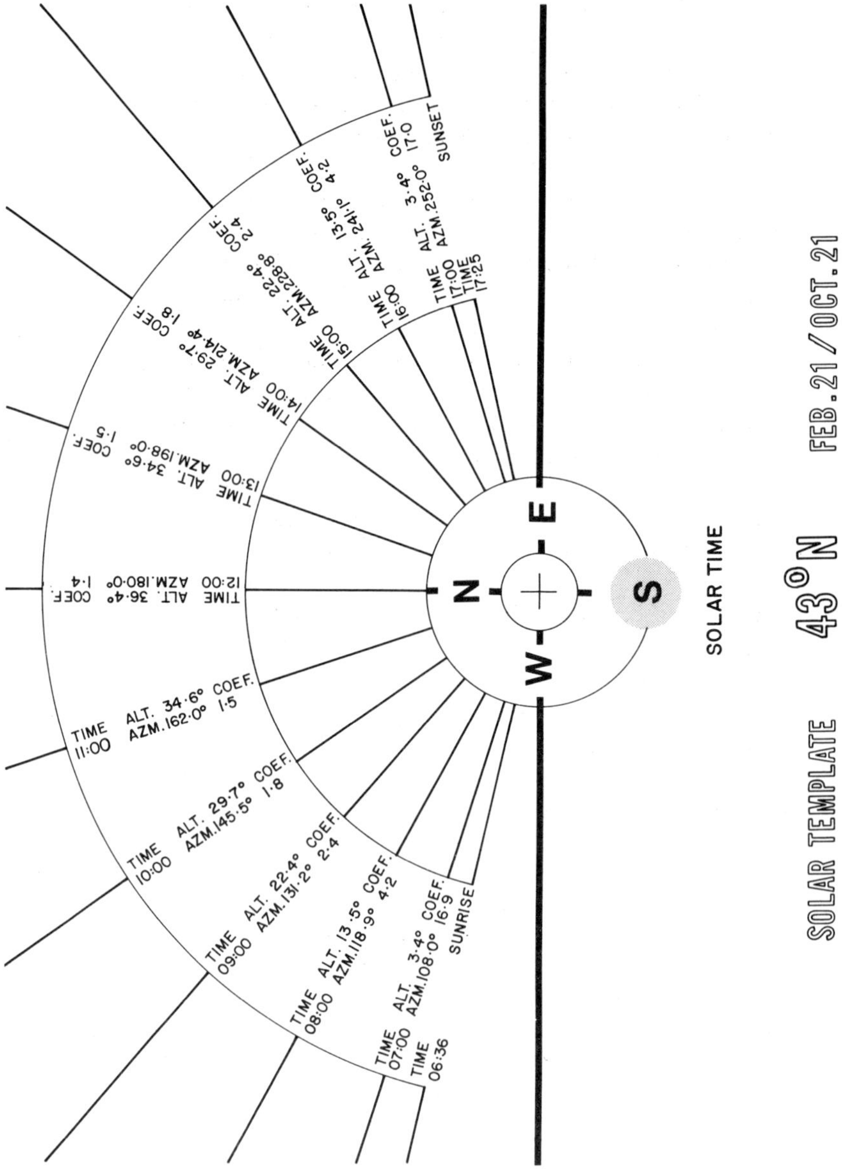

8-6. Solar Template #3—February 21/October 21.

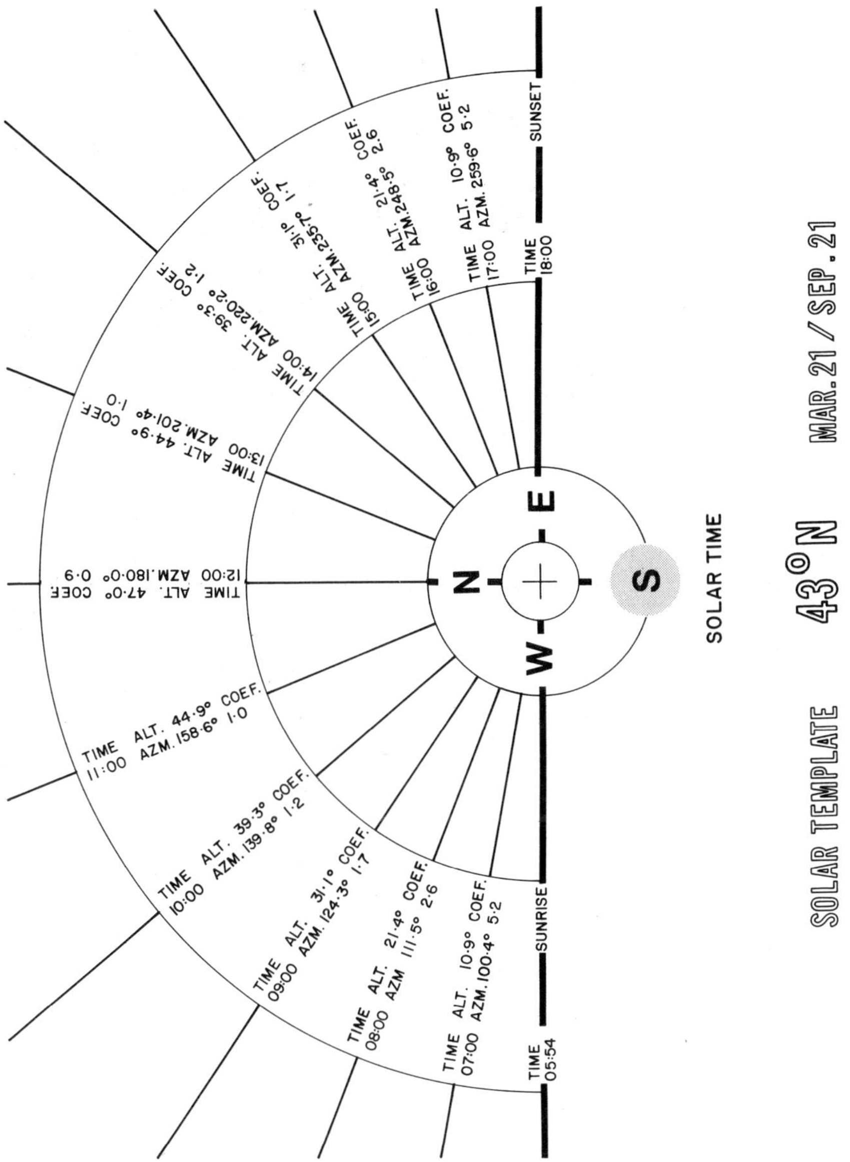

8-7. Solar Template #4—March 21/September 21 (spring and fall equinox).

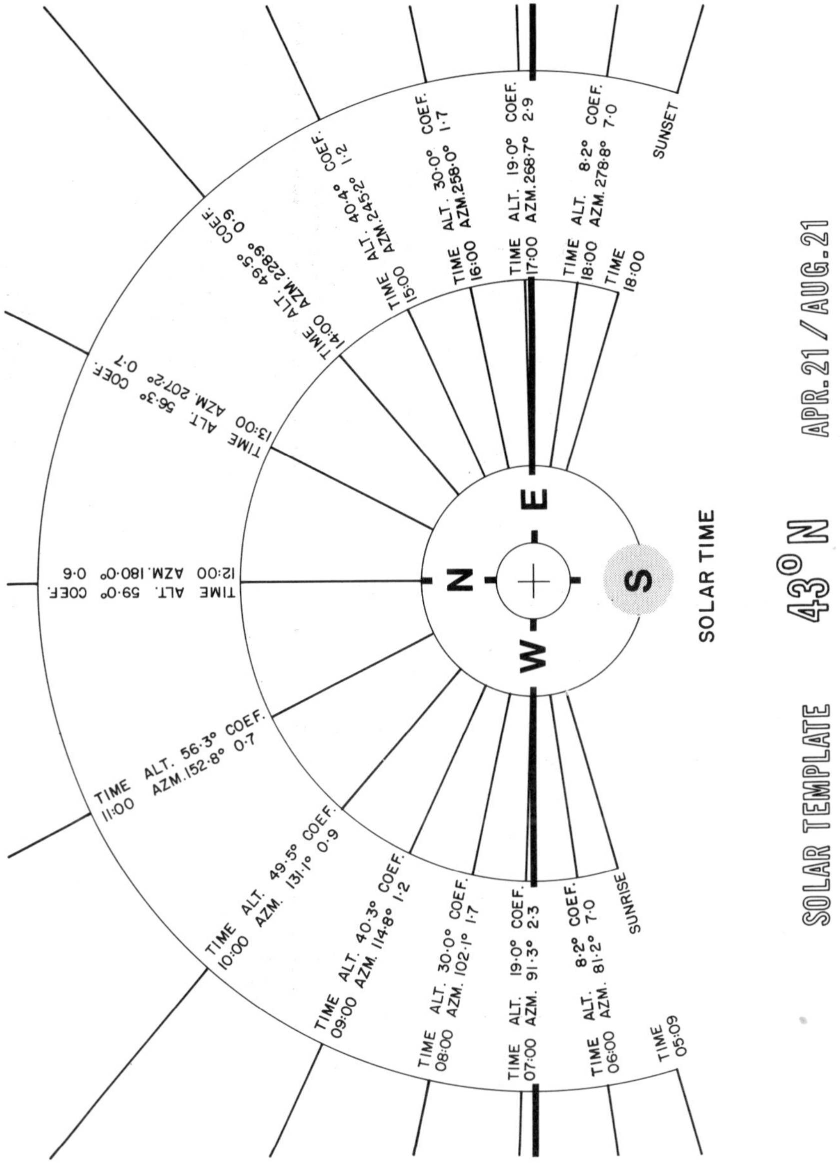

8-8. Solar Template #5—April 21/August 21.

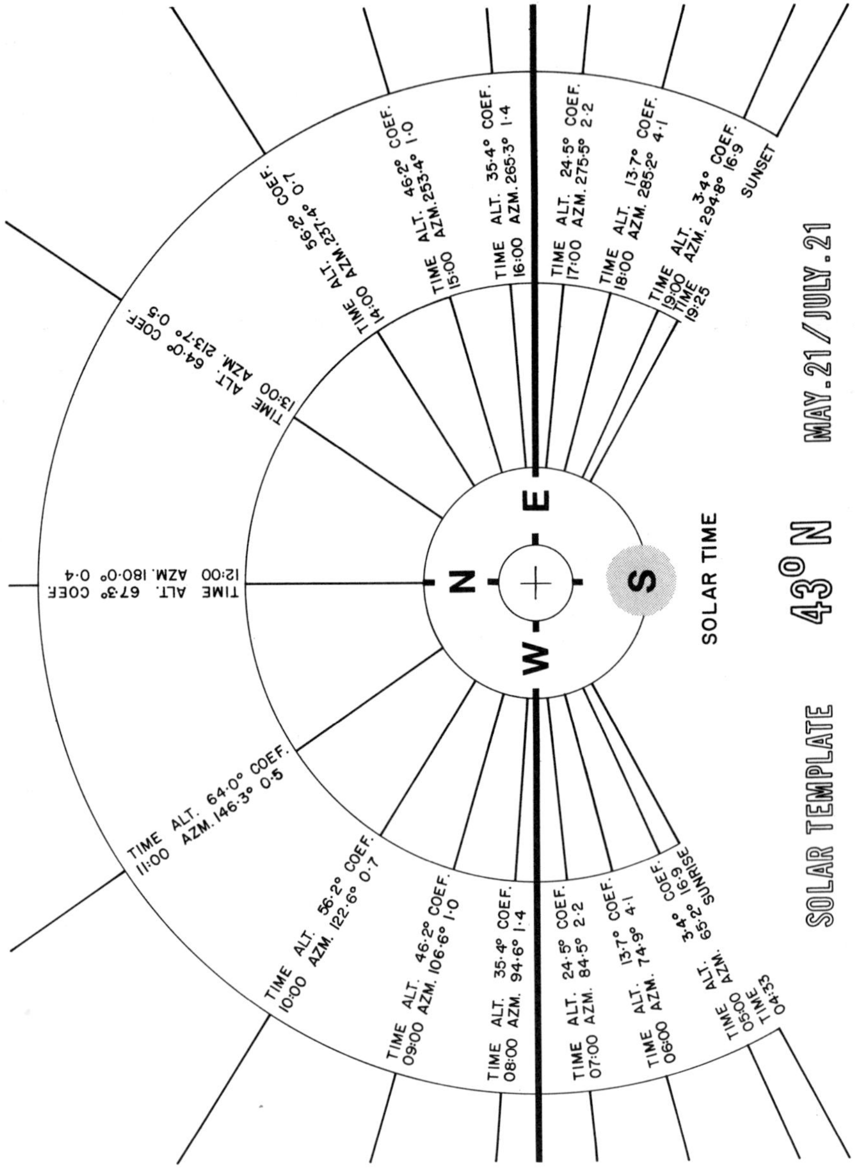

8-9. Solar Template #6—May 21/July 21.

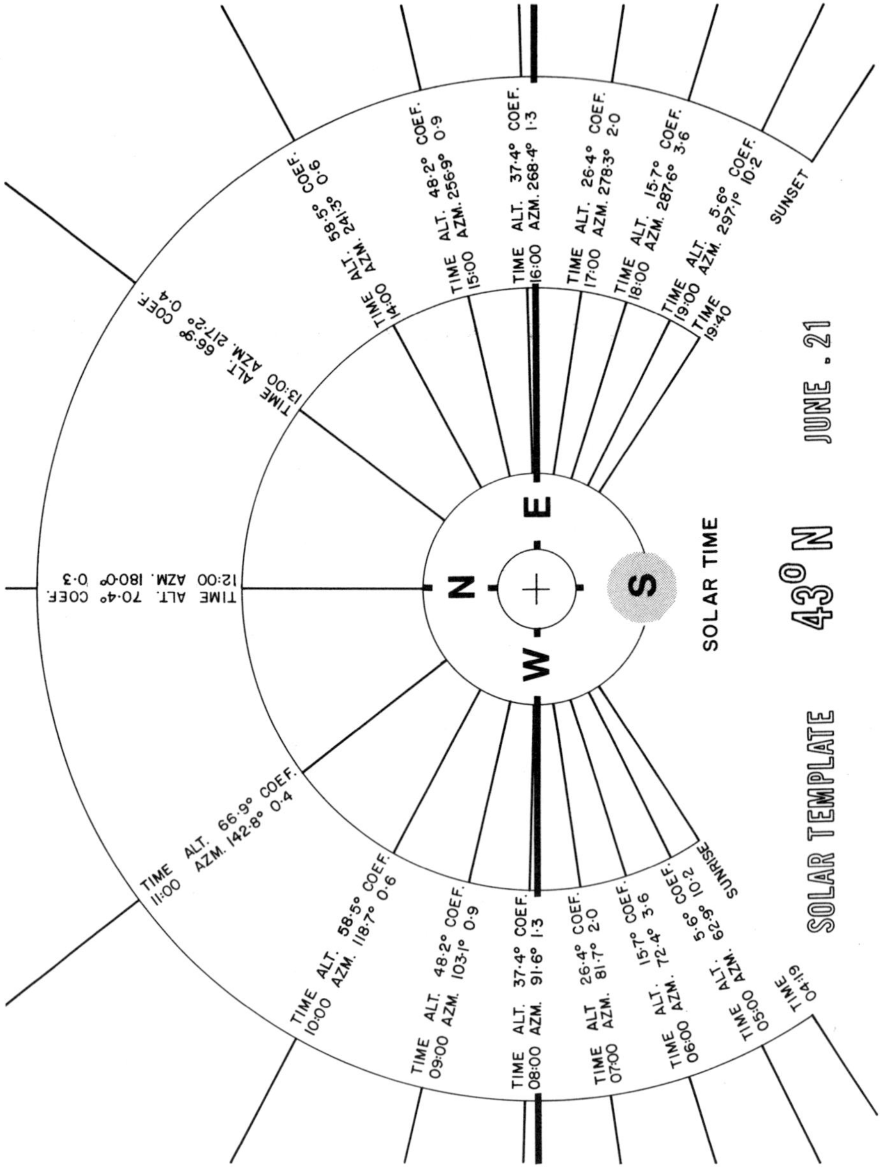

8-10. Solar Template #7—June 21 (the longest day of the year in the northern hemisphere).

TORONTO SHADOW TEMPLATE SET IS APPLICABLE FOR ANY CITY LOCATED BETWEEN 42°N AND 44°N LATITUDE

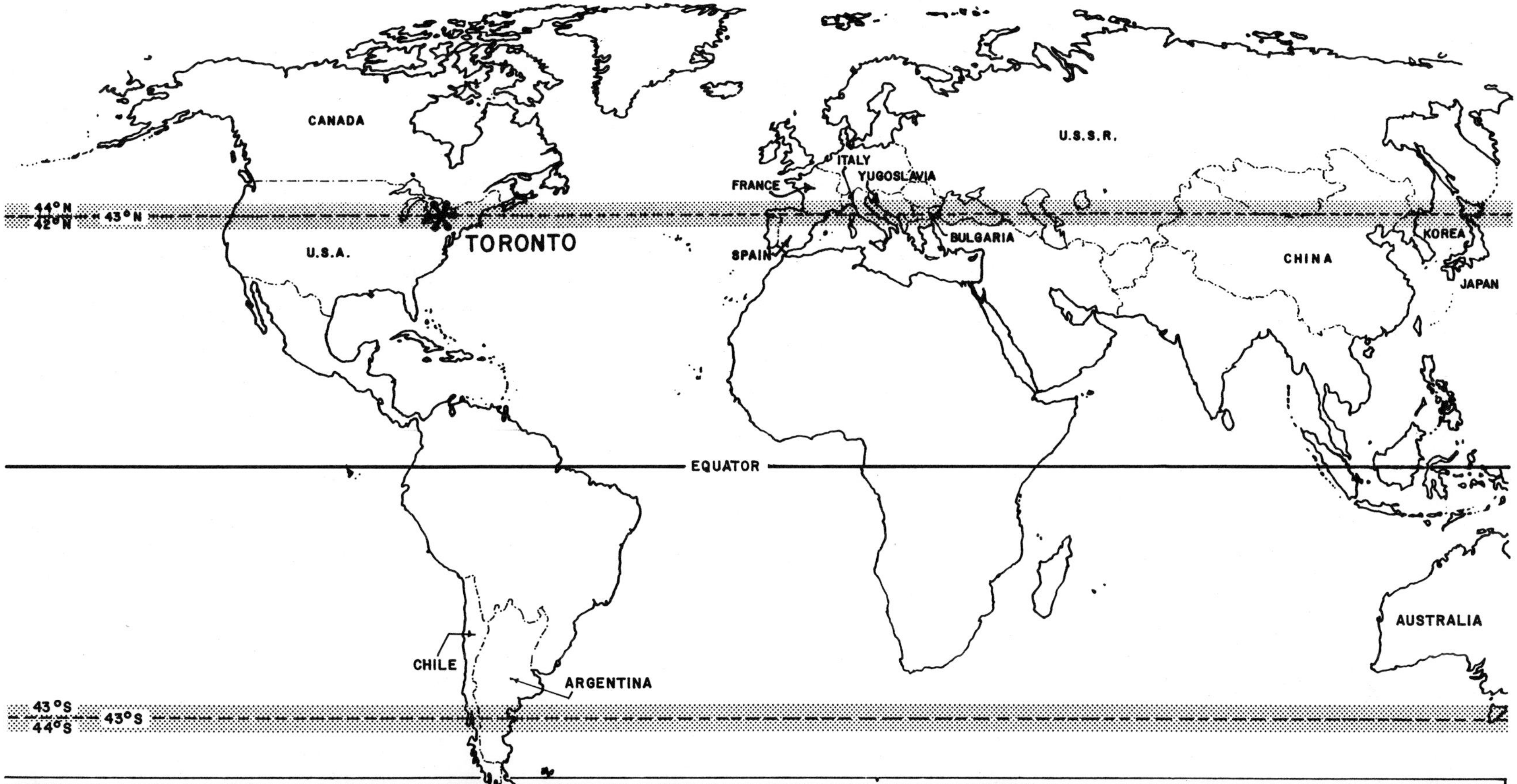

CANADA		U.S.A.		SPAIN	FRANCE	ITALY	YUGOSLAVIA	BULGARIA	U.S.S.R.		CHINA	N. KOREA	JAPAN
HAMILTON	GUELPH	BOSTON	DETROIT	BILBAO	NICE	ROME	SPLIT	SOFIJA	GROZNYJ	USSURIJSK	SHENYANG	CH'ONGJIN	SAPPORO
TORONTO	KITCHNER	CAMBRIDGE	WARREN	LA CORUNA	TOULOUSE	FIRENZE	SARAJEVO	RUSE	FRUNZE	SUCHUMI	FUSHUN		ASAHIKAWA
ETOBICOKE	WATERLOO	WORCESTER	LIVONIA	VIGO	MARSEILLE	FOLIGNO		VARNA	ALMA-ATTA	POTI	CHANGCHUN		OTARU
NORTH YORK	CAMBRIDGE	SPRINGFIELD	DEARBORN	OVIEDO	MONTPELLIER	LIVORNO			VLADIVOSTOK	MOZDOK	JILIN		MURORAN
SCARBOROUGH	BRANTFORD	ROCHESTER	FLINT	GIJON	NIMES	PRATO			SOCI	GORI	YINING		HAKODATE
NIAGARA FALLS	LONDON	BUFFALO	LANSING	SANTANDER	TOULON				KUTAISI	KIZL'AR	WULUMUQI		OBIHIRO
ST. CATHARINES	ST. THOMAS	ALBANY	GRAND RAPIDS	SAN SEBASTIAN					NAL'CIK		FUXINSHI		KUSHIRO
WELLAND	CHATHAM	UTICA	CHICAGO	PAMPLONA					ORDZONIKIDZE		SIPING		
BURLINGTON	WINDSOR	SYRACUSE	ROCKFORD						MACHACKALA		LIAOYUAN		
OAKVILLE	SARNIA	NIAGARA FALLS	MILWAUKEE						CIMKENT				
BRAMPTON		ERIE	MADISON						DZAMBUL				

8-11. Solar Templates constructed for 43°N latitude can be used within the 42°N and 44°N latitude strip.

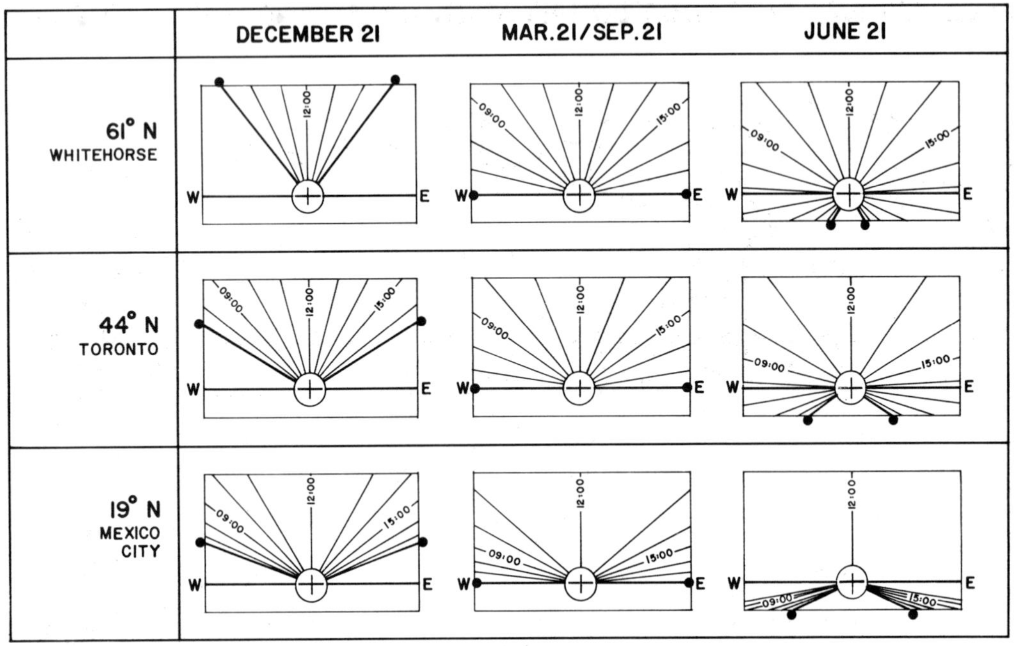

8-12. Comparison chart of selected Solar Templates.

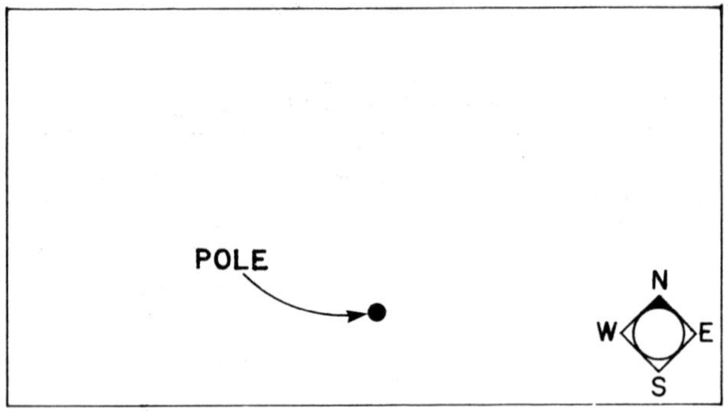

8-13. The plan.

ficients. However, the dates must be reversed. For example, the solar data shown for December 21 in northern latitudes would be equivalent to those for June 21 in southern latitudes.

Figure 8-12 compares selected Solar Templates. The direction of shadows from sunrise to sunset also indicates the level of solar exposure in a given time and place. There are profound variations between different places and different times of the year in terms of characteristics of shadows cast as well as solar energy receivership.

THE COMPOSITE SHADOW PATTERN OF A POLE

The simple task of constructing the shadows of a pole would proceed as follows:

GIVEN DATA

- Object: a pole, height = 3 units
- Date: December 21
- Time frame: 09:00, 10:00, 11:00, 12:00, 13:00, 14:00, and 15:00 hours (solar time)
- Latitude: 43°N
- Slope: 0° (a horizontal, flat plane)
- Direction of true north (solar azimuth at 12:00 hours solar time)
- Set of transparent Solar Templates designed for 43°N

STEP 1. Select the corresponding template, for example, December 21 (see fig. 8-4).

STEP 2. Overlay the center of the template on the center of the pole. Ascertain that the north direction on the template is parallel with the direction of true north on the plan (figs. 8-13, 8-14).

STEP 3. Transfer the required (seven) shadow directions from the template to the plan (fig. 8-15).

STEP 4. Determine the shadow length for each hour. Use the formula, shadow length = pole height × coefficient. The length of the shadow at each of the stipulated times is as follows:

09:00 = 3×4.829 = 14.48 units
10:00 = 3×3.078 = 9.23 units
11:00 = 3×2.463 = 7.39 units
12:00 = 3×2.289 = 6.86 units (solar noon, shortest shadow length, and axis of symmetry)
13:00 = 3×2.463 = 7.39 units
14:00 = 3×3.078 = 9.23 units
15:00 = 3×4.829 = 14.48 units

Note: In figure 8-4, coefficients are shown to two decimal places only.

STEP 5. Transfer the resultant shadow lengths to the drawing (fig. 8-16).

Note: The sum of the pole's projected shadows (the east/west movement of the pole's shadow) represents a composite shadow pattern (the impact area) of the pole within a stipulated time frame (09:00 to 15:00 hrs.).

THE COMPOSITE SHADOW PATTERN OF SOLID OBJECTS

Any object can be represented by a set of vertical poles. Figure 8-17 illustrates the transition from four poles to a solid object. The construction of the composite shadow of a small house is illustrated in figure 8-18. The proper translation of a solid object into a set of poles requires a certain amount of practice as does the identification of poles that are critical to the construction of a required shadow pattern.

In summary, the construction of a composite shadow pattern of a solid object requires three basic steps.

1. Abstract the object into a number of relevant poles.
2. Construct the impact area for each pole (these are partial impact areas).
3. Connect the outer edges of the partial impact areas to arrive at the total composite shadow pattern of a given object.

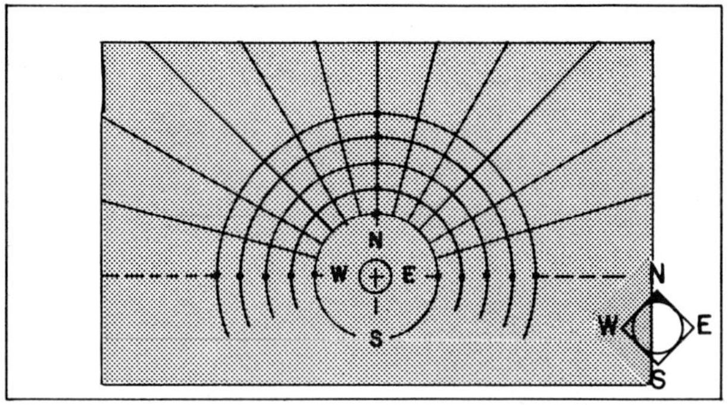

8-14. A template superimposed on the plan.

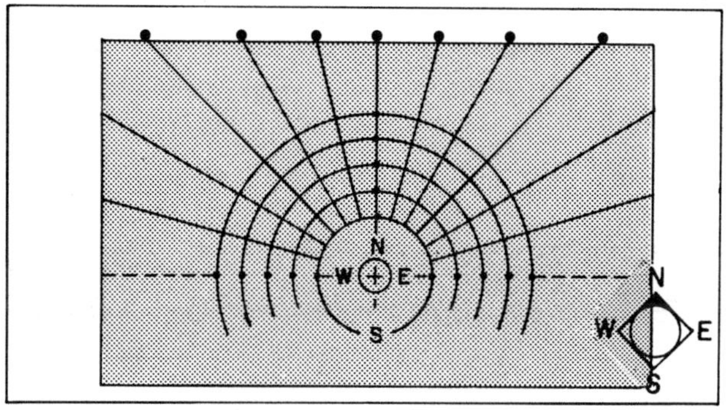

8-15. The seven shadow directions required are transferred from the template to the plan.

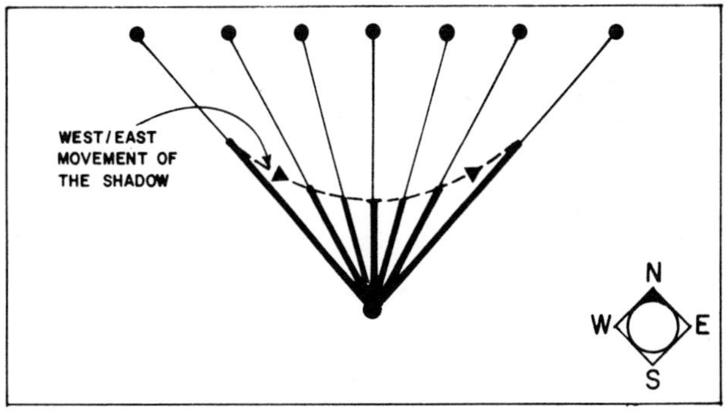

8-16. The actual length of shadows.

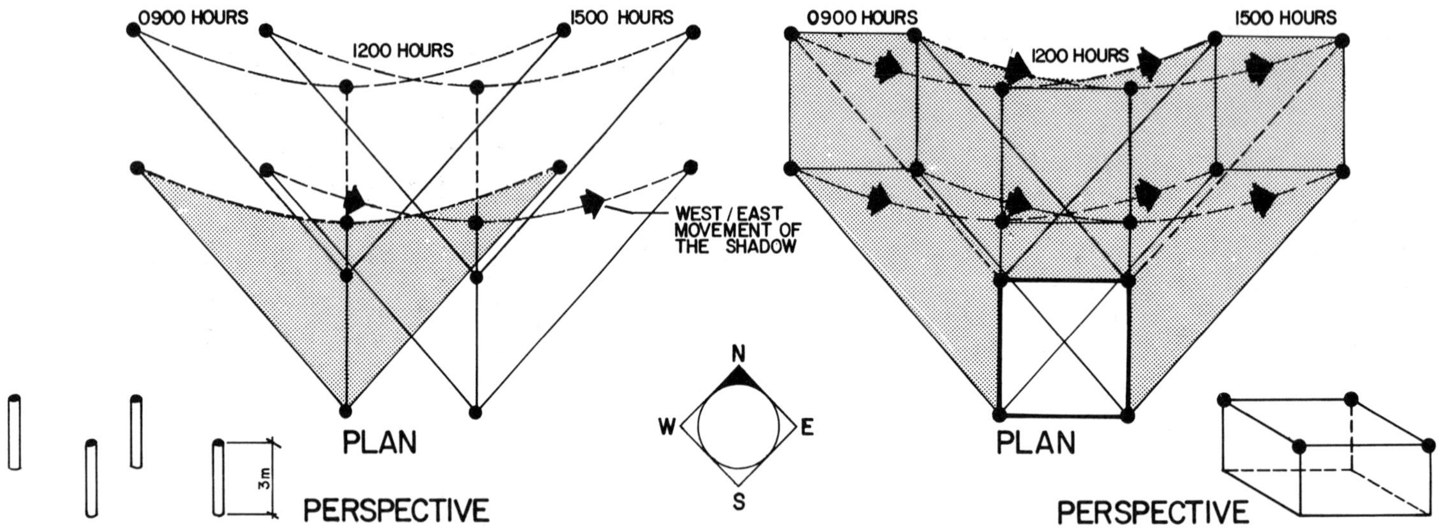

8-17. Four poles and a solid.

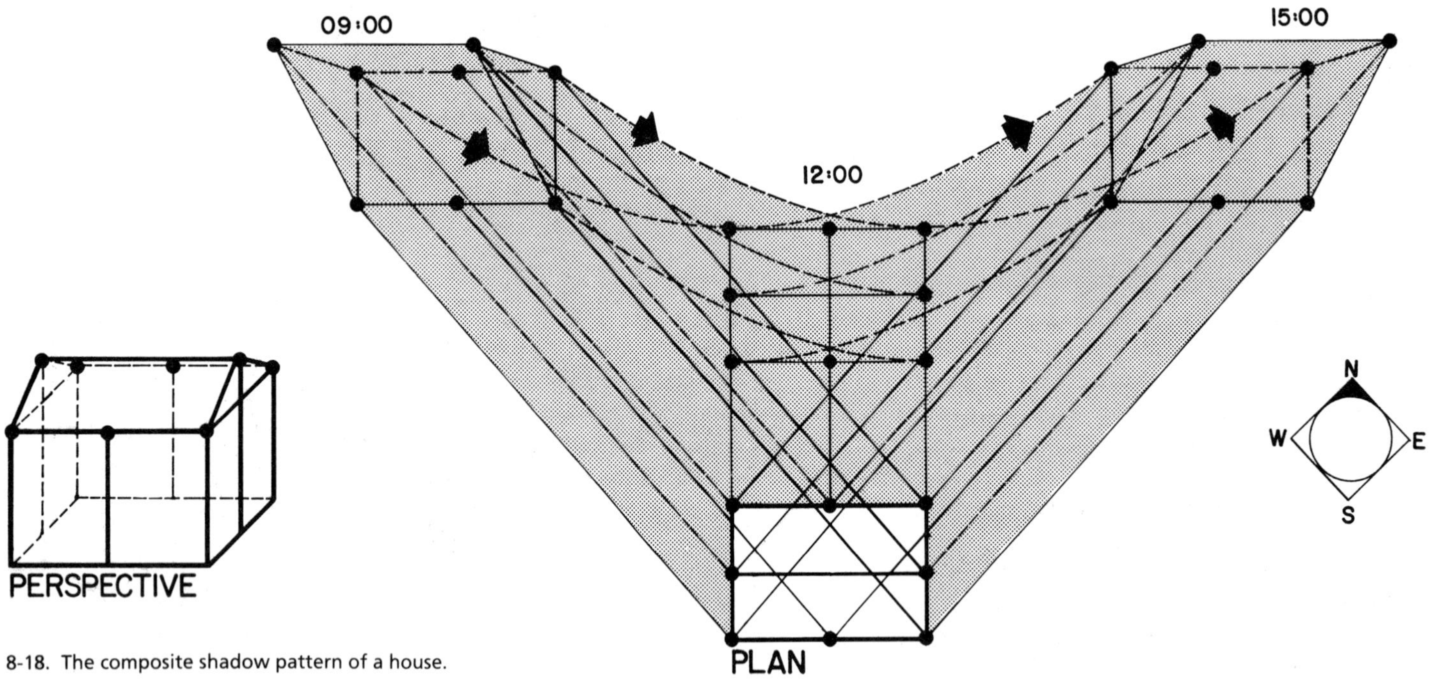

8-18. The composite shadow pattern of a house.

CONSTRUCTION OF SHADOW PATTERNS OVER INCLINED PLANES

In practice, flat, horizontal surfaces are rare. The following exercise will demonstrate the construction of composite shadow patterns on flat, uniform slopes. Figure 8-19 shows a nominal hill that has four slopes (northern, eastern, western, and southern) and a flat top. The conventional contours are uniformly distributed. Any straight line that is parallel with the contours would be exactly horizontal. All other lines are inclined. The degree of slope they represent is directly related to the angle of deviation from the direction of contours. In other words, a line parallel with the contours has a 0° slope; a gradual deviation from this direction represents various degrees of slope, and the line reaches the maximum degree of slope, 5°, when it is perpendicular to the contours. Radial lines superimposed over the nominal hill plan are shown in figure 8-20. This relationship does not change with the hill's orientation (fig. 8-21).

It is important to understand that when the Solar Template is superimposed over the plan of an inclined surface, each shadow direction has its own specific slope degree. The exceptions are shadow directions parallel with the contours (0° slope). The other extreme is shadow directions perpendicular to the contours (in our case, a 5° slope).

In addition, it must be determined which shadows are cast in an uphill direction (generally shorter), and which ones are cast in a downhill direction (generally longer) (fig. 8-22). For that purpose, an additional transparent template (fig. 8-23) can be prepared and used in conjunction with any Solar Template (figs. 8-24, 8-25). The method and formula for calculating coefficients are shown in figure 8-26.

The construction of a shadow pattern on inclined planes is described in the following example:

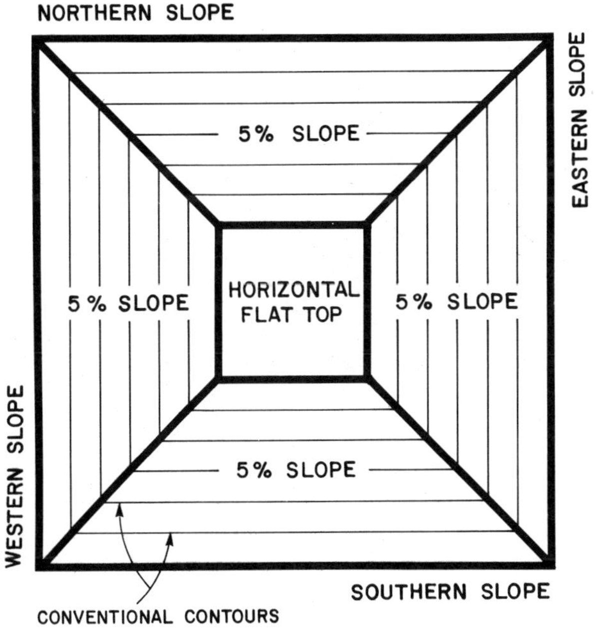

8-19. A nominal hill.

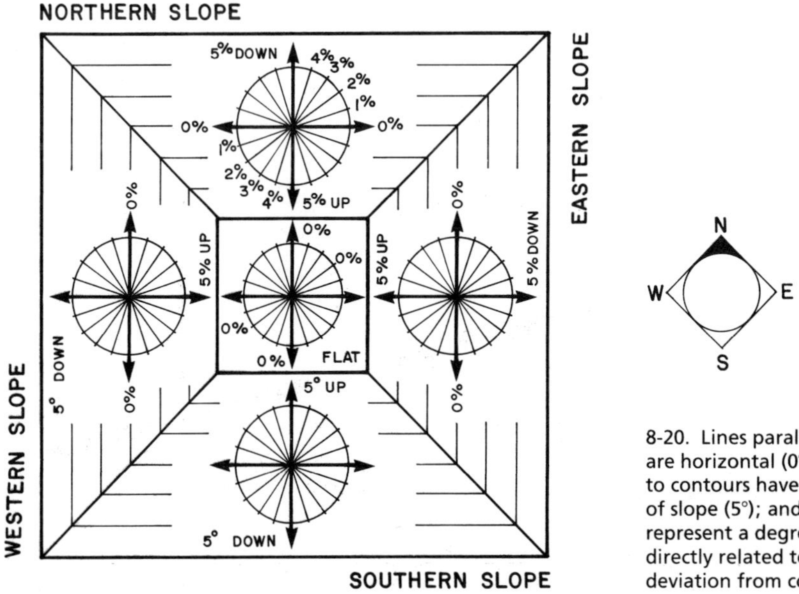

8-20. Lines parallel with the contours are horizontal (0°); lines perpendicular to contours have the maximum degree of slope (5°); and all other lines represent a degree of slope that is directly related to the angle of deviation from contours.

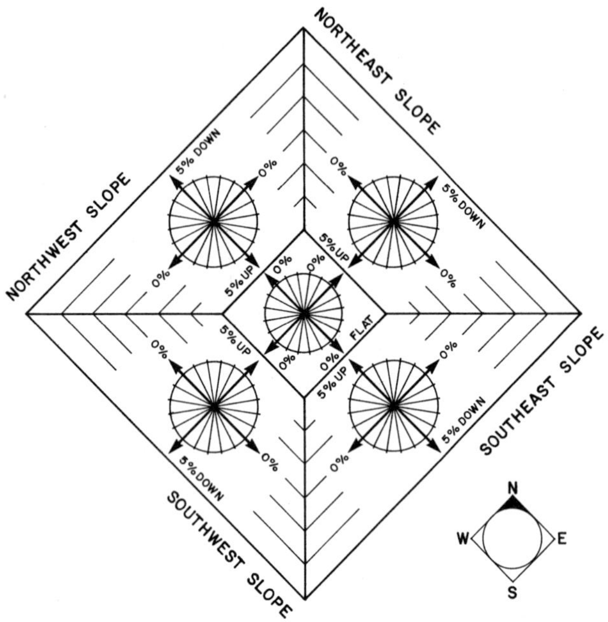

8-21. The principles described in figure 8-20 apply to a slope of any orientation.

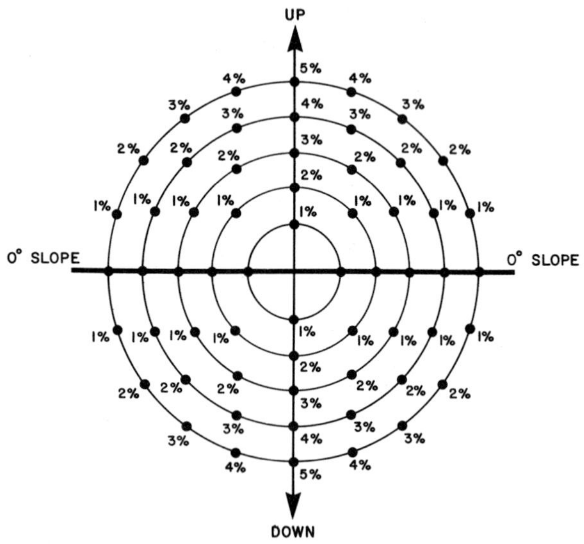

8-23. A template indicating the degree of slope (slope gradient) for different radial lines.

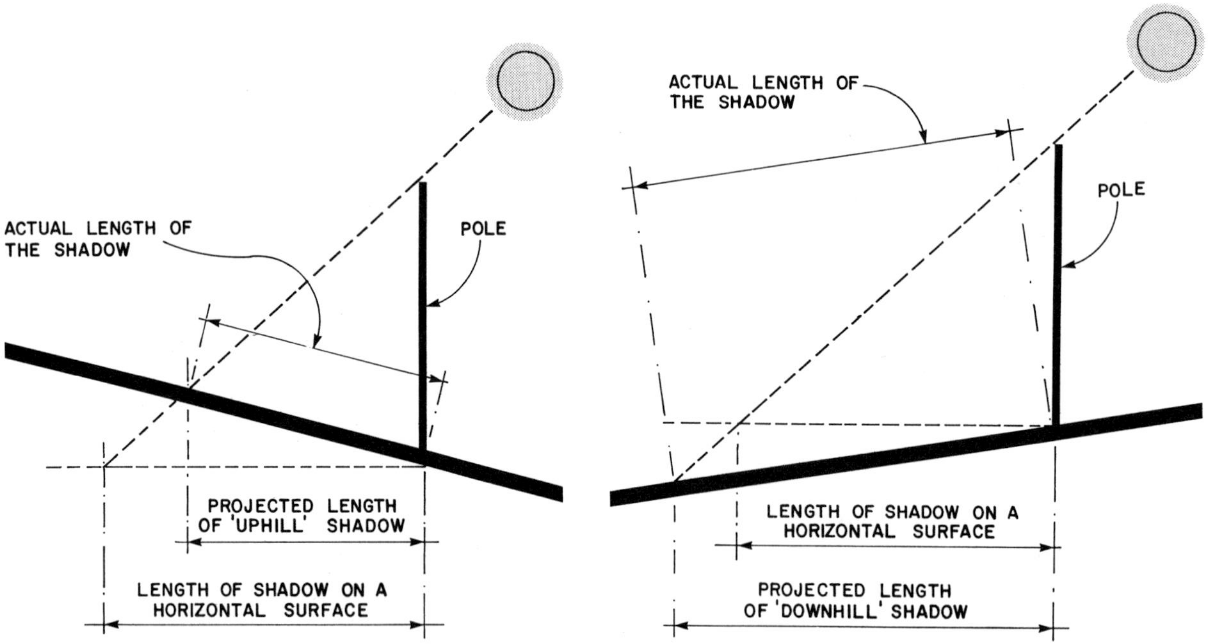

8-22. Shadows cast in uphill and downhill directions.

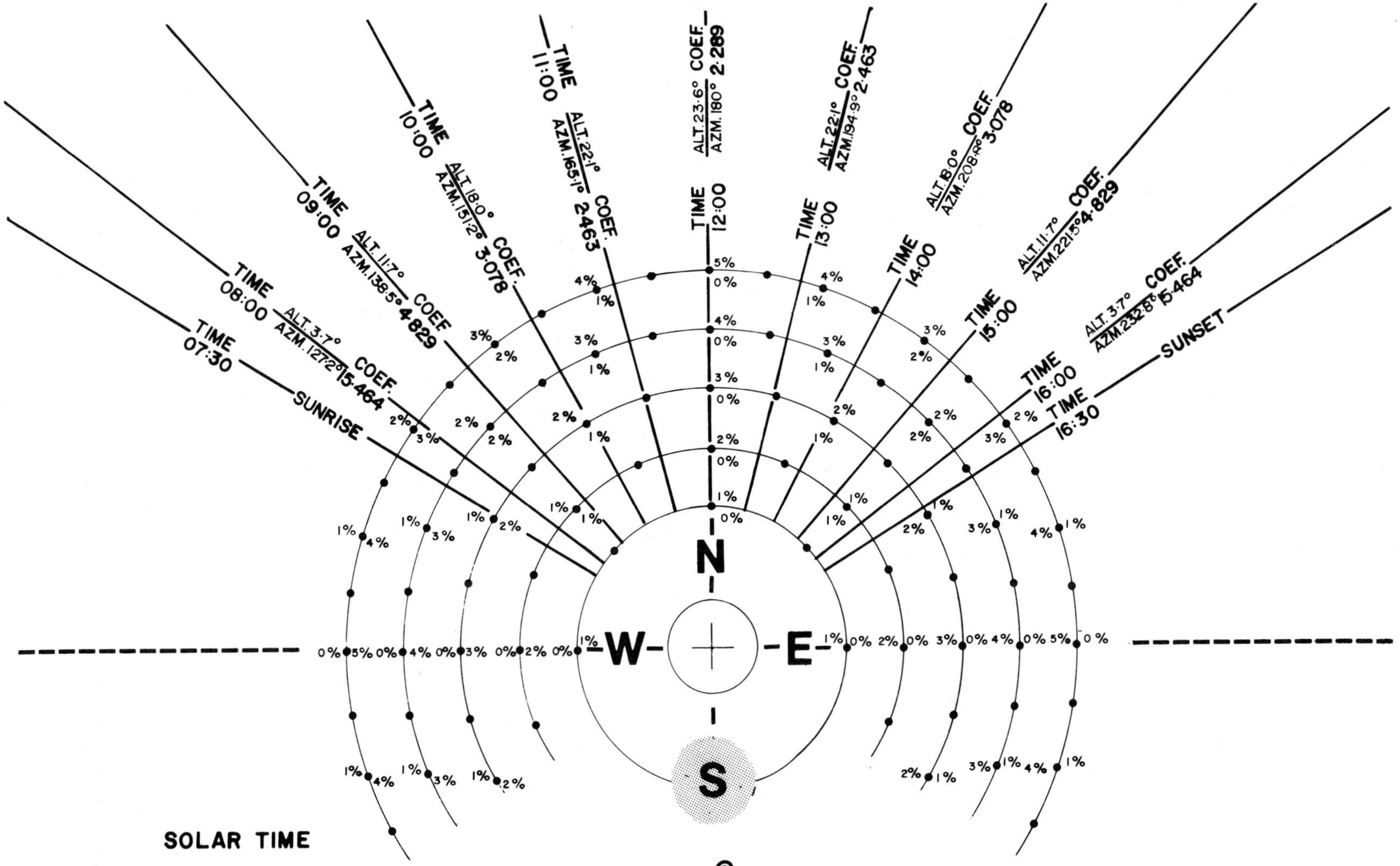

43° N **DEC. 21**

SHADOW LENGTH = HEIGHT x COEF. HEIGHT = SHADOW LENGTH ÷ COEF.

SOLAR TIME

SLOPE IN % AND DIRECTION		05:00 19:00 HOURS	06:00 18:00 HOURS	07:00 17:00 HOURS	08:00 16:00 HOURS	09:00 15:00 HOURS	10:00 14:00 HOURS	11:00 13:00 HOURS	12:00 HOURS
1%	UP				13·395	4·607	2·986	2·403	2·238
	DOWN				18·289	5·014	3·176	2·524	2·342
2%	UP				11·815	4·404	2·900	2·347	2·189
	DOWN				21·745	5·344	3·280	2·590	2·398
3%	UP				10·562	4·218	2·818	2·293	2·142
	DOWN				28·863	5·647	3·392	2·659	2·458
4%	UP				9·552	4·047	2·741	2·242	2·097
	DOWN				40·570	5·986	3·511	2·732	2·520
5%	UP				8·723	3·890	2·669	2·193	2·054
	DOWN				68·075	6·365	3·638	2·808	2·584

8-24. A Solar Template for December 21 with slope gradients for different times of day.

SLOPE IN % AND DIRECTION		05:00 19:00 HOURS	06:00 18:00 HOURS	07:00 17:00 HOURS	08:00 16:00 HOURS	09:00 15:00 HOURS	10:00 14:00 HOURS	11:00 13:00 HOURS	12:00 HOURS
1 %	UP				13·395	4·607	2·986	2·403	2·238
	DOWN				18·289	5·014	3·176	2·524	2·342
2 %	UP				11·815	4·404	2·900	2·347	2·189
	DOWN				21·745	5·344	3·280	2·590	2·398
3 %	UP				10·562	4·218	2·818	2·293	2·142
	DOWN				28·863	5·647	3·392	2·659	2·458
4 %	UP				9·552	4·047	2·741	2·242	2·097
	DOWN				40·570	5·986	3·511	2·732	2·520
5 %	UP				8·723	3·890	2·669	2·193	2·054
	DOWN				68·075	6·365	3·638	2·808	2·584

8-25. Coefficients for a variety of slope gradients to be used in conjunction with the December 21 Solar Template.

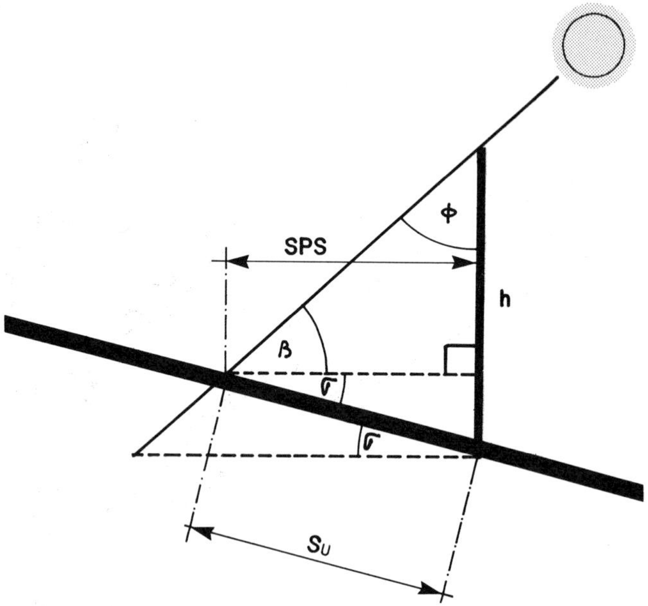

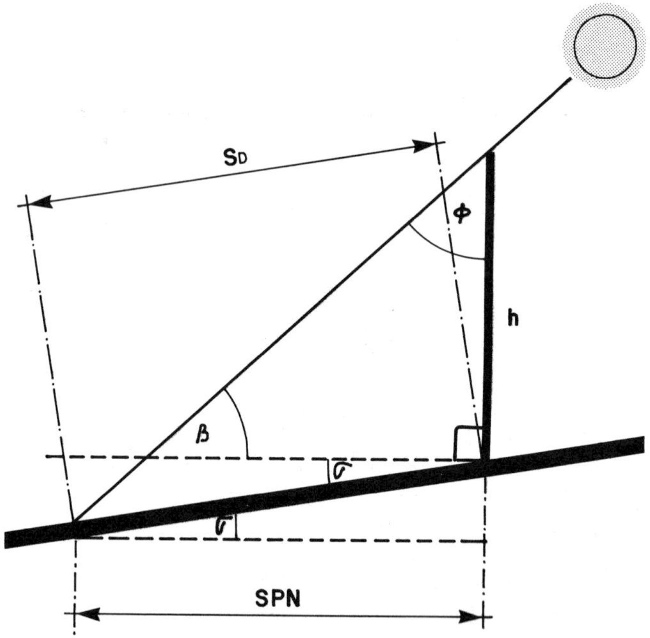

8-26. *(Right)* The formula for calculating the length of horizontal projections of uphill and downhill shadows. If h equals one, the result is a number that may be used as a multiplier (coefficient) for a variety of pole heights.

KNOWN:
Solar Altitude "β"

Pole Height "h"

Slope Gradient "γ"

Formula for SPS: $SPS = \dfrac{h \cos \gamma \cos \beta}{\sin (\beta + \gamma)}$

Formula for SPN: $SPN \dfrac{h \cos \gamma \cos \beta}{\sin (\beta - \gamma)}$

TASK

To construct the composite shadow pattern of a pole.

GIVEN DATA

- Object: a pole, height = 3 units
- Date: June 21
- Time frame: 06:00, 07:00, 08:00, 09:00, 10:00, 11:00, 12:00, 13:00, 14:00, 15:00, 16:00, 17:00, 18:00 hours (solar time)
- Latitude: 43°N
- Slope: 5% north (a uniform, inclined plane)
- Direction of true north (solar azimuth at 12:00 hrs. solar time)
- Set of solar templates designed for 43°N

Note: The procedure is the same as in the construction of the shadow pattern of a pole on a flat, horizontal surface, with one additional step.

STEP 1. Select the corresponding template, for example, June 21 (see fig. 8-10).

STEP 2. Overlay the center of the template onto the center of the pole. Ascertain that north on the template is parallel with true north on the plan.

STEP 3. Transfer the required (thirteen) shadow directions from the template to the plan.

STEP 4 (the additional step). Determine whether each shadow line is cast uphill or downhill and the percent of its specific slope (see the northern slope in fig. 8-20).

STEP 5. Determine the projected shadow length for each hour, using the formula, shadow length = pole height × coefficient. In this case, each shadow direction requires its own coefficient (see figs. 8-20, 8-23). Table 8-1 shows the calculation of shadow lengths for different times and slope gradients.

STEP 6. Transfer the resultant shadow lengths to the drawing and connect the ends of the shadows cast.

TABLE 8-1
Calculation of Shadow Lengths

TIME	% SPECIFIC SLOPE	HEIGHT × COEFFICIENT		LENGTH OF SHADOW
06:00	1% up	3 × 3.399	=	10.197 units
07:00	½% up	3 × 1.994	=	5.982 units
08:00	0%	3 × 1.308	=	3.924 units
09:00	¾% down	3 × 0.9001	=	2.700 units
10:00	1½% down	3 × 0.6185	=	1.855 units
11:00	3% down	3 × 0.432	=	1.296 units
12:00	5% down	3 × 0.363	=	1.089 units
13:00	3% down	3 × 0.432	=	1.296 units
14:00	1½% down	3 × 0.6185	=	1.855 units
15:00	¾% down	3 × 0.9001	=	2.700 units
16:00	0%	3 × 1.308	=	3.924 units
17:00	½% up	3 × 1.994	=	5.982 units
18:00	1% up	3 × 3.399	=	10.197 units

The construction of shadow patterns over inclined surfaces is a fairly simple drafting exercise provided that the relationships between time, shadow direction, shadow length, and slope gradient are understood. On flat horizontal planes, shadow direction and shadow length change with time. On inclined planes, not only shadow direction and shadow length but also slope gradient changes with time, that is, the shadow may be climbing uphill or running downhill.

Solar Templates must always be aligned with the true north/south direction regardless of the actual slope orientation (figs. 8-27, 8-28).

CONSTRUCTION OF SHADOW PATTERNS OVER IRREGULAR TOPOGRAPHY

Flat horizontal and flat inclined surfaces are usually human-made. The application of the above techniques is therefore limited to carefully landscaped areas and to an urban paved environment. But a vast number of

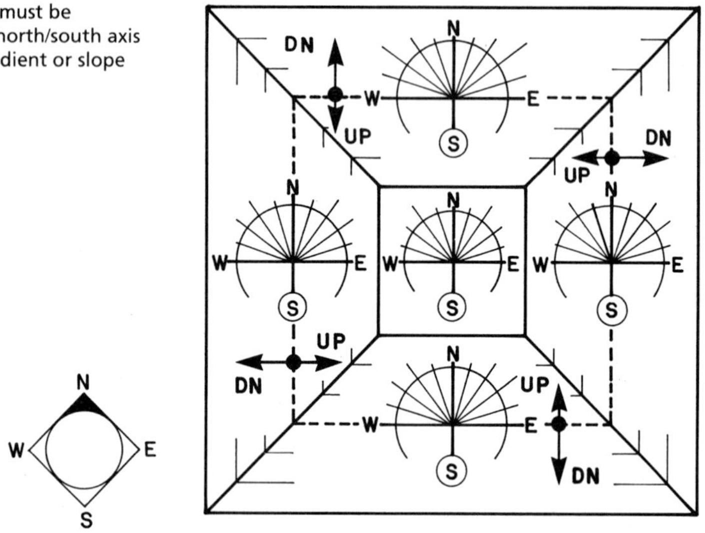

8-27. Solar Templates must be oriented along a true north/south axis regardless of slope gradient or slope orientation.

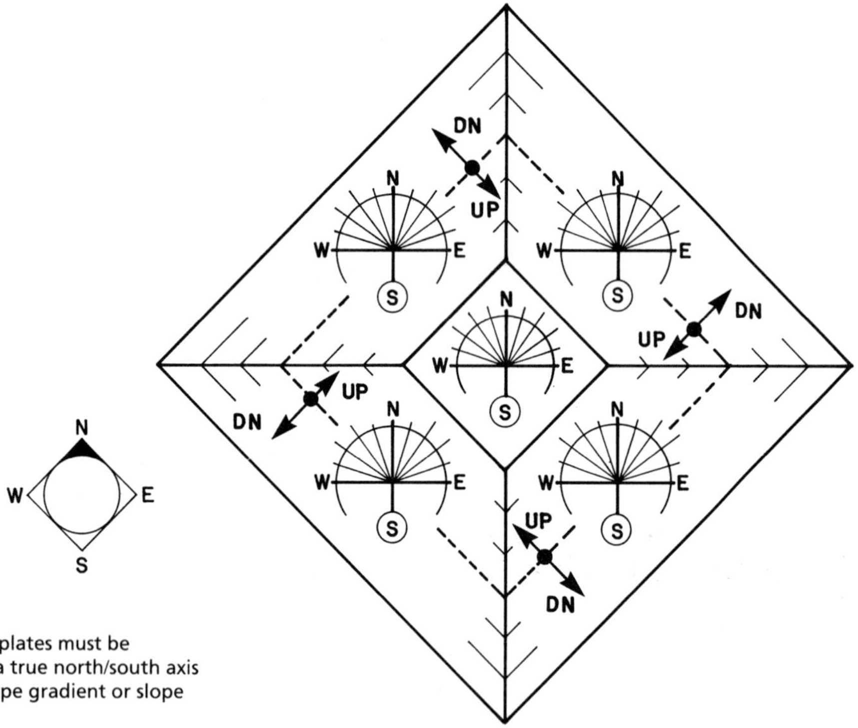

8-28. Solar Templates must be oriented along a true north/south axis regardless of slope gradient or slope orientation.

tasks require the construction of cast shadow patterns over an irregular topography. Normally, this is a cumbersome, complex exercise. The following method is very fast and efficient, because shadow patterns can be constructed from plans only, with no sections required.

Figure 8-29 shows a pole (height = 10 units) and topography described by conventional contours. It also shows the composite shadow pattern of the pole as if it were projected on a flat, horizontal plane (shaded area), and the actual shadow pattern cast over a given topography (the dotted line). The challenge is to determine where the sunbeam intersects with the ground and to avoid the time-consuming construction of sections in each shadow direction. This can be achieved by assigning a scale to the sunbeam.

Suppose each contour represents a drop of five units in elevation. The pole is situated on a 0 level and is 10 units high. The first step is to determine the length of the shadow as if it were cast on flat ground. This is done using a Solar Template as previously described. The end of the shadow represents the intersection of the sunbeam and the horizontal plane. It is called *point 0*. Since the top of the pole represents point +10, the sunbeam would intersect a theoretical +5 plane at point +5. The +5 point is located exactly in the middle, between points +10 and 0. This interval can be plotted along the direction of the sunbeam (the direction of the shadow). These additional points would represent intersections with horizontal planes −5, −10, −20, and so on (see fig. 8-30). (Note that the elevation is included for explanatory purposes only—the whole exercise would be performed on a plan.)

In figure 8-31, a calibrated (scaled) beam intersects the given topography. Since each contour represents the edge of a horizontal plane, for example, contour −10 indicates the very end of level −10, the meeting point between the sunbeam and the topography is simple to locate. It must be on one of the points of the calibrated sunbeam. The intersection occurs where the point of the sunbeam and the level have the same number. In the example shown in figure 8-31, it must be point −15.

The process of constructing shadow patterns over an uneven topography can be summarized in a few steps.

1. Construct the impact area of a pole as if cast on a flat horizontal surface.
2. Calibrate each sunbeam.
3. Determine the lengths of shadows.
4. Draw the composite shadow pattern.

CONSTRUCTION OF SOLAR ENVELOPES

A Solar Envelope describes the maximum developable volume. It represents the shape of the largest possible building with a permissible impact area. Any building constructed within the limits of the Solar Envelope would have the same or smaller impact area.

Therefore, the construction of Solar Envelopes must be derived from impact areas (fig. 8-32). The generic form of a Solar Envelope is shown in figure 8-2 (see the axonometry). It resembles a three-sided figure, hence the name *pyramid*. The base is formed by the impact area, and the height is determined by the height of the pole. If a building were shaped to fit this envelope exactly, such a building would not cast any shadow between 09:00 and 15:00 hours. The volume of a Solar Pyramid can be manipulated in one of two ways.

1. Alter the height of the pole (to accommodate larger and smaller developments).
2. Move the pole forward, backward, or to the side (to accommodate longitudinal building forms, such as row houses on east/west and north/south streets) (see chapter 4).

The Solar Pyramid and its variants can thus assist designers in planning new developments and in determining distances between buildings, sun access, and the allocation of solar harnessing devices.

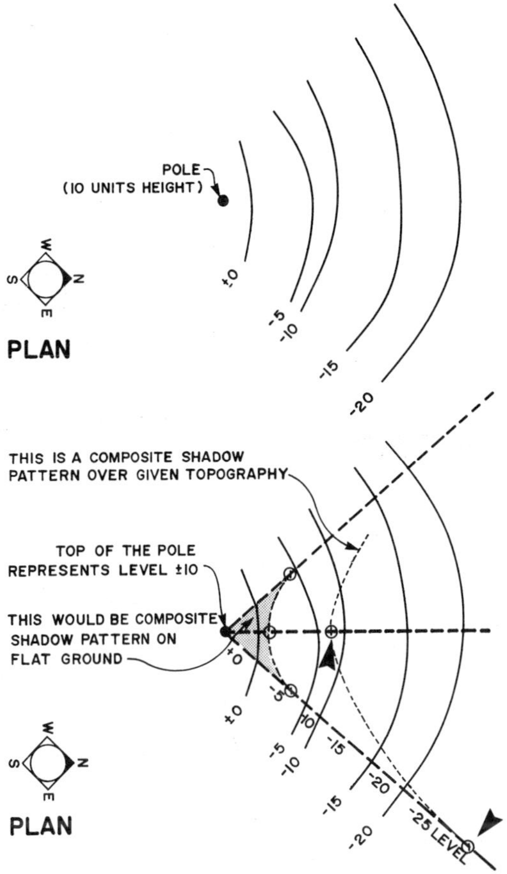

8-29. The construction of a composite shadow pattern over irregular topography.

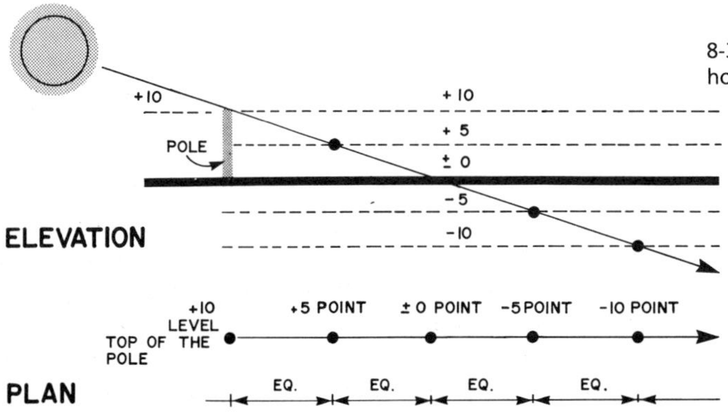

8-30. A sunbeam intersecting a horizontal plane.

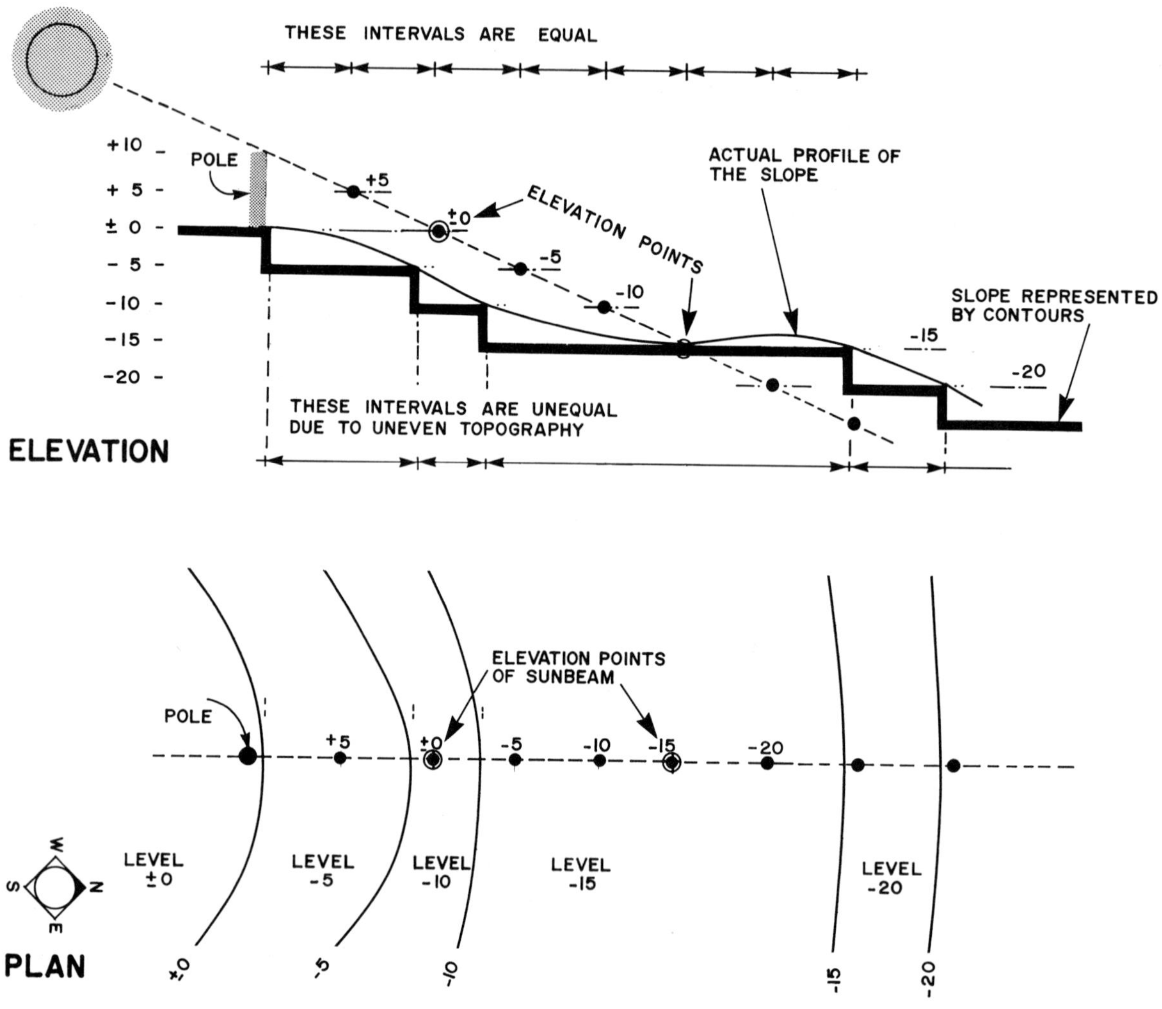

8-31. A sunbeam intersecting a given topography.

Very frequently, however, the Solar Envelope is used to help determine the size of a development to be built in an already developed area (fig. 8-32). The first step is to determine the extent of the impact area. Then, by using the formula, pole height = shadow length ÷ coefficient, the individual poles of the Solar Envelope are constructed. In principle, the construction of a Solar Envelope is the reverse of the construction of the composite shadow pattern of a three-dimensional (solid) object:

object → poles → impact area
impact area → poles → object

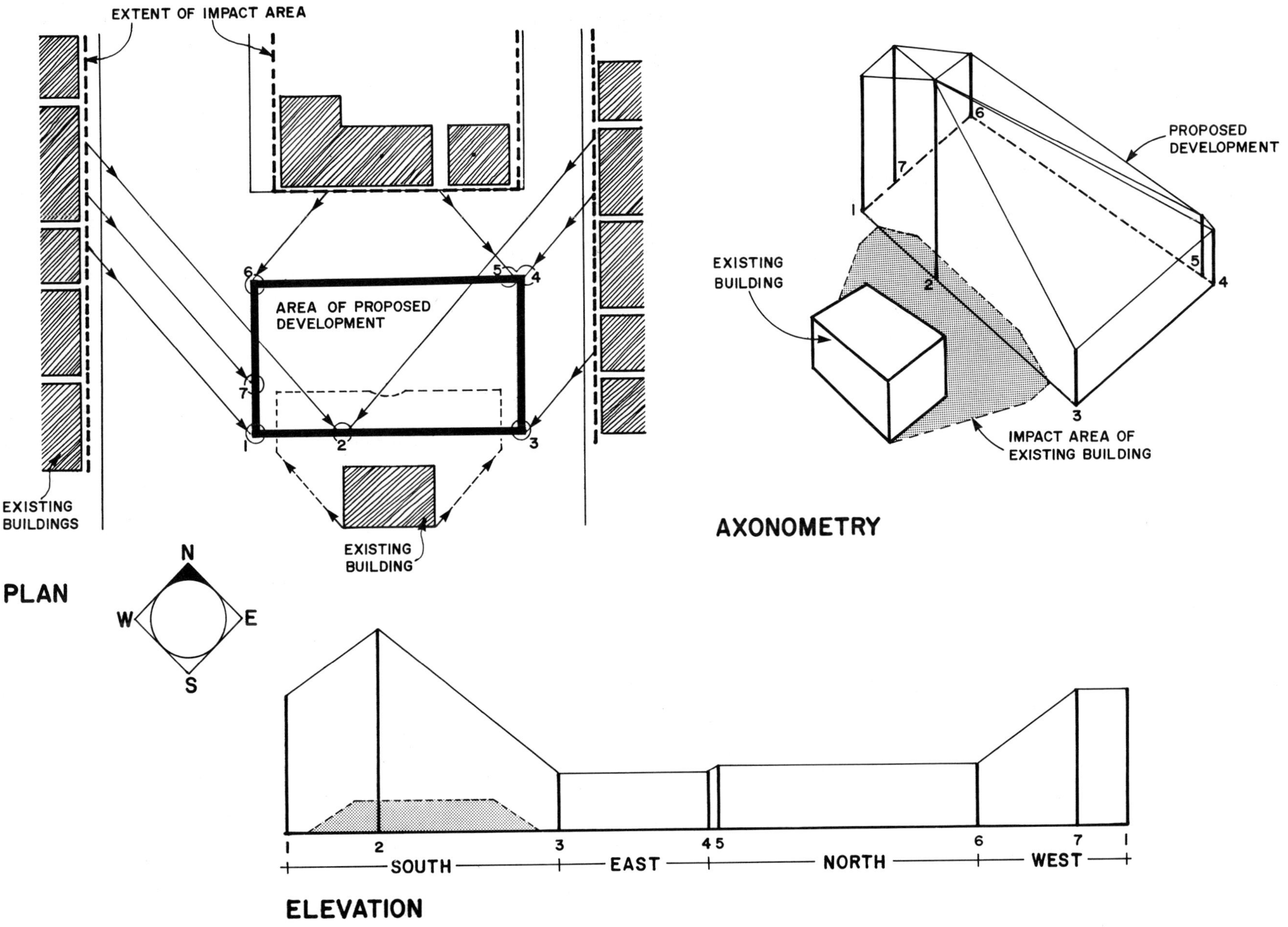

8-32. The construction of a Solar Envelope.

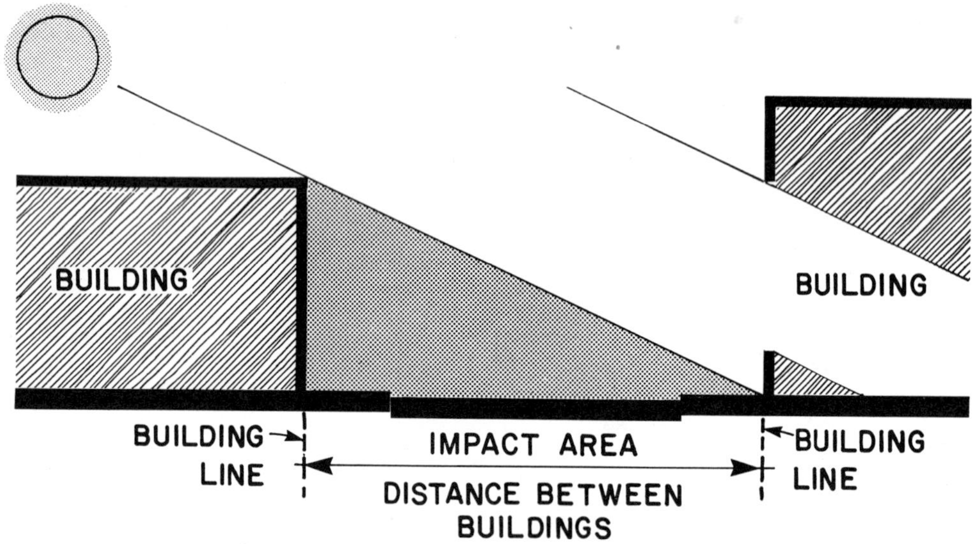

8-33. The distance between buildings equals the extent of the impact area.

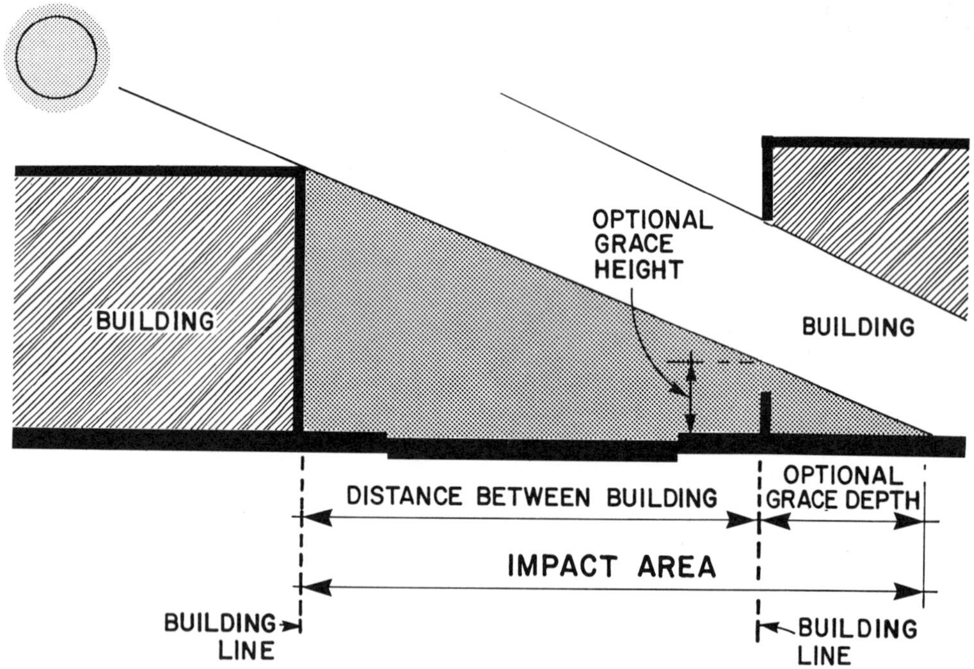

8-34. The distance between buildings equals the extent of the impact area minus the optional grace depth.

Certain existing structures may occasionally cast shadows on the site of a proposed development. These impact areas should be recorded (see fig. 8-32) to provide the architect with a guide for fenestration design and the allocation of solar harnessing systems (sun spaces, greenhouses, solar collectors).

Solar Envelopes can be constructed not only over flat horizontal ground but also over inclined planes (uniform slopes) and over a complex topography.

THE SPATIAL RELATIONSHIPS OF BUILDINGS

Ideally, one building should not be constructed within the impact area of another building (fig. 8-33). Therefore, the distance between two buildings should be equal to the depth of the impact area. The concepts of optional grace height and optional grace depth are introduced in cases where it is necessary to compromise this rule (fig. 8-34). In this case, the distance between buildings should be equal to the depth of the impact area minus the optional grace depth. The optional grace height, that is, the maximum permissible level of southern wall overshadowing, is determined by the building designers. It would therefore follow that:

$$\text{optional grace depth} = \text{optional grace height} \times \text{coefficient}$$

This is the same equation as:

$$\text{shadow length} = \text{pole height} \times \text{coefficient}$$

SOLAR IMPREGNATION OF INTERNAL SPACES

The sunlit areas of internal surfaces, both horizontal and vertical, can be outlined by a simple method using Solar Templates. Figure 8-35 shows a transparent vertical wall with a solid rectangular panel in the center. Constructing a shadow of this vertical panel is an elementary task. In figurc 8-36, the figure is reversed: a solid vertical wall has a rectangular opening, a window. The projection of the window on the horizontal plane (the floor) is constructed using the same method as in figure 8-35.

The depth of penetration varies during the year (fig. 8-37). The daily variations, a composite sun penetration pattern, can be constructed using the same procedure as in constructing the composite shadow patterns of solid objects.

A construction method illustrating window projection on vertical surfaces (walls) is shown in figure 8-38.

Projection on a wall = floor extension ÷ coefficient

Note that this is the same formula as:

pole height = shadow length ÷ coefficient

Note also that:

floor extension =
maximum projection on the floor – room depth

A designer can manipulate the amount of direct sunlight using a number of methods, such as orientation of the room, size and shape of the opening, and fixed and movable horizontal and vertical louvres. In all cases, performance can be determined with the help of Solar Templates.

1. For protection against a high sun, a horizontal element is attached to the outside face of the building, above the opening (fig. 8-39). In reality, it could be an overhang, balcony, or just a slab with no other function. The width of the horizontal element and the height above the opening are two variables to consider when adjusting the amount of direct sunlight. In northern latitudes, this method applies mainly to facades with a southern exposure.
2. For protection against a low sun, a vertical element is placed in front of the opening on the outside (fig. 8-40). In reality, this could be a solid fence, a wall, or a suspended panel in front of the opening. The height and the distance from the opening are two variables available to the designer. This approach may be used on facades with an eastern or western exposure. One must carefully block the low sun without blocking too much of the view.

CONSTRUCTION OF SHADOW PATTERNS ON VERTICAL SURFACES

An Overhang

A simple task is to construct the shadow cast by an overhang on the wall of a house at a given time (fig. 8-41). In order to use Solar Templates, this task must be reformulated by following two steps (fig 8-42).

1. Abstract what is given (the overhang) and what is to be determined (the extent of the shadow).
2. Calculate using the formula, pole height = shadow length ÷ coefficient.

Note: As in the previous cases, the task can be reversed. If the maximum extent of the shadow cast by the overhang is given, then the size of the overhang is calculated from this equation:

shadow length = pole height × coefficient

A Horizontal Peg

A horizontal peg assists in determining shadow patterns on vertical walls cast by various architectural features and elements (fig. 8-43). As in the case of the overhang, reformulation must take place (fig. 8-44). Then the center of the selected Solar Template, say December 21, is placed onto the end of the peg. Again, north on the template is parallel with true north on the plan (fig. 8-45). Dimension "a" is determined by the solar azimuth before and after noon. At solar noon, "a"

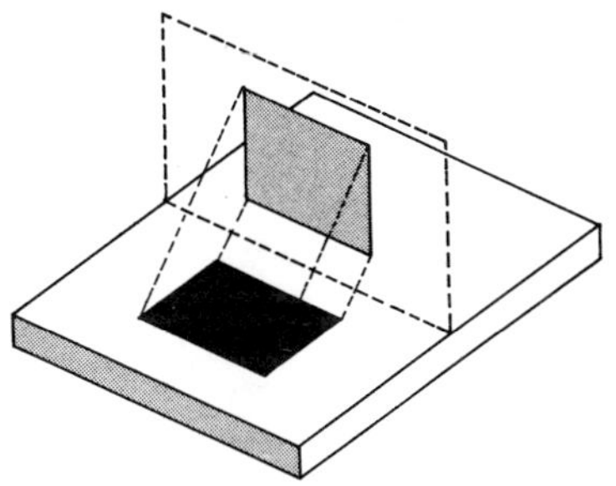

8-35. The shadow of a vertical panel.

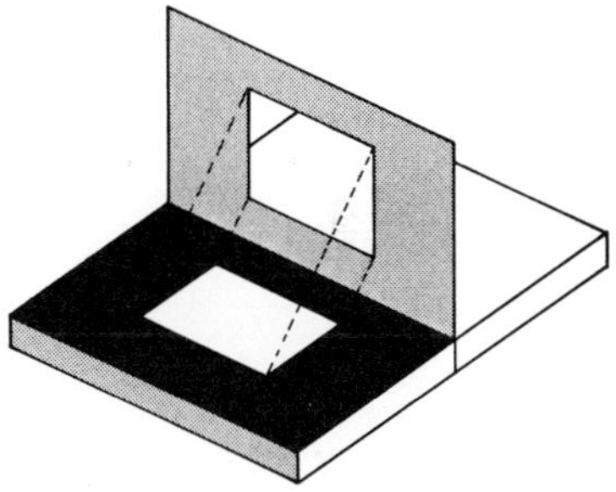

8-36. The shadow of a vertical wall with an opening.

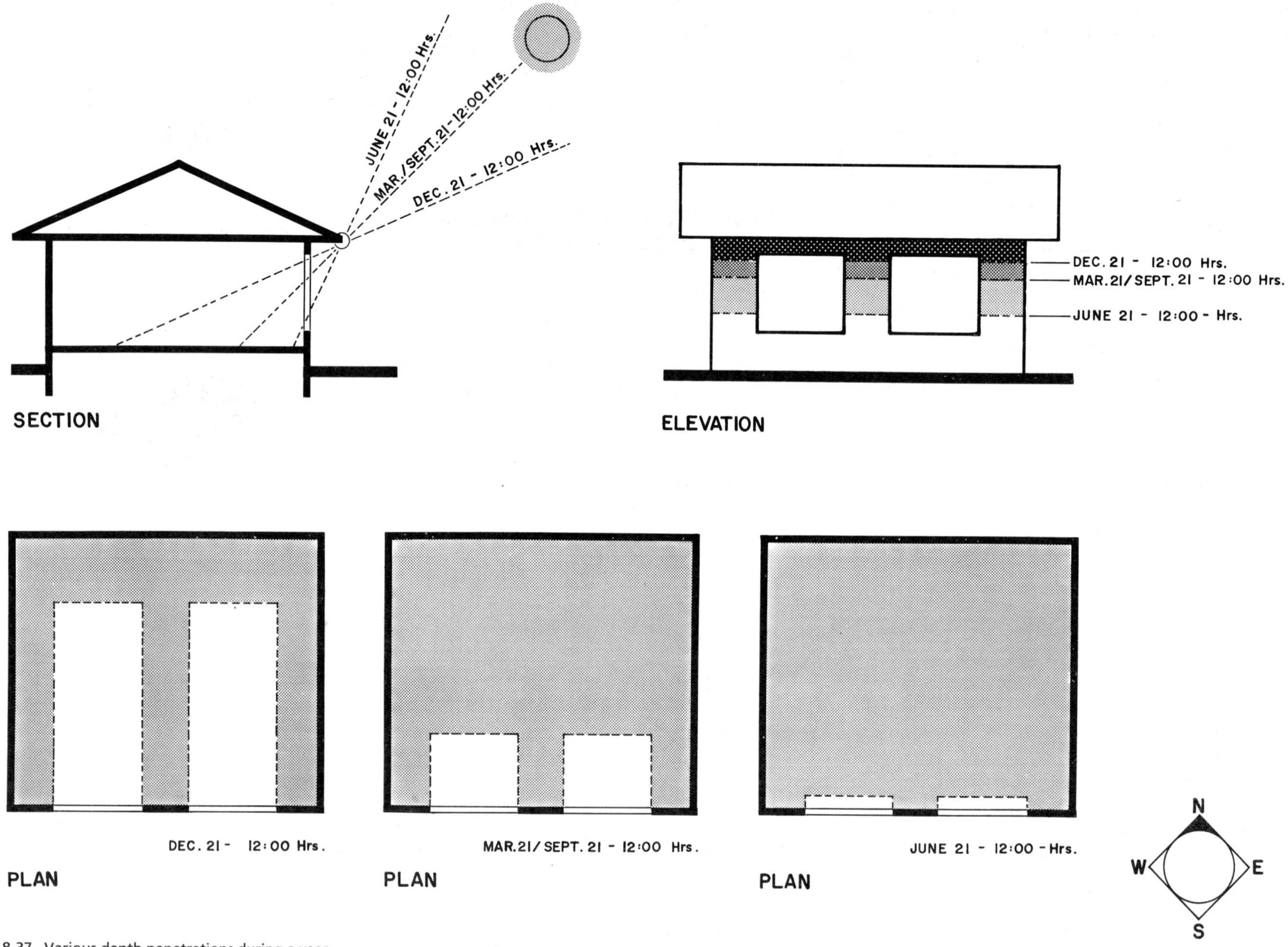

8-37. Various depth penetrations during a year.

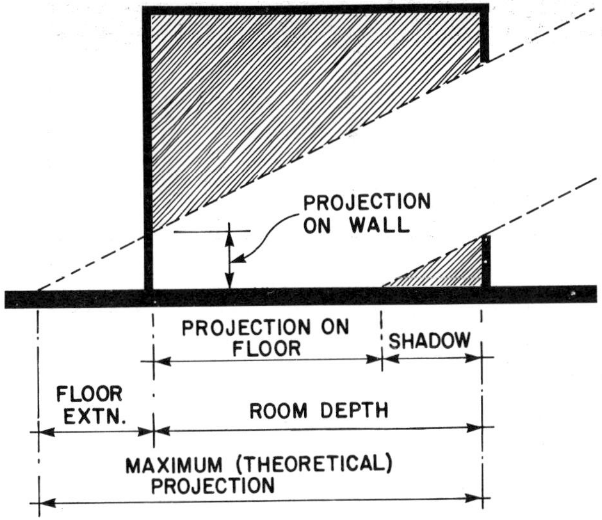

8-38. A window projection on a floor and wall.

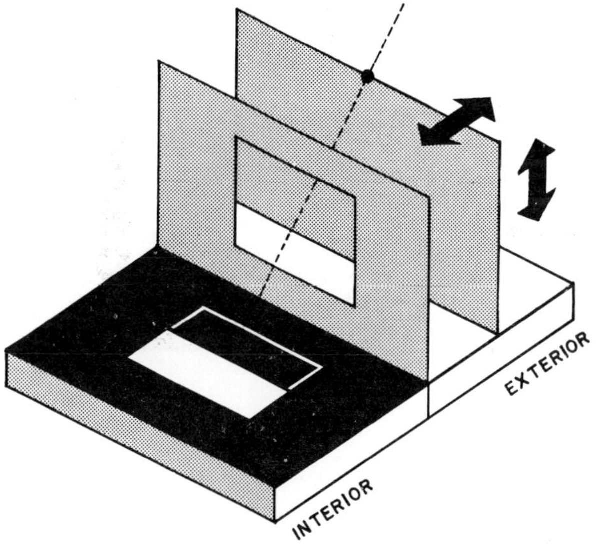

8-40. Protection against a low sun.

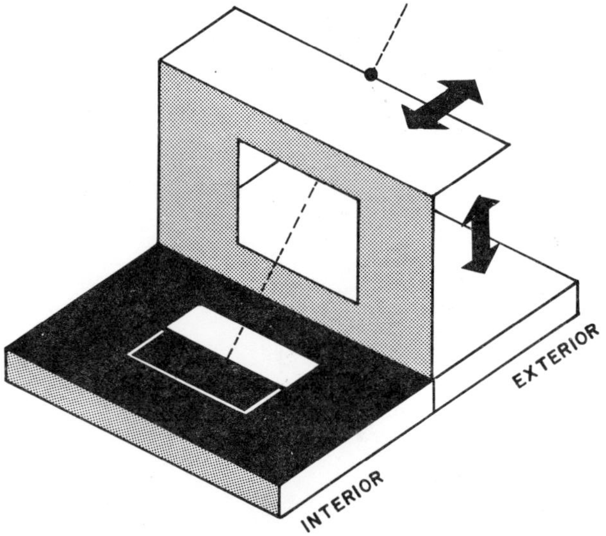

8-39. Protection against a high sun.

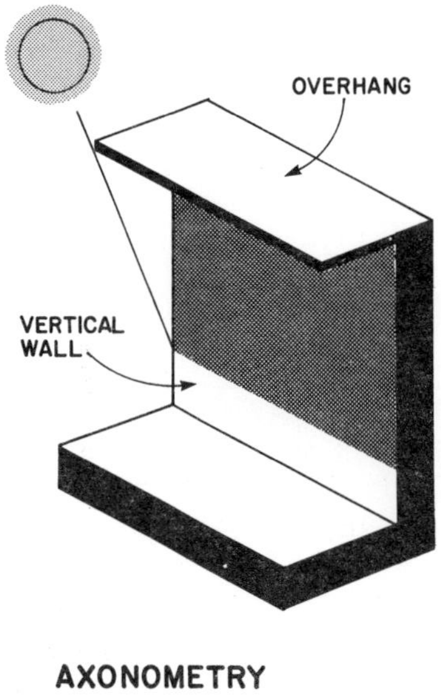

8-41. The construction of an overhang to cast a shadow on a wall oriented due south.

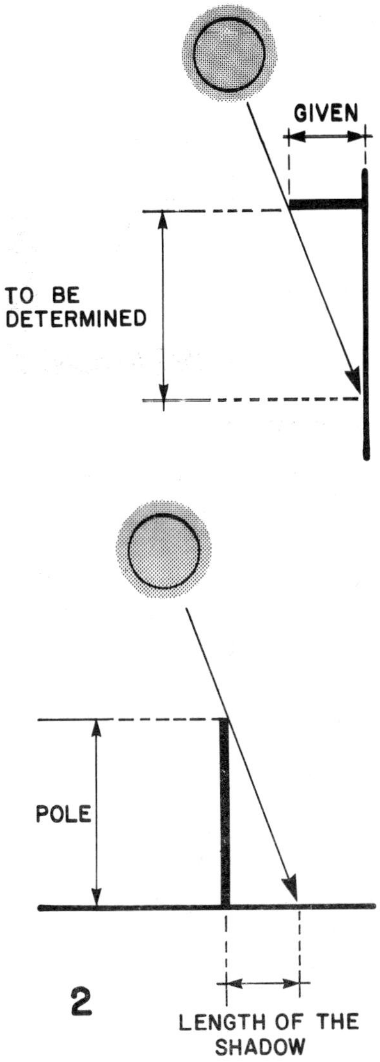

8-42. A section showing a vertical wall, an overhang, and a sunbeam at 12:00 hours. The size of the overhang is given, and the task is to determine the extent of the shadow cast. To use Solar Templates directly, apply the formula: height of pole = length of shadow − coefficient.

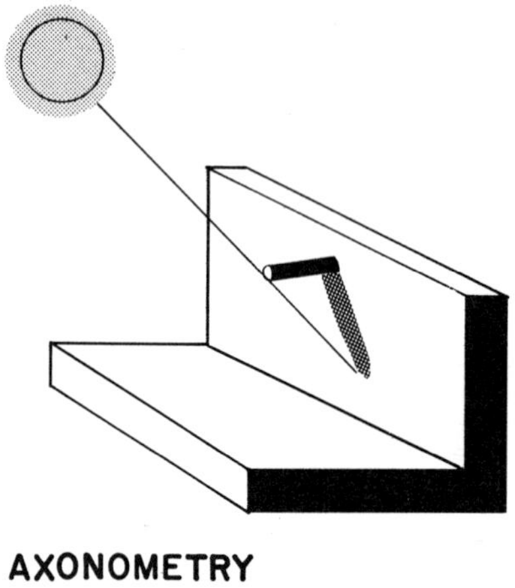

8-43. A horizontal peg casts a shadow on a vertical wall.

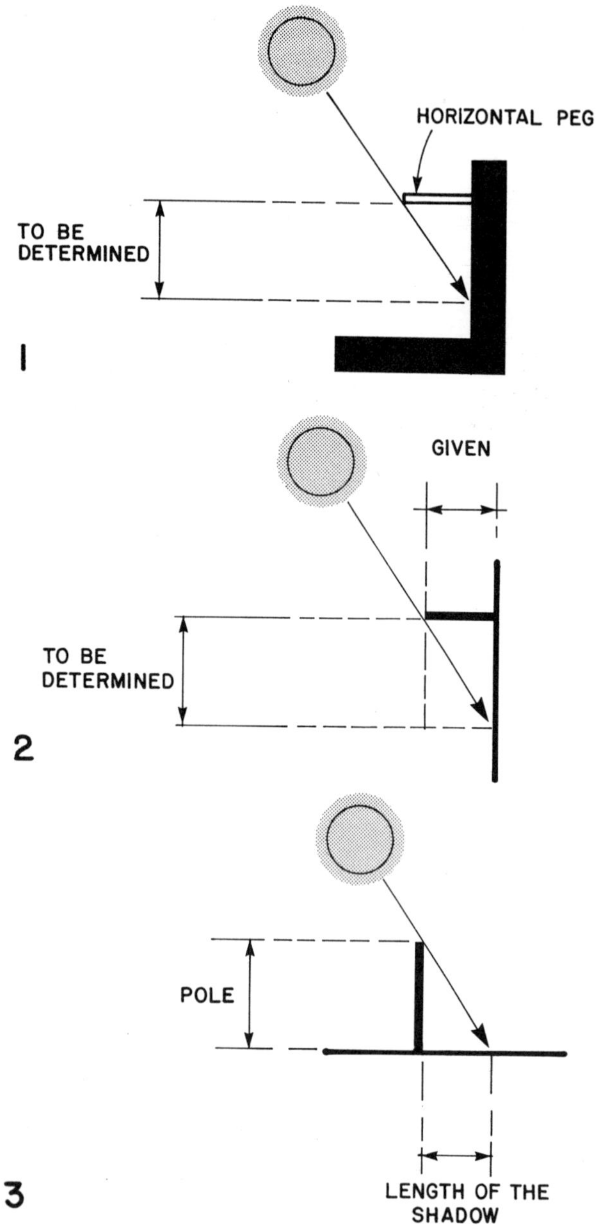

8-44. A section showing a vertical wall, horizontal peg, and sunbeam at a given time.

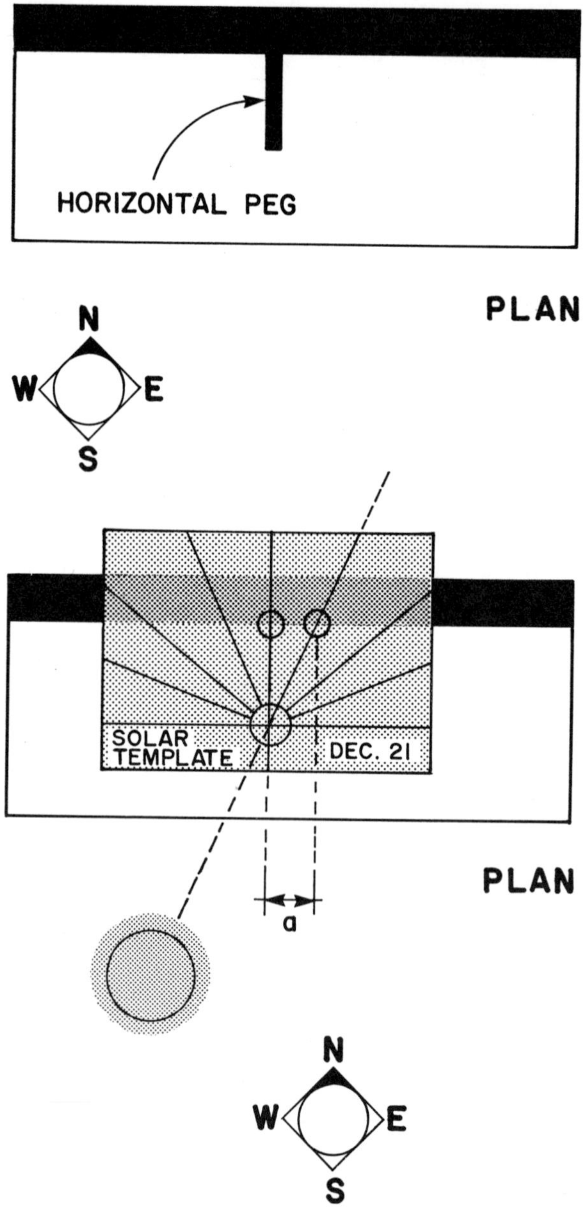

8-45. The construction of a peg to cast shadow.

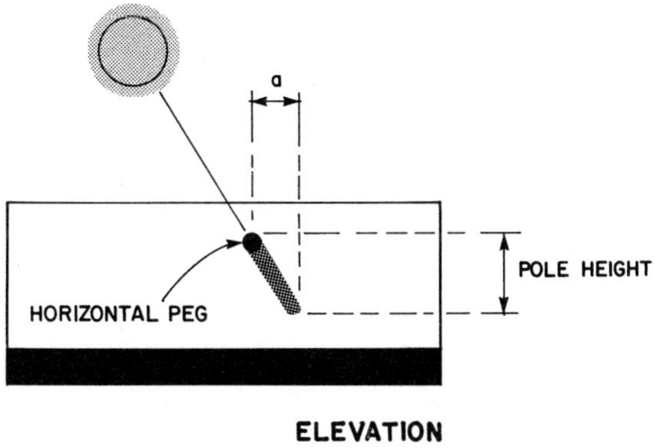

8-46. The elevation of a vertical wall with a shadow cast by a peg.

is equal to zero. Next the pole height is calculated:

pole height = shadow length ÷ coefficient

The actual shadow is drawn on the elevation (fig. 8-46).

CONSTRUCTION OF REFLECTED SUNLIGHT GEOMETRIES

Sun Reflection from Vertical Planes

Solar Templates can be used to determine the position of sun images on reflecting (mirrored) facades and to construct reflected sunbeams. Figure 8-47 illustrates the basic geometries pertaining to sun reflection. The introduction of a world behind the mirrored surface, a phantom world, helps to simplify the task to the level of an elementary exercise.

The direction of a reflected sunbeam is determined in the following example:

TASK

To construct a reflected sunbeam for November 21 at 09:00 and 15:00 hours.

GIVEN

- The plan of an east/west street (fig. 8-48)
- True north
- Location of mirrored facade
- Location of reflection point (Rp)
- Latitude 43°N

SOLUTION

STEP 1. Select the corresponding template, November 21 (see fig. 8-5).

STEP 2. Overlay the template; the center of the template must be placed over the Rp. North on the template must point in the direction of true north on the plan (fig. 8-49).

STEP 3. Rotate the template counterclockwise until the 09:00 hours direction coincides with the mirrored facade line. West on the template represents the direction of the reflected sunbeam at 09:00 hours (fig. 8-50).

STEP 4. Rotate the template clockwise until the 15:00 hours direction coincides with the mirrored facade line. East on the template represents the direction of the reflected sunbeam at 15:00 hours (fig. 8-51).

The Ip and Rp are constructed by using already familiar formulas:

shadow length = pole height × coefficient
pole height = shadow length ÷ coefficient

In this case, the Rp represents the top of the pole and the Ip represents the end of the shadow. Obviously, one point must be given in order to construct the other one (see fig. 8-47).

1. An observer can now identify where the reflected sun image will be at a given time of day and year when facing a mirrored vertical plane (facade).
2. It can also be determined where a given part of a mirrored vertical plane (facade) will cast reflected sunlight at a given time of day and year.

8-47. The basic geometry of a mirror image.

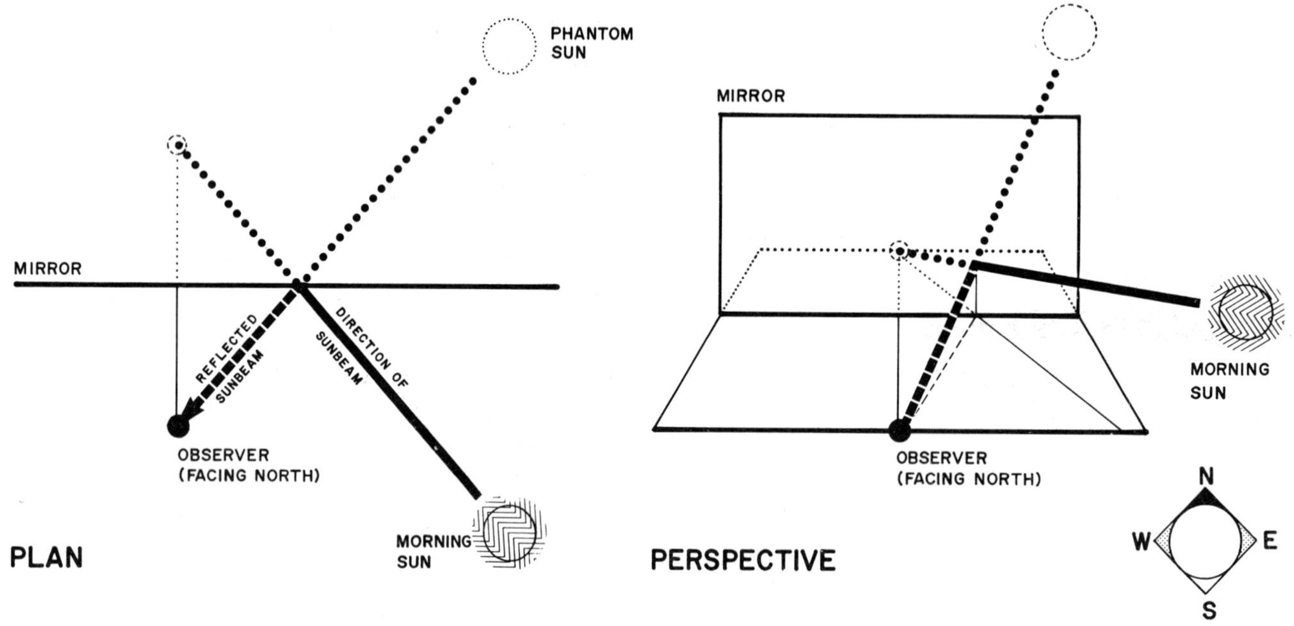

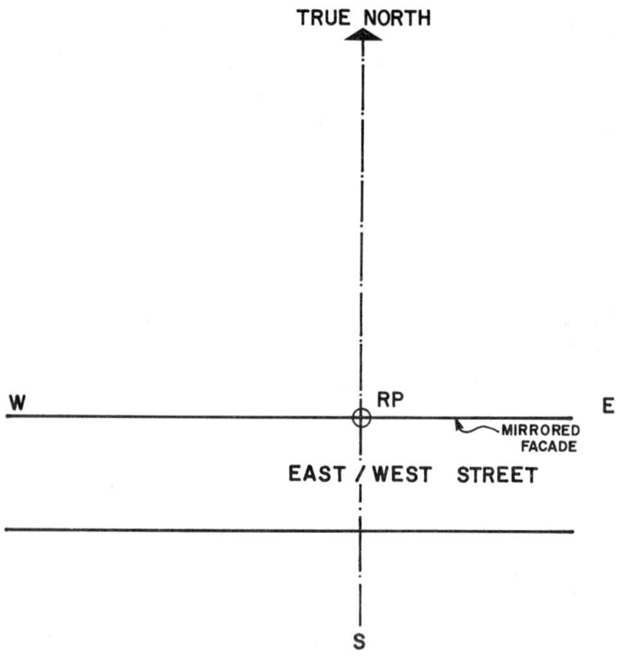

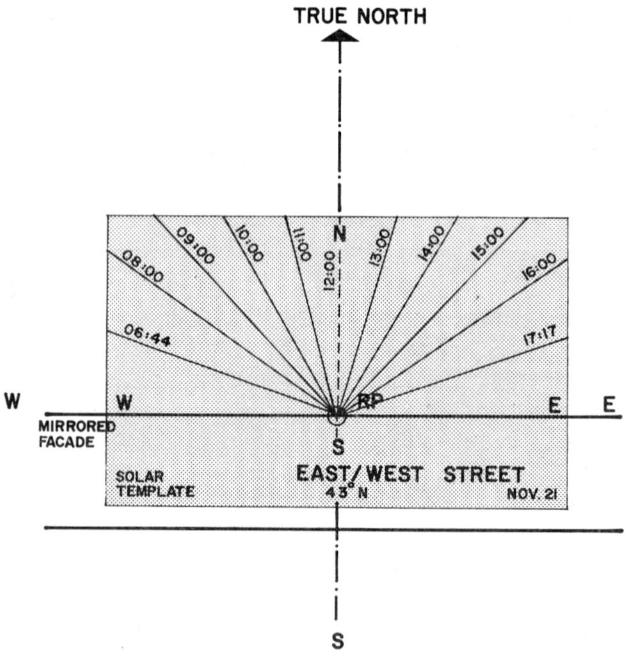

8-48. *(Left)* An east/west street with a mirrored facade facing south.

8-49. *(Right)* A Solar Template over a reflection point.

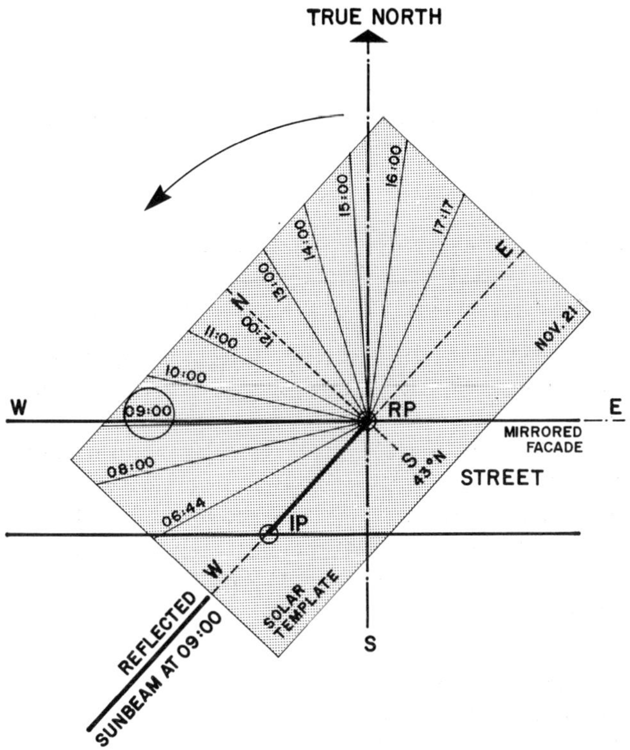

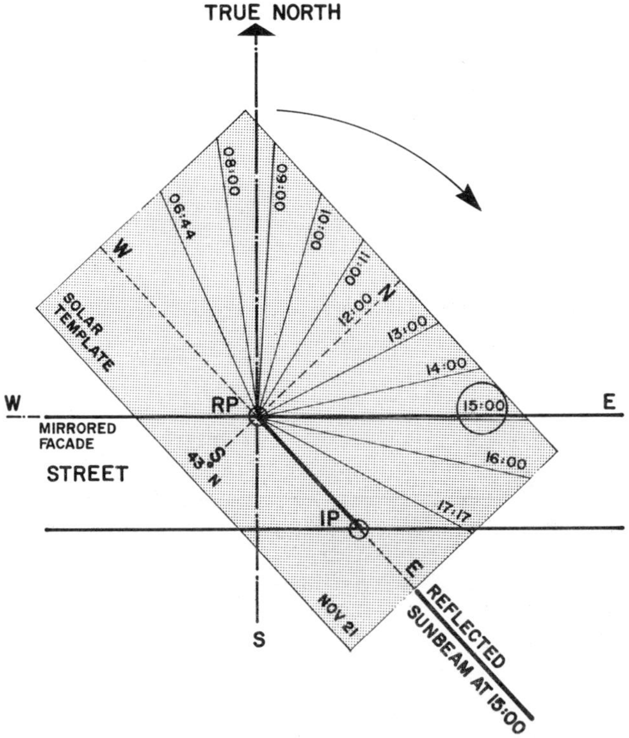

Sun Reflection from Horizontal Planes

The Ip may be located on vertical surfaces, such as building facades and interior walls or on horizontal surfaces, such as the underside of pedestrian walkways and the ceilings of internal spaces as shown in figure 8-52.

The location of the Ip does not affect the calculation. In figure 8-53(1), a case of typical sun reflection is shown. Again, it has to be reformulated to take advantage of the simple pole equation. Since the angle of incidence is equal to the angle of reflection (fig. 8-53[2]), it can be assumed that dimension "b" represents the pole height and "a" represents the shadow length. To make the use of Solar Templates even easier, the task is simplified as shown in figure 8-53(3). Once the reformulation process is understood, the direct use of Solar Templates is very simple.

TASK NO. 1 Find the location of the Ip on May 21 at 14:00 hours. The location of the Rp is known. The distance between the reflection point and the vertical surface is "c" (fig. 8-54). Assuming the Rp represents a hypothetical pole that casts a shadow toward a vertical wall, using the Solar Template is simple (fig. 8-55).

In order to calculate the height of the pole, the length of the shadow must be known. Figure 8-56 shows distance "a" as the length of the shadow; this can be either directly scaled or computed from the following equation:

$$a = c \div \cos \text{ of solar azimuth}$$

8-50. *(Top)* The Solar Template is rotated counterclockwise to find the direction of the impact point (Ip) at 09:00 hours.

8-51. *(Bottom)* The Solar Template is rotated clockwise to find the direction of the (Ip) at 15:00 hours.

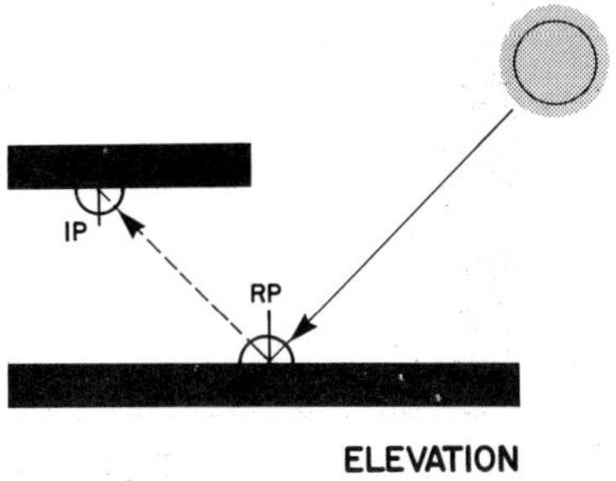

8-52. The Ip is located on the underside of horizontal surfaces.

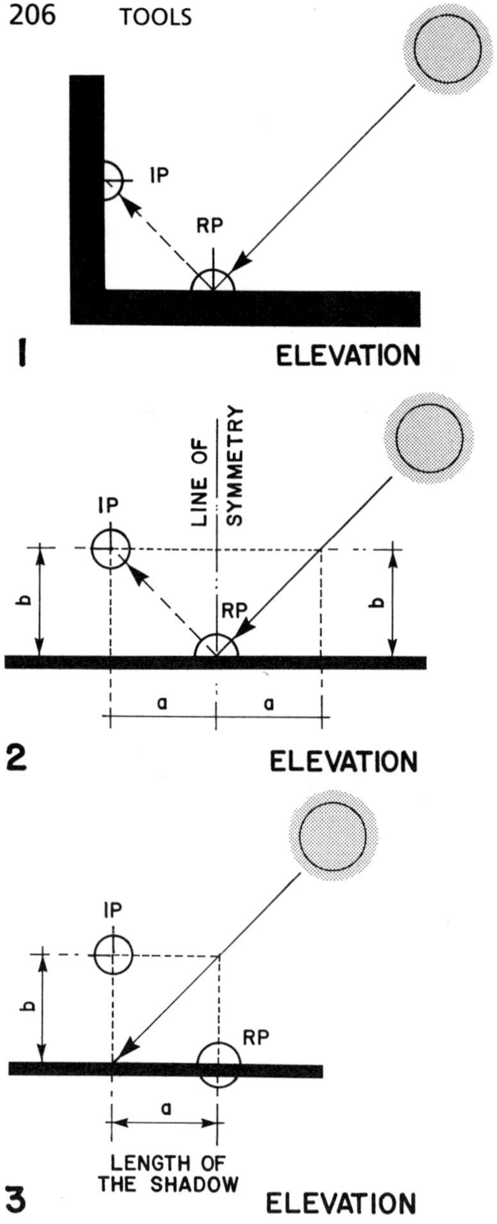

8-53. The Ip is located on the vertical wall. The task of establishing the height of the Ip is reformulated to a simple pole/shadow task.

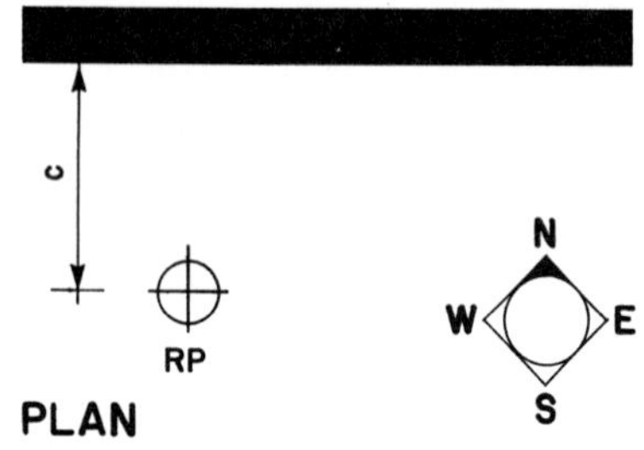

8-54. The Reflection Point (Rp) in front of a vertical wall.

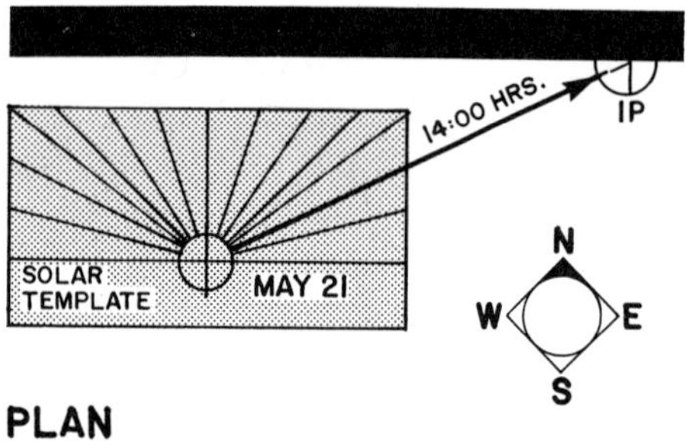

8-55. A Solar Template superimposed on the Rp. The sunbeam is aimed toward the Ip at 14:00 hours.

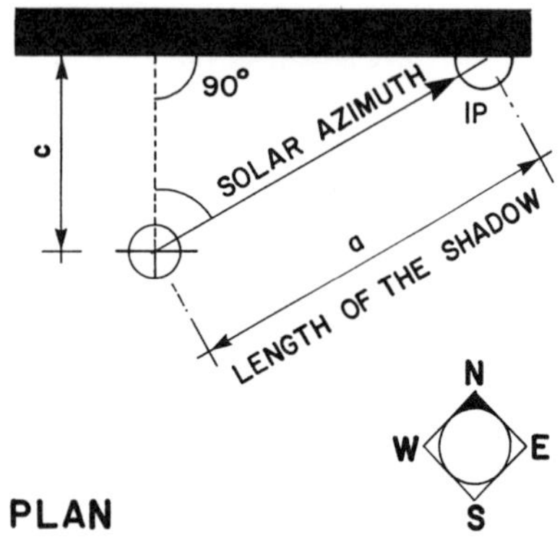

8-56. Distance "a" can be considered the shadow length.

Now the height of the pole can be calculated from the equation:

$$\text{pole height} = \text{shadow length} \div \text{coefficient}$$

The task is thus completed, since the height of the pole is equivalent to the height of the Ip above the horizontal surface (height of Ip $= a \div$ coefficient) (see fig. 8-53).

TASK NO 2. Find the location of the Rp on April 21 at 15:00 hours. The Ip is 15 units above the horizontal plane (fig. 8-57).

The first step is to determine the solar azimuth using the appropriate template (figs. 8-58, 8-59). Since the Ip location is known, the position of the Rp (fig. 8-60) can be calculated from the familiar equation, shadow length $=$ pole height $\times$ coefficient. In this case, $a = 15 \times$ coefficient.

METHOD OF CALCULATION OF APPARENT SOLAR TIME

North America is divided into six time zones: Newfoundland, Atlantic, Eastern, Central, Mountain, and Pacific. The central longitudes (reference lines) of these zones are as follows:

52.5° longitude—Newfoundland Standard Time
60° longitude—Atlantic Standard Time
75° longitude—Eastern Standard Time
90° longitude—Central Standard Time
105° longitude—Mountain Standard Time
120° longitude—Pacific Standard Time

Only solar time can be used in conjunction with Solar Templates. The precise calculation of solar time is relatively complex. The following method, however, is sufficiently accurate for the purposes of architects, planners, and design professionals.

STEP 1. During daylight saving time, subtract one hour from local time to arrive at the standard time.

STEP 2. Add four minutes to the standard time for every degree of longitude if the site is east of the central longitude (reference line), or subtract four minutes for every degree of longitude if the site is west of the central longitude (reference line) (fig. 8-61).

STEP 3. Correct seasonal variations in the earth-sun geometry by adding or subtracting the times listed in the table below.

	MINUTES	SECONDS
January 21	−11	23
February 21	−13	39
March 21	−07	12
April 21	+01	19
May 21	+03	30
June 21	−01	40
July 21	−06	19
August 21	−03	04
September 21	+06	57
October 21	+15	22
November 21	+14	07
December 21	+01	53

The following is an example showing how to calculate solar time:

TASK
To calculate the solar time for Hamilton, Canada, at 12:00 hours (noon) local time on June 21.

GIVEN
- Local time in Hamilton
- Daylight saving time is in effect from the last Sunday in April until the last Sunday in October
- Hamilton's longitude is 80°W, thus Hamilton is located within the eastern standard time zone
- The eastern standard time zone has its central longitude (reference line) at 75°, thus,
- Hamilton is located 5° west of the central longitude (reference line)
- The seasonal variation for June 21 is −1:40 (subtract one minute, forty seconds).

STEP 1. In June daylight saving time is in effect, so one hour must be subtracted from the local time:

12:00 (local time) − 01:00 = 11:00 (standard time)

STEP 2. Subtract four minutes times the number of degrees of longitude that Hamilton is west of central longitude (75°) from 11:00 hours:

80° − 75° = 5°
5 × 4 minutes = 20 minutes
11:00 hours − 20 minutes = 10:40 (10 hours, 40 minutes)

STEP 3. Subtract the seasonal variation for June 21:

10:40 − 00:01:40 = 10:38:20
(10 hours, 40 minutes, minus 1 minute and 40 seconds equals 10 hours, 38 minutes, and 20 seconds)

Consequently, on June 21 in Hamilton, Canada, Noon (12:00 hrs. local time) = 10 hours, 38 minutes, and 20 seconds solar time.

Note that the difference between local time and solar time is 1 hour, 21 minutes, and 40 seconds. This is the constant difference throughout the 21st day of June. Thus, solar noon would take place at 13:21:40 hours local time (when the clock shows 1 hour, 21 minutes, and 40 seconds P.M.). In this instance, June 21 happened to be the longest day of the year; thus, the shortest shadow of the 21st day of June would also be the shortest shadow of the year.

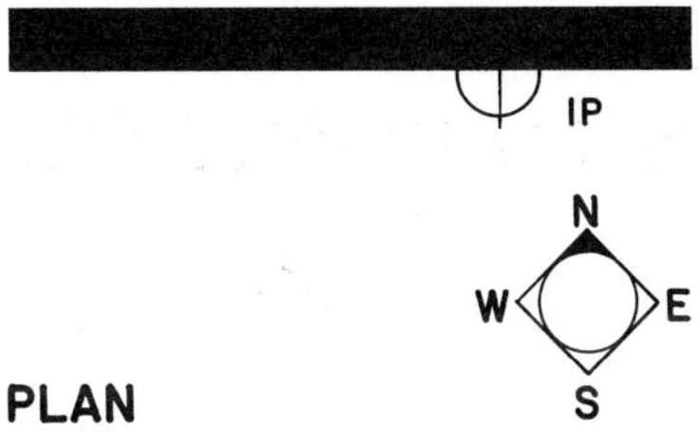

8-57. The Ip is located on a vertical wall.

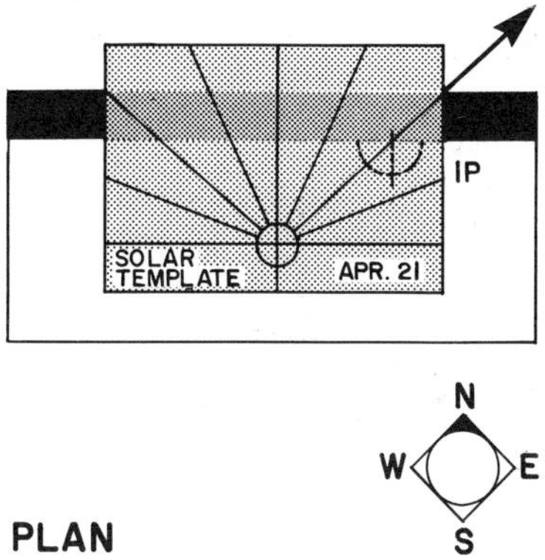

8-58. The Solar Template is superimposed on the Ip.

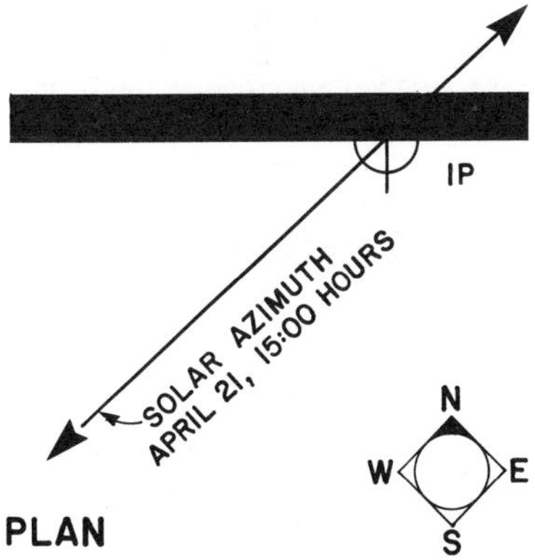

8-59. The solar azimuth is projected over the Ip.

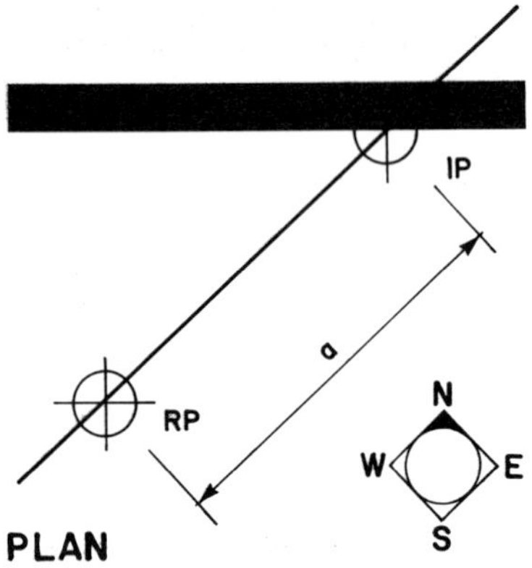

8-60. The Rp is located at distance "a" from the Ip.

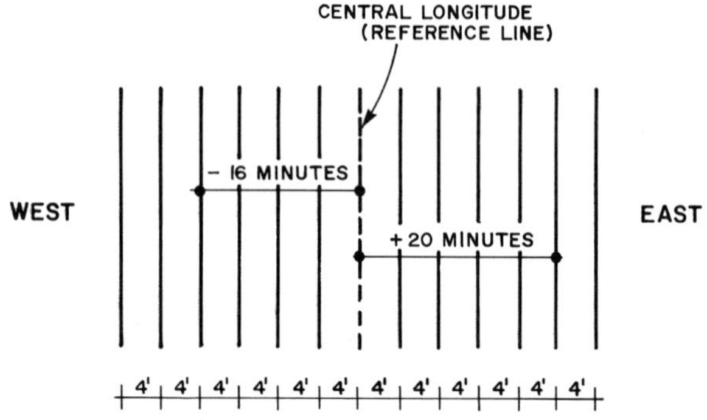

8-61. The central longitude of a time zone; every degree of longitude equals four minutes.

8-62. A pole in the ground.

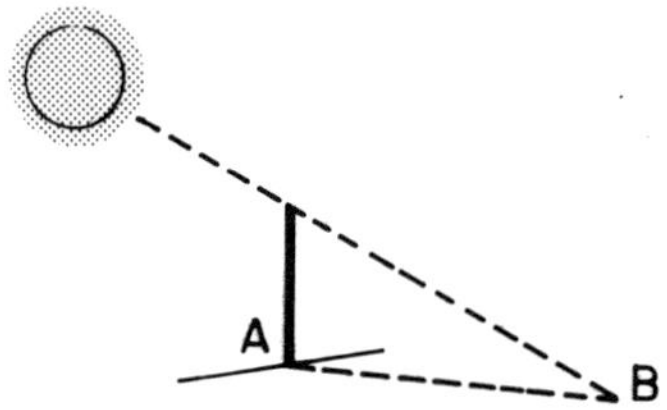

8-63. The morning shadow cast by the pole.

CONSTRUCTION OF TRUE NORTH

Only true north can be used in conjunction with Solar Templates. Magnetic north, used by surveyors, is the direction shown by a magnetic compass. All maps intended for general use show magnetic north. On occasion, either true north or the angle of deviation of magnetic north from true north is included for a given locality.

If neither true north nor the angle of deviation from true north is indicated, the most reliable method for determining true north is the *field* or *Vitruvian* method, shown in the following figures.

STEP 1. Place a pole upright in the ground (fig. 8-62).

STEP 2. Mark the end of the pole's shadow in the morning (sometime before noon). Points A and B represent the shadow of the pole at that time (A.M.) (fig. 8-63).

STEP 3. Using point A as a center, draw a circle as shown in figure 8-64.

STEP 4. In the afternoon, when the end of the shadow touches the circle again, mark point C (fig. 8-65).

STEP 5. Divide the distance between B and C in half to arrive at point D (fig. 8-66).

STEP 6. Join point A and point D. This line represents true north (fig. 8-67).

True north as shown in the previous figures is, in fact, the sun azimuth at solar noon, that is, the direction of the sun's beams at the moment when the shadow is shortest. This method can be performed directly on the ground at the site, or on a map, provided that the map is first properly oriented toward the magnetic pole. Although this method was introduced by the Roman architect, Vitruvius, in the first century B.C., it was most likely known and used by the ancient Babylonian and Egyptian astronomers as well. Solar time and true north are prerequisites for the construction and use of Solar Templates.

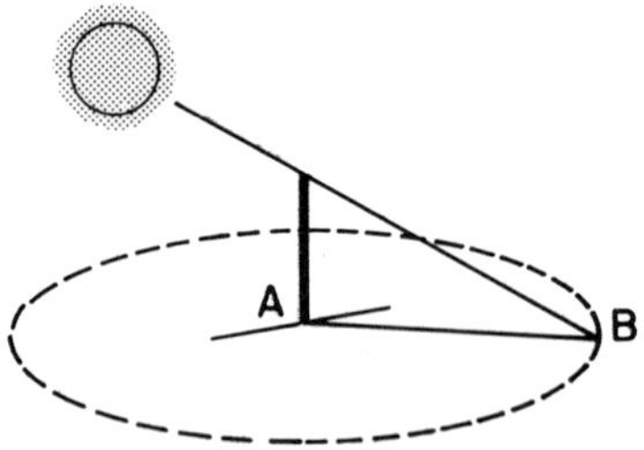

8-64. A circle of AB radius.

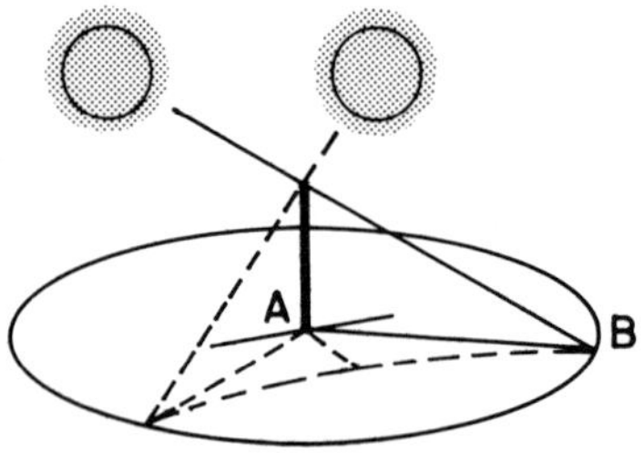

8-65. The afternoon shadow cast by the pole.

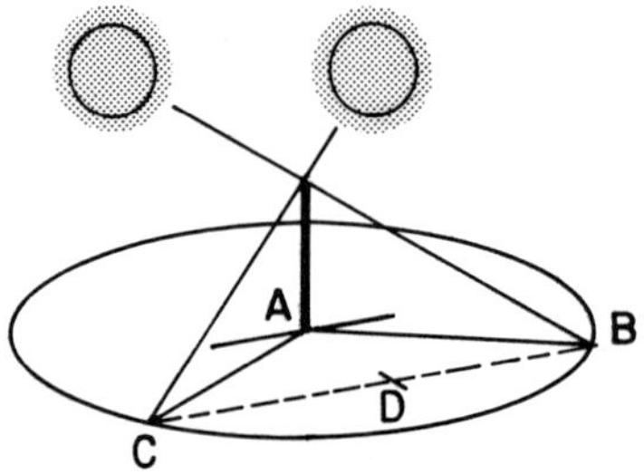

8-66. The distance between point C and B divided.

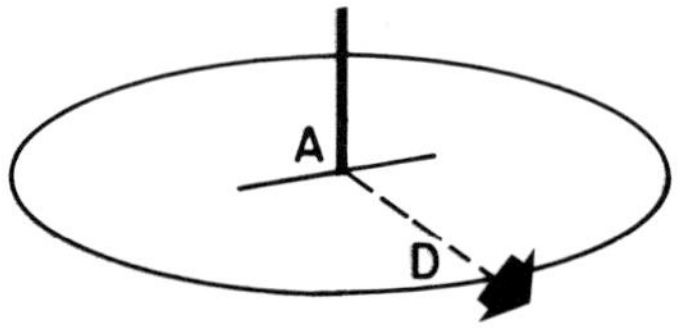

8-67. A line going from point A to point D indicates the true north direction.

BIBLIOGRAPHY

Adamenki, V. N. *Possibilities of Improving Winter Micrometeorological Conditions in Existing Planned Urban Areas.* Downsview, Ontario, Canada: Atmospheric Environment Service, 1970.

Arens, E. "Outdoor Comfort of Pedestrians in Cities." *In Proceedings of the SUNY/USFS/AMS Conference.* Syracuse, New York: Metropolitan Physical Environment, 1975.

Aronin, J. E. *Climate and Architecture.* New York: Reinhold Press, 1956.

Atwater, M. A., and J. P. Pandolfo. "Tundra Environment Changes Induced by Urbanization." *Climate of the Arctic.* Anchorage: University of Alaska Press (1975): 312–15.

Auer, A. H. Jr., and R. A. Dirks. "Contributions to an Urban Meteorological Study." *Bulletin of the American Meteorological Society* 55 (1974): 106–10.

Bach, W. "Seven Steps to Better Living in the Urban Heat Island." *Landscape Architecture* 61 (no. 2): 136–41.

Baird, P. D. *The Polar Wind.* New York: Wiley Interscience, 1964.

Baker, D. G., J. W. Erz, and H. J. Paulus. "Frequency, Duration, Commencement Time and Intensity of Temperature Inversion at St. Paul, Minneapolis." *Journal of Applied Meteorology* 8 (1969): 747–53.

Barry. R. G., and R. J. Chorley. *Atmosphere, Weather and Climate.* New York: Holt, Rinehart and Winston, 1970.

Bazajanac, V. "Energy Analysis: One Year Late." *Progressive Architecture* 4 (April 1981): 154–56.

Benson, C. S., and S. A. Bowling. "The Sub-Arctic Heat Island as Studied at Fairbanks Alaska." *In Climate of the Arctic.* Anchorage: University of Alaska Press, 1975.

Berjand, M. E. "Meteorological Factors Determining the Dispersion of Atmospheric Pollutants under Urban Conditions." *Urban Climate, Technical Note #8.* Geneva: Secretariat of the World Meteorological Organization, 1970.

Berlyand, M., M. I. Budyko, and K. Kondratyev. "Urban Climate and the Problems of Variation of the Global Climate." *Meteorology and Hydrology* 1 (1973): 3–14.

Bernatzky, A. "Climate Influences of the Greens and City Planning." *Journal of Arboriculture* 3 (1977): 121–26.

Bolin, B. *Energy and Climate.* Stockholm: Secretariat for Future Studies, 1975.

Bornstein, R. D. , A. Lorenson, and D. Johnson. "Recent Observations of Urban Effects on Wind and Temperatures In and Around New York City." In *Preprints Conference on Urban Environment and Second Conference on Biometeorology.* Philadelphia: American Meteorological Society, 1972.

Braham, R. R. Jr., "Ice Nucleus Measurements in an Urban Atmosphere." *Journal of Applied Meteorology* 13 (1974): 940–45.

Braham, R. R., Jr. "Overview of Urban Climate." In *Proceedings of SUNY/USFS/AMS Conference, Metropolitan Physical Environment, 1975.* Syracuse, New York.

Briggs, A. *The Fitness of Man's Environment.* New York: Harper & Row Publishers, 1968.

Broberg, P. "Regional Urbanisms." *The Teilhard Review* 13 (no. 1), Spring 1978.

Burke, J. *Connections.* London: Macmillan London Ltd., 1978.

Burton, A. C., and O. G. Edholm. *Man in a Cold Environment.* New York, London: Hafner Publishers, 1955.

Canada Mortgage and Housing Corporation. *An Introduction to Energy Conservation in a Residential Development*. Ottawa: Queen's Printer, 1983.

Canadian Housing Design Council and the University of Manitoba. *The Winter City Book: Proceedings of a Conference on Winter Cities, Winter 1982*. Ottawa: Canadian Housing Design Council, 1982.

Carborn, J. M. *Shelterbelts and Windbreaks*. London: Faber and Faber Publishers, 1965.

Carter, N. "Urban Environment and the Quality of Life." In *The Winter City, Proceedings of a Conference on Winter, Winnipeg, Manitoba 1982*. Ottawa: Canadian Housing Design Council, 1982.

Casper, W. "Introduction to the Discussion on Climatological Data for Regional and Town Planning." In *Teaching the Teachers on Building Climatology*. Stockholm: Swedish Institute for Building Research, 1973.

Center for Energy Policy and Research. *The Energy Efficient Community: A Model for the Nation*. New York: New York Institute of Technology, 1977.

Chaine, P. M., and P. Skeates. "Ice Accretion Handbook (Freezing Precipitation)." In *Industrial Meteorology Study VI*. Downsview, Ontario: Atmospheric Environment Service and Environment Canada, 1974.

Chandler, T. J. "Urban Climates: Inventory and Prospect." In *Urban Climates, Technical Note # 108*. Geneva: Secretariat of the World Meteorological Organization, 1970.

Chandler, T. J. "Urban Climates and the Natural Environment." *Journal of Biometeorology* 20 (1976): 128–38.

Chandler, T. J. "Urban Climates." *Special Environment Report #2*. Geneva: Secretariat of the World Meteorological Organization, 1979.

City of Brampton Planning Department. *Energy Conservation in Land-Use Planning*. Brampton, Ontario: City of Brampton, 1979.

City of New York Planning Commission. *Plazas for People, Streetscape and Residential Plazas*. New York: New York City Planning Department, 1976.

City of Ottawa, Planning Branch. *Consideration of Wind Problems in the Downtown Core*. Report prepared by P. Chen, and M. Hart. Ottawa: Planning Branch, Department of Community Development, 1977.

City of Ottawa, Planning Branch. *Ottawa Central Area: Sunlight, Wind and Snow Studies*. Report prepared by P. Chen. Ottawa: Planning Branch, Department of Community Development, January 1983.

City of Toronto Planning and Development Department. *Sunlight, Wind Study for the Central Core*. Toronto: City of Toronto, June 1980.

Claiborne, R. *Climate, Man and History*. New York: Norton and Company, 1970.

Clarke, J. F., and W. Bach. "Comparison of the Comfort Conditions in Different Urban and Suburban Micro-environments." *International Journal of Biometeorology* 15 (1971): 41–54.

Clarke, J. F., and J. T. Peterson. "The Effect of Regional Climate and Land Use on the Nocturnal Heat Island." In *Preprints Conference on Urban Environment and Second Conference on Biometeorology*. Philadelphia: American Meteorological Society, 1972.

Conrads, L. A. *Observations of Meteorological Urban Effects: The Heat Island of Utrecht*. Utrecht: University of Netherlands, 1976.

Craig, C. D., and W. P. Lowry. "Reflection on the Urban Albedo." In *Building Climatology, Technical Note #109*. Geneva: Secretariat of the World Meteorological Organization, 1970.

Crowe, R. B. "The Importance of Climate to Recreation-Tourism Facility Planning." Paper presented to 9th Congress. Vancouver: Canadian Meteorological Society, May 1975.

Crowther, R. L. "Climate and Architecture." *Bulletin of the American Meteorological Society* 58 (1977): 595–96.

Culjat, B. "Climate and the Built Environment in the North." Ph.D. diss., Royal Institute of Technology, Stockholm, 1975.

Culjat, B., and R. Erskine. *Climate Responsive Social Space: Scandinavian Perspective*. Stockholm: Institute for Architecture, 1980.

Curtis, F. A., and M. W. Schlosser. "Modelling the Rykmans [sic] Neighbourhood Subdivision." *Plan Canada* 26 (no. 10) (January 1987): 282–90.

Curtis, F., ed. *Considerations and Opportunities for Energy Conservation in Urban and Regional Planning: Conference Proceedings*. Kingston, Ontario: Queen's University, 1979.

Davenport, A. G. "An Approach to Human Comfort Criteria for Environmental Wind Conditions." In *Preprint, WMO—CIB Colloquium on Teaching the Teachers on Building Climatology 1*. Stockholm: National Institute for Building Research, 1973.

DeFreitas, C. R. "Estimation of the Disruptive Impact of Snowfalls in Urban Areas." *Journal of Applied Meteorology* 14 (1975): 1166–73.

Egan, M. D. *Concepts in Thermal Comfort*. Englewood Cliffs, New Jersey: Prentice-Hall, 1975.

Energy, Mines and Resources Canada. *Passive Solar Remodelling: A Feasibility and Design Study to Assess Methods of Residential Passive Solar Energy*. Ottawa: Energy, Mines and Resources Canada, Conservation and Renewable Energy Branch, 1981.

Environment Canada, Lands Directorate. *Energy Conservation through Land Use Planning: A Synthesis of Discussions at a Symposium in Montreal, Working Paper No. 6*. Edited by W. R. D. Sewell, and H. D. Foster. Ottawa: Environment Canada, August 1980.

Federer, C. A. "Trees Modify the Urban Micro-climate." *Journal of Arboriculture* 2 (1976): 121–27.

Fitch, J. M. *American Building: The Environmental Forces that Shape It*. Boston: Houghton Mifflin, 1972.

Fitch, J. M. and D. P. Branch. "Primitive Architecture and Climate." *Scientific American* 208 (1960): 134–44.

Folk, G. E. *Textbook of Environmental Physiology*. Philadelphia: Lea and Febiger, 1974.

Foster, G. E. *Landscaping that Saves Energy Dollars*. New York: David McKay, 1980.

Gajzago, L. "Outdoor Microclimate and Human Comfort." In *Preprints, WHO-CIB Conference on Teaching the Teachers on Building Climatology*. Stockholm: The National Swedish Institute for Building Research, 1973.

Gandemer, J. *La Protection Contre Le Vent*. Paris: Centre Scientifique & Technique du Batiment, 1981.

Gandemer, J. *Discomfort Due to Wind Near Buildings: Aerodynamic Concepts and Method of Wind Integration in Built-up Areas Design*. Nantes, France: Centre Scientifique & Technique du Batiment, 1983.

Garnier, B. J., and A. C. Lafleur. "A Study of Human Comfort in Montreal, Canada." In *Preprints, Conference on Urban Environment and Second Conference on Biometeorology*. Philadelphia: American Meteorological Society, 1972.

Gates, D. M. *Man and His Environment: Climate*. New York: Harper and Row Publishers, 1972.

Givoni, B. *Man, Climate and Architecture*. New York: Elsevier Publishing Company, Ltd., 1969.

Givoni, B. "Urban and Building Biometeorological Research: an Overview." *International Journal of Biometeorology* 5 (no. 16) (1973): 85–87.

Givoni, B. "Effect of Town Planning on the Urban Climate." In *Preprints, WHO-CIB Conference on Teaching the Teachers on Building Climatology*. Stockholm: The National Swedish Institute for Building Research, 1973.

Givoni, B. *Architectural Design Based on Climate*. New York: Van Nostrand Reinhold Company, Inc., 1981.

Givoni, B., and M. Paciuk. "Airflow Around High-Rise Buildings." In *Preprint No.19, Teaching the Teachers: Colloquium on Building Climatology*. Stockholm: The National Swedish Institute for Building Research, September 1972.

Goldstein, K. M. "Weather, Mood and Internal-External Control." *Perceptual and Motor Skills* 35 (1972): 786.

Griffiths, J. F. *Applied Climatology*. London: Oxford University Press, 1976.

Griffiths, J. F. *Climate and the Environment: The Atmospheric Impact on Man*. London: Paul Elek, 1976.

Grillo, J. N., and J. Spar. "Rain-snow Mesoclimatology of the New York Metropolitan Area." *Journal of Applied Meteorology* 10 (1971): 56–61.

Grinnell, G. "Thermodynamics and World History." *The Energy Newsletter* 1(no. 2) Hamilton, Canada: McMaster University (October 1980): 37–43.

Gutheim, F. "A Livable Winter City." *Architectural Record (no. 165)*(February 1979): 111–16.

Hamilton-Wentworth Planning and Development Department. *Energy Consideration in Urban and Regional Planning*. Hamilton, Canada: Regional Municipality of Hamilton-Wentworth, January 1978.

Hamilton-Wentworth Planning and Development Department. *Performance Standards—An Alternative Land Use Control Mechanism*. Hamilton, Canada: Regional Municipality of Hamilton-Wentworth, 1978.

Hamilton-Wentworth Planning and Development Department. *Planning for Low Density—a Solar Approach*. Hamilton, Canada: Regional Municipality of Hamilton-Wentworth, 1980.

Hamilton-Wentworth Planning and Development Department. *Design Criteria for Energy Efficient Neighbourhood Planning, Stage One*. Hamilton, Canada: Regional Municipality of Hamilton-Wentworth, September 1982.

Hamilton-Wentworth Planning and Development Department. *Design Criteria for Energy Efficient Neighbourhood Planning, Stage Three*. Hamilton, Canada: Regional Municipality of Hamilton-Wentworth, March 1983.

Heschong, L. *Thermal Delight in Architecture*. Cambridge, Massachusetts: The MIT Press, 1979.

Hix, J. "Energy Conservation and the Architect." *The Canadian Architect* 2 (February 1977): 22–36.

Hough, M. *City Form and Natural Process*. New York: Van Nostrand Reinhold Company, Inc., 1984.

Hough, M., and Suzanne Barrett. *People and City Land-*

scapes. Toronto: Conservation Council of Ontario, 1987.

Huckuby, J. L., and L. Roach. "The Microclimatic Impact of Reflective Glass Buildings in an Urban Environment." In *Proceedings of the SUNY/USFS/AMS Conference: Metropolitan Physical Environment*. Syracuse, New York, 1975.

Huntington, E. *Principles of Human Geography*. New York: John Wiley and Sons, 1951.

Idso, S. B. "Transmittance of Atmosphere for Solar Radiation on Individual Clear Days." *Journal of Applied Meteorology* 9 (1970): 239–40.

Joyner, S., and R. Reiter. "Microclimatic Aspects of Developmental Suitability of the Metropolitan Landscape." In *Proceedings of SUNY/USFS/AMS Conference: Metropolitan Physical Environment*. Syracuse, New York, 1975.

Knowles, R. L. *Energy and Form: An Ecological Approach to Urban Growth*. Cambridge, Massachusetts: The MIT Press, 1974.

Knowles, R. L. *Sun Rhythm Form*. Cambridge, Massachusetts: The MIT Press, 1981.

Kung, R. "Improvement of Pedestrian Level Wind Problems." Bachelor thesis, University of Waterloo, Canada, 1984.

Laconte, P., ed. *Environment of Human Settlements*. Oxford: Pergamon Press, 1976.

Lam, W. M. C. *Sunlighting as Formgiver for Architecture*. New York: Van Nostrand Reinhold Company, Inc., 1986.

Landsberg, H. E. *Weather and Health: An Introduction to Biometeorology*. New York: Doubleday Anchor Books, 1969.

Landsberg, H. E. "Climate of the Urban Biosphere." *International Journal of Biometeorology* 5 (1973): 71–83.

Landsberg, H. E. "Inadvertent Atmospheric Modification through Urbanization." In *Weather and Climate Modification*. Edited by W. N. Hess. New York: John Wiley and Sons, 1974.

Landsberg, H. E. "Micrometeorology." *Architectural Forum* 83 (no. 3) (1974): 114–19.

Levine, R., and N. Brining, eds. *The Cold Weather Catalogue*. New York: Dolphin Books/ Doubleday, 1977.

Los, S. "The Design of Urban Outdoor Space: A Bioclimatic Approach." In *Proceedings of the Third International P.L.E.A. Conference*. London: Pergamon Press, 1984.

Matus, V. "Habitation in Hostile Environments." Master's thesis, University of Toronto, 1969.

Matus, V. "The Energy Argument—Use and Misuse." *The Energy Newsletter* 1 (no. 2) (October 1980): 27–36.

Matus, V. "Beyond Conservation—Prospect for Planning." In *Proceedings of the 1982 National Conference of the Canadian Institute of Planners*. Toronto, 1982.

Matus, V. "Climate Sensitive Urban Planning and Design." In *Proceedings of the Third International P.L.E.A. Conference*. London: Pergamon Press, 1984.

Matus, V. "Solar Access Geometries." In *Proceedings of the 10th Annual National Conference of the Solar Energy Society of Canada, SESCI '84*. Winnipeg, Canada: Solar Energy Society of Canada, Inc., 1984.

McBoyle, G., ed. *Climate in Review*. Boston: Houghton Mifflin Company, 1973.

Melaragno, M. P. *Wind in Architectural and Environmental Design*. New York: Van Nostrand Reinhold Company, Inc., 1982.

Milne, M., and B. Givoni. *Energy Conservation through Building Design*. New York: McGraw-Hill, 1979.

Ministry of Agriculture and Food. *Heat Units for Corn in Southern Ontario* by Brown, D. M. Toronto: Ontario Ministry of Agriculture and Food, Order No. 78-063, June 1978. Reprinted May, 1981.

Nash, J. E. "Human Relations in Frozen Places." *Architecture Minnesota*. (Special issue: An Architecture for Winter—The Idea of the North) 9 (no. 1) (January/February 1983): 28–31.

National Research Council of Canada. *Tables of Solar Altitude, Azimuth, Intensity and Heat Gain Factors for Latitudes from 43 to 55 Degrees North: Technical Paper No. 243*. Ottawa: National Research Council of Canada, April 1967.

Oke, T. R. "City Size and the Urban Heat Island." *Atmospheric Environment* 7 (1973): 769–79.

Oke, T. R., and F. G. Hanell. "The Form of the Urban Heat Island in Hamilton, Canada." In *Urban Climates, Technical Note No. 108*. Geneva: Secretariat of the World Meteorological Organization, 1970.

Olgyay, V. *Design with Climate: Bioclimatic Approach to Architectural Regionalism*. Princeton: Princeton University Press, 1963.

Oliver, J. E. *Climate and Man's Environment*. New York: John Wiley and Sons, 1973.

Ontario Ministry of Energy. *Perspectives on Access to Sunlight*. Toronto: Ministry of Energy, 1978.

O. P. A. Consulting Ltd. *Sparks Street Economic Study*. Ottawa: O. P. A. Consulting Ltd., August 1983.

Persinger, M. A. *The Weather Matrix and Human Behaviour*. New York: Praeger Publishers, 1980.

Phillips, D. W. "Understanding Winter Climate." Paper presented at the Winter Cities Forum '86, Edmonton, Alberta, 1986.

Phillips, D. W., and R. B. Crowe. *Climate Severity Index for*

Canadians. Downsview, Ontario, Canada: Atmospheric Environment Service, Environment Canada, 1984.

Pihlak, M. "Philosophy, Principles and Practices of Northern Latitude Urban Landscape Design." Master's thesis, University of California, Berkeley, 1983.

Pressman, N., and X. Zepic. *Planning in Cold Climates: A Critical Overview of Canadian Settlement Patterns and Policies*. Winnipeg, Canada: Institute of Urban Studies, 1986.

Regional/Urban Design Assistance Team. *Livable Winter Cities*. Edmonton, Canada: RAIC, AIA, and R/UDAT, 1986,.

Rogers, W. *The Winter City Book*. Edina, Minnesota: Dorn Books, 1980.

The Royal Architectural Institute of Canada. *Energy Conservation: Design Resource Handbook*. Toronto: Carswell Printing Company, 1979.

The Royal Botanical Gardens. *Annual Meteorological Summary 1983: Hamilton, Ontario*. Edited by W. J. Lamoureux. Burlington, Ontario, Canada, 1983.

Royle, J. C. "How to Make Cities More Livable in Winter." *Canadian Geographic* (February/March 1979): 21–27.

Royle, J. C. "Livable Winter City—Coming Faster Than You Think." *Housing Ontario* 21 (no. 9) (November/December 1980): 12–15.

Secretariat General du Groupe Central des Villes Nouvelles. *Integration du Phenomene Vent dans la Conception du Milieu Bati*. Paris: Secretariat General du Groupe Central des Villes Nouvelles, 1976.

Siddiqi, A. A. "Buildings, Energy and Urban Morphology." In *Proceedings of the Third International P.L.E.A. Conference*. London: Pergamon Press, 1984.

Statistics Canada. *Causes of Death by Province, Age, and Sex*. Ottawa: Statistics Canada, 1984.

Statistics Canada. *Mortality Differences in Canada*. Ottawa: Statistics Canada, 1984.

Terjung, W. H. "World Patterns of the Distribution of the Monthly Comfort Index." *International Journal of Biometeorology* 12 (1968): 119–51.

Tighe, D. "Climate Considerations in Urban Development: A Discussion Paper." Peterborough Planning Department, Peterborough, Ontario, Canada, December 1978. Photocopy.

United States Department of Energy. *Site Planning for Solar Access*. Washington, D. C.: Government Printing Office, September 1979.

Welch, J. D. "The Effects of Sun, Wind and Snow on Urban Design." In *Proceedings of The Winter City Conference: Winter, 1982*. Ottawa: Canadian Housing and Design Council, 1982.

Werier, V. "The Livable Winter City." *Habitat* 26 (no. 3) Ottawa: Canada Mortgage and Housing Corporation, 1983.

Winter Cities Conference Corporation. *Winter Cities Forum '86: Symposium Proceedings*. Edmonton, Canada: University of Alberta, 1986.

Wise, A. F. E., D. E. Sexton, and M. S. T. Lillywhite. "Urban Planning Research: Studies of Airflow around Buildings." *Architect's Journal* 141 (May 1965): 1185–89.

World Meteorological Organization. *Proceedings of the Symposium on Urban Climates and Building Climatology: Technical Notes No. 108*. Geneva: Secretariat of the World Meteorological Organization, 1970.

World Meteorological Organization. *Proceedings of the Symposium on Urban Climates and Building Climatology: Technical Notes No. 109*. Geneva: Secretariat of the World Meteorological Organization, 1970.

INDEX

INDEX